Materials and structures

D1423627

Materials and structures

2nd edition

R. Whitlow
C.Eng., M.I.Struct.E., F.G.S.
Senior Lecturer in Geotechnics
and Foundation Engineering,
Bristol Polytechnic

 LONGMAN

The CHARTERED
INSTITUTE OF
BUILDING

Pearson Education Limited
Edinburgh Gate, Harlow,
Essex CM20 2JE, England
and Associated Companies throughout the world.

Co-published with The Chartered Institute of Building through

Englemere Limited
White House, Englemere,
Kings Ride, Ascot,
Berkshire, SL5 8BJ
England

© Longman Group UK Limited 1973, 1991

First published 1973
Second edition 1991
Second impression 1992
Third impression 1995
Fourth impression 1996
Fifth impression 1998
Seventh impression 1999
Eighth impression 2000

British Library Cataloguing in Publication Data
Whitlow, R. (Roy)
    Materials and structures. — 2nd ed.
    1. Materials
    I. Title
    620.11

Set in Compugraphic Times 10/11 pt

Printed in Malaysia, GPS

Contents

Preface

The purpose of this book is to provide in a single volume an adequate
coverage of those principles of strength of materials and elementary
theory of structures that are usually included, in one form or another, in
courses followed by students of architecture, building, civil engineering,
structural engineering and allied disciplines. The topics discussed should
prove to be equally relevant to the requirements of both full-time and
part-time students engaged in study leading to the award of various
certificates, diplomas and degrees.

The choice and presentation of the subject matter reflects the author's
central philosophy that the qualitative aspects of the behavioural
characteristics of structures are of primary importance. Understanding
structural behaviour is the key to understanding structural calculations and
design methods. There should be no mystique and very little needs to be
taken for granted. The student should take every opportunity offered in
laboratory and other practical periods to examine, especially from a
qualitative point of view, the behaviour of various structural elements
under load. Careful examination of phenomena observed in the
laboratory, both expected and unexpected, will often provide a valuable
insight into the theoretical reasoning which attempts to explain the same
phenomena quantitatively. Lecturers and teachers should make liberal use
of classroom demonstrations and of impromptu experiments.

It is intended that this book be used as a direct teaching text, thus
obviating the need for single-sheet handouts and the copying out of

assignment exercises. The author has assumed, quite rightly, that lecturers and teachers are perfectly capable of providing detailed explanations and will be able to answer queries, help with difficulties, etc. The text therefore sets out to provide the essential theories, formulae and principles in as simple and straightforward a manner as is consistent with clarity and ease of reading. Extensive use is made of worked examples in the text, the solutions being accompanied by explanatory remarks. At the end of each chapter practice exercises are given to cover the whole of the work done in the chapter. Answers to these exercises are provided at the end of the book. In this, the first edition, it is expected that a number of inadvertent errors will come to light. If readers who discover numerical errors would kindly write to him, the author will be most appreciative.

R. Whitlow
Bristol 1972

Preface to Second Edition

In the eighteen or so years since I wrote the first edition of *Materials and Structures* very little has changed in the body of knowledge it represents. Most of the subject is basic theory and fundamental principles and, as such, is unalterable.

Nevertheless, there have been changes in both the student population and in the nature of the knowledge that they are required to learn. A technological revolution has occurred in the form of the development of computers, especially with respect to personal computers. In their general education, students will become 'computer literate' to a high degree and one of the tasks in further and higher education will be to enhance these skills. Also, the processes of analysis, design, testing and manufacture have themselves been revolutionized. As a result, the practical applications of the long-established disciplines entailed in structural mechanics and strength of materials have to be learned in different ways. Long-winded hand-calculation methods, the bane of engineering students for many years, are a thing of the past. However, the new era of split-second calculations also demands a deeper understanding of material and structural behaviour. Mechanical calculations and engineering judgement (based on expertise and experience) are separated and many problem-solving approaches must be redesigned. The methods of learning and teaching these subjects have changed dramatically: more examples can be worked through, but a keener and a more discerning understanding of underlying principles is essential.

In the second edition, therefore, some of the material involving lengthy hand-drawing or hand-calculation routines has been removed and more relevant material has been added. Frequent reference is made to using computer-aided learning techniques (see Section 1.14) and computer solutions to selected examples are given.

A number of readers have suggested ideas for strengthening or extending certain topic areas. This I have attempted to do, without widening the scope too much and making the book too large. The treatment of loading in Chapter 1 now reflects definitions and guidelines given in British Standards. The concept concerning limit states is introduced in Chapter 3 and appears also later in the text. A section has been added to Chapter 4 on principal axes and principal second moments of area, and the corresponding work on asymmetric bending in Chapter 6 strengthened. Also in Chapter 6, the sections on plastic bending theory and design have been rewritten. A section on Macaulay's method has been added to Chapter 7 to enable lecturers particularly to make use of computers to aid learning. The work on slender columns and struts in Chapter 8 has been extended to include consideration of recommended buckling formulae.

My aims and intentions remain unchanged: to assist both teachers and students in a study which, for me, continues to be absorbing and fascinating. I should like to thank all those readers who have kindly written to me with advice and suggestions (and corrections!) and I hope that this will continue. My thanks also go to my colleagues at Bristol Polytechnic for their help and support. Finally, I should like to thank my wife Marion for her unstinting help and understanding.

Roy Whitlow
Bristol 1990

1 Forces and reactions

Symbols

a = acceleration
d = lever arm of couple
E = equilibrant
g = acceleration due to gravity ($9 \cdot 81$ m/s^2)
H = horizontal force
l = lever arm of moment
M = moment of force
m = mass
N = direct force in a member
P = applied force
P_x = force component in the x-direction
P_y = force component in the y-direction
Q = applied force, shearing force
Q_x = force component in the x-direction
Q_y = force component in the y-direction
R = reaction force, resultant
R_x = resultant component in the x-direction
R_y = resultant component in the y-direction
T = torque or twisting moment, tension force
t = elapsed time
V = vertical force
v = velocity

W = applied force or load
$\bar{x}$ = distance to centroid or resultant
x = Cartesian co-ordinate, distance in the x-direction
y = Cartesian co-ordinate, distance in the y-direction
z = Cartesian co-ordinate, distance in the z-direction
θ = angle

1.1 Mass, force and gravity

Force is described in the *Oxford English Dictionary* as that 'measurable and determinable influence tending to cause motion of a body'. In order to cause a body to move or to cause a change in its movement, a FORCE must act upon that body. Sir Isaac Newton defined force quantitatively as the *product of the mass of the body and the rate of change of velocity* caused by the application of the force:

$$P = ma$$

where: P = applied force
$\quad\quad\ m$ = mass of the body
$\quad\quad\ a$ = acceleration caused by the application of the force

$$= \frac{v_2 - v_1}{t}$$

(where v_1 = initial velocity, v_2 = final velocity, t = time elapsing during the change).

The application of a force on a body will cause the body to accelerate. Similarly, if a body of mass m tends to be accelerated but is restrained from doing so, then the restraining force will be given by:

$$P = ma$$

Now all objects in the proximity of the Earth tend to be accelerated towards the centre of the Earth's mass due to gravitational attraction. At mean sea-level this acceleration, i.e. *the acceleration due to gravity*, has a magnitude of $9 \cdot 81$ m/s^2. Hence all objects resting on the surface of the Earth are held down or 'weighed down' by a force (often referred to as the *weight* of the object) which has a magnitude of $9 \cdot 81$ times the mass of the object.

The basis of constructional engineering is to devise structural systems by means of which a variety of objects may be supported and their gravity-forces transmitted to the crustal rocks and soils of the Earth.

1.2 Force is a vector

A SCALAR quantity may be defined by a single item of data, e.g. a number; whereas a VECTOR quantity requires at least *two*, sometimes *three*, items of data. A force has to be specified in terms of its *magnitude* and also in terms of its *direction*, therefore force is a vector quantity. For example, if a certain force is said to act at a particular point in a structure, it is necessary, in addition to knowing how *big* the force is, to know in which *direction* it is being applied. It could be vertical, horizontal, or even at some oblique angle; this must be taken into account in calculations involving the force.

Forces which lie in the same plane are said to be COPLANAR and are vector quantities capable of definition by the minimum of two items of data (i.e. magnitude and two-dimensional direction). Non-coplanar forces have to be related to three-dimensional space and therefore require *two* items of directional data in addition to magnitude.

It is usual to relate directional data to specific reference directions, such as horizontal or vertical, or the longitudinal axis of the member itself may be used (Fig. 1.1). In order to simplify the mathematics, the conventional Cartesian axes are employed; x and y for coplanar systems and x, y, z for three-dimensional space systems (Fig. 1.2).

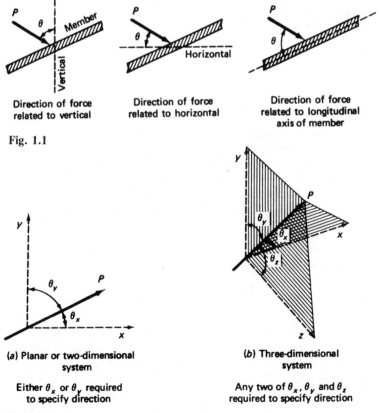

Direction of force related to vertical

Direction of force related to horizontal

Direction of force related to longitudinal axis of member

Fig. 1.1

(*a*) Planar or two-dimensional system

Either θ_x or θ_y required to specify direction

(*b*) Three-dimensional system

Any two of θ_x, θ_y and θ_z required to specify direction

Fig. 1.2

4

1.3   Vector components

In analysis and calculations it is sometimes convenient to consider the
effects of a force in a given direction other than the actual direction of
the force itself. In particular, the effects of forces along the Cartesian
axes are often required. The force-effects along these axes are called
VECTOR COMPONENTS. A single force may be fully specified by its
vector-components.

In Fig. 1.3 the force P has vector-components P_x and P_y,

where: $\quad P_x = P\cos\theta_x$ $\qquad\qquad\qquad$ [1.1a]

$\qquad\quad\; P_y = P\cos\theta_y = P\sin\theta_x$ $\qquad\qquad$ [1.1b]

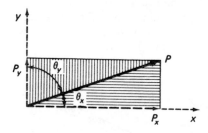

Fig. 1.3

1.4   Coplanar forces and resultants

Most of the work in this chapter will be related to systems of coplanar
forces. An important point to be considered is how to deal with several
coplanar forces, all of which pass through the same point; this is called a
system of CONCURRENT COPLANAR FORCES.

First, consider two concurrent forces, P and Q (Fig. 1.4). These two
forces may be replaced by a single force R which produces the same
effect. Now forces are vector quantities and, as such, cannot be added
together directly as in the case of scalar numbers. However, *vector-
components referred to the same axis* can be added directly giving the
vector-components of the resultant,

$$P_x + Q_x = R_x \qquad\qquad [1.2]$$

$$P_y + Q_y = R_y \qquad\qquad [1.3]$$

Since the addition is counterchangeable, i.e. $P + Q = Q + P$, a
parallelogram construction results (Fig. 1.5) which gives rise to what may
be termed the PARALLELOGRAM RULE. The PARALLELOGRAM
RULE states that *the resultant vector R of two vectors P and Q is the
diagonal of the parallelogram in which P and Q are the adjacent sides.*

When more than two coplanar forces comprise the system, the resultant
may be found by repeated applications of the parallelogram rule to pairs

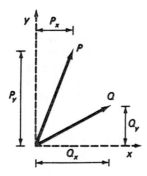

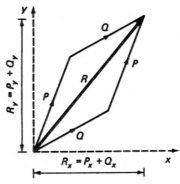

Fig. 1.4

Fig. 1.5

of forces. Solutions to problems involving coplanar forces may be easily found using a graphical method based on the parallelogram rule.

EXAMPLE 1.1 Obtain the resultant of the coplanar force system shown in Fig. 1.6: (a) using a graphical method; (b) using vector-components.

(a) The graphical solution is shown in Fig. 1.7.
By scaling off the length of the diagonal:
$$R = 435 \text{ kN (Answer)}$$

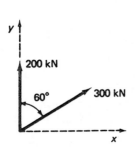

Fig. 1.6

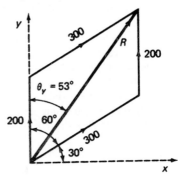

Fig. 1.7

(b) The reference axes are given in Fig. 1.7.
Then $R_x = 0 + 300 \cos 30°$
$= 260 \text{ kN}$
and $R_y = 200 + 300 \sin 30°$
$= 200 + 150$
$= 350 \text{ kN}$
from which $R = \sqrt{(260^2 + 350^2)}$
$= \sqrt{(67\ 600 + 122\ 500)}$
$= \sqrt{(190\ 100)}$
$= 436 \text{ kN (Answer)}$

1.5 Force triangles and polygons

If Fig. 1.5 is looked at again, it will be seen that the parallelogram consists of two triangles. The resultant R could have been found by drawing either one of these triangles. In the upper triangle, vector P is drawn first, followed by vector Q, which starts at the point where vector P finishes. In the lower triangle, Q is drawn first, followed by P, the same answer being obtained.

Consider now how the resultant of the coplanar force system shown in Fig. 1.8 may be found. Vector S is drawn first and then vector T is drawn, starting from the end of vector S (Fig. 1.9). The resultant is the straight line drawn from the end of vector T back to the start of vector S.

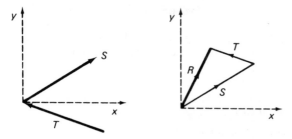

Fig. 1.8   **Fig. 1.9**

The TRIANGLE OF FORCES RULE states that *the resultant vector R of two vectors P and Q is the third side of a triangle in which vectors P and Q are the other two sides*. This is of course very similar to the parallelogram rule (Section 1.4), and may also be used to determine the vector-components of a given force. This will be evident if Fig. 1.3 is re-examined; the shaded areas are in effect triangles of forces.

Where the system consists of three or more forces, the *triangle* of forces is extended to become a *polygon*. Consider the system shown in Fig. 1.10. Vector P is drawn first, followed by vectors Q, S and T in

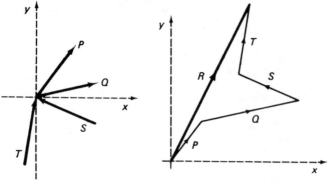

Fig. 1.10   **Fig. 1.11**

that order. It is important to note that the lines representing forces in the polygon are drawn in the *direction and sense of each force*. The resultant *R* is then drawn between the start of the first vector and the end of the last vector (Fig. 1.11). Any number of concurrent coplanar forces can be dealt with in this way. The same answer is obtained irrespective of the order in which the vectors are drawn. However, it is usual to work around the point in a clockwise direction. The student should understand that in complicated cases such methodical treatment is necessary if errors are to be avoided.

1.6 Moments of forces

The MOMENT of a force is defined as the *turning effect* of that force about a given point. In Fig. 1.12(*a*) the application of force *P* will cause *AB* to rotate about the hinge *A*. The product of the magnitude of the force and the radius of its potential rotation gives a measure of the turning effect and it is this product that is termed the MOMENT of the force. The moment of force *P* about point *A* is equal to *Pl*.

$$M_A = Pl$$

If the joint at *A* is fixed so that rotation is prevented, (Fig. 1.12(*b*)) the turning effect (moment) is constrained by a *reaction moment* in the support. This reaction moment will be equal in magnitude and opposite in rotational direction to M_A.

The units of a moment will be those of force × distance, e.g. newtons × millimetres (N mm) or kilonewtons × metres (kN m). These of course are also units of work, which should be understandable since the application of a force tends to cause work to be done.

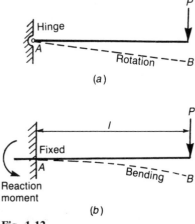

(a)

(b)

Fig. 1.12

1.7 Equilibrium and equilibrants

A body is said to be in a state of *equilibrium* when there is no tendency for it to be disturbed from its existing state of rest (or state of uniform motion). A system of forces acting on a body (such as a component of building structure) which is in equilibrium must therefore itself be in equilibrium and its resultant be zero.

If the equilibrium of a body is to be maintained, both of the following conditions must be satisfied:

1. The algebraic sum of the vector-components of all the *applied forces*, referred to any given direction, must be zero.
2. The algebraic sum of the *moments* of all the applied forces, about any given point and in any given plane, must be zero.

These two statements are often referred to as the LAWS OF STATIC EQUILIBRIUM. For coplanar systems they may be summarized as follows:

Sum of vertical forces or components = 0, i.e. $\Sigma (V)$ = 0
Sum of horizontal forces or components = 0, i.e. $\Sigma (H)$ = 0
Sum of moments about any point = 0, i.e. $\Sigma (M)$ = 0

The equilibrium of a body to which a single force is applied may be maintained by the application of a second force, which is equal in magnitude and direction, but opposite in sense, to the first force. This second force, since it restores equilibrium, is called the EQUILIBRANT. In Fig. 1.13 the body is in equilibrium only when:

(i) $P = Q$,
(ii) P and Q lie along the same directional line, and
(iii) Q is opposite in sense to P.

When a body is acted upon by two or more forces, the equilibrant has to be *equal and opposite** to the *resultant* of the system. In Fig. 1.14 forces P, Q and S act upon the body as shown and it is required to find the equilibrant. This may be done simply by drawing the force polygon,

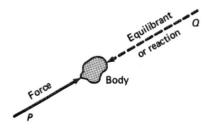

Fig. 1.13

* This is a shortened way of saying 'equal in magnitude and direction, and opposite in sense'.

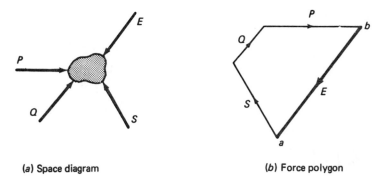

(a) Space diagram   (b) Force polygon

Fig. 1.14

from which the length *ab* gives the magnitude and direction of the resultant and therefore the magnitude and direction of the equilibrant. The sense of the direction of the equilibrant *E* is opposite to that of the resultant and is therefore from *b* towards *a*. The force polygon is said to be *closed* by the equilibrant; the final vector *ba* terminates at *a*, which was the starting point. A *closed* force polygon therefore represents a system of forces which is in equilibrium.

EXAMPLE 1.2 Determine the magnitude and direction of the equilibrant for the force system shown in Fig. 1.15.

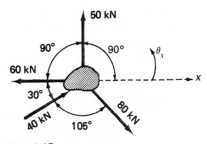

Fig. 1.15

The force polygon (Fig. 1.16) is drawn starting at point *a* with the 80 kN force and working around the body in a clockwise direction. The line representing the 50 kN force terminates at *b*; the equilibrant is then represented by the line *ba*. By scaling off the length of *ba* and measuring the angle with a protractor, the magnitude and direction of equilibrant may be found.

$$E = 34 \cdot 2 \text{ kN}$$
$$\theta = 66 \cdot 7°$$

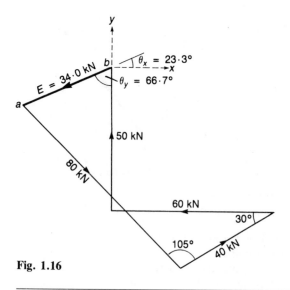

Fig. 1.16

1.8 Reactions

Bodies such as structural components are usually held in equilibrium by being secured to rigid fixing points; these are often other parts of the same structure. The tendency of the applied forces (loads) is to cause the members to move. The fixing points or supports react against this tendency and so the forces generated in the supports are called REACTIONS.

In general, a structural member has to be held or supported at at least two points (an exception to this is the cantilever, which will be explained later). Anyone who has tried 'balancing' a long pole or something similar will realize that, although only one support is theoretically necessary, two are required to give satisfactory stability. A further explanation is illustrated in Fig. 1.17. The body is shown in exactly balanced equilibrium in diagram (*a*), where *P* is an applied force (and could be the resultant of several forces) and *Q* is a single reaction from a rigid support. It is immediately apparent that a small movement will cause a misalignment of *P* and *Q*, producing an *out-of-balance* situation. Rotation will take place under the influence of the turning moment generated by the couple *P* × *d* (which also equals *Q* × *d*). In view of the variable nature of loading in buildings, a single point of support without other means of restraint does not provide for adequate stability.

In Fig. 1.18 a body is shown pinned to rigid supports at *A* and *B*. A measure of the forces generated in the supports is required. Since the body must be in equilibrium the *laws of static equilibrium* will apply:

$$\Sigma \ (V) \ = \ 0$$

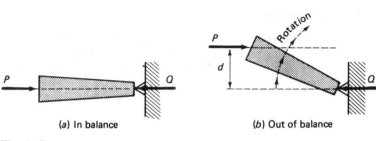

| (a) In balance | (b) Out of balance |

Fig. 1.17

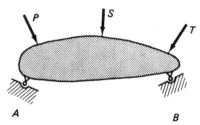

A   B

Fig. 1.18

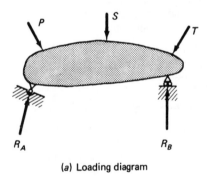

(a) Loading diagram

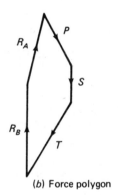

(b) Force polygon

Fig. 1.19

$$\Sigma\ (H)\ =\ 0$$
$$\Sigma\ (M)\ =\ 0$$

Thus, three equations could be established. There are, however, FOUR unknown items to be determined, i.e. the magnitude and direction of each of the two reactions. Mathematically therefore a solution is impossible unless either the magnitude or the direction of one or the other of the two reactions is known beforehand.

Let it be assumed that there is a horizontal roller at B (Fig. 1.19). In this case, the reaction at B will be purely vertical. The solution can now be found since only three unknown quantities remain. Numerical examples of this type of problem will be found in Chapter 2.

1.9 Parallel forces

Earlier in this chapter it was shown that two or more forces could be replaced by a single resultant. Consider then two *parallel* forces and their resultant (Fig. 1.20). *P* and *Q* are vertical parallel forces and from the *laws of static equilibrium* it can be established that:

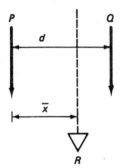

Fig. 1.20

(i)   the total vertical force remains constant and equals $P + Q$,
(ii) the total horizontal force remains constant and equals zero
(iii) the total moment about any point remains constant

In order to satisfy (i) and (ii), $R = P + Q$ and must be vertical.
In order to satisfy (iii), R must be positioned such that

$$Q \times d = R \times \bar{x}$$

The resultant of any number of parallel forces may be found in this way, as in the following example.

EXAMPLE 1.3 Determine the magnitude and position of the resultant of the system of parallel forces shown in Fig. 1.21.

The resultant must produce exactly the same effects as the system it replaces:

(i)   the total horizontal force remains constant and is equal to zero; then R must be purely vertical.

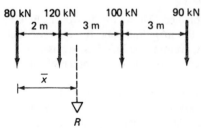

Fig. 1.21

(ii) the total vertical force remains constant; then

$$R = 80 + 120 + 100 + 90$$
$$= 390 \text{ kN}$$

(iii) the total turning moment about any point must remain constant; consider therefore the turning moment about the extreme left-hand load (80 kN):

$$\Sigma (M) = 80 \times 0 + 120 \times 2 + 100 \times 5 + 90 \times 8$$
$$= 0 + 240 + 500 + 720$$
$$= 1\,460 \text{ kNm}$$

The resultant R must also produce a turning moment of this magnitude about the same point; therefore:

$$R \times \bar{x} = 1\,140 \text{ kNm}$$

Hence
$$\bar{x} = \frac{1\,460}{R}$$

$$= \frac{1\,460}{390} = 3 \cdot 74 \text{ m (Answer)}$$

(*Note*: It will be seen that the lever arm of the 80 kN force is zero and therefore the moment about this point due to this force is zero.)

1.10 Beam reactions

If the forces shown in Fig. 1.21 were applied to a horizontal beam, then theoretically the beam could be held in equilibrium by the application of a single force (reaction) which is equal and opposite to the resultant R (Fig. 1.22). However, as explained in Section 1.7, *two* reactions are required to ensure the necessary stability and a more likely arrangement is that shown in Fig. 1.23. The reactions R_A and R_B must both be vertical, since there is no horizontal force component. Furthermore, in order to satisfy the condition of static equilibrium that the sum of the vertical forces must be zero, the sum $R_A + R_B$ must be equal to the sum of the downward-acting forces,

$$R_A + R_B = 390 \text{ kN}$$

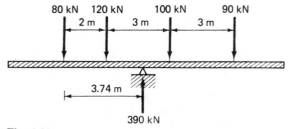

Fig. 1.22

14

The magnitude of the reactions may now be found by the application of the third law of static equilibrium, *i.e. the algebraic sum of the moments of the forces about any point must be zero.* This is demonstrated in the following example.

EXAMPLE 1.4   Calculate the magnitude of the reactions in the case of the loaded beam shown in Fig. 1.23.

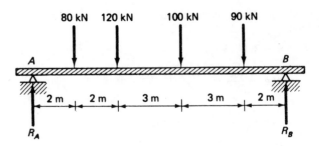

Fig. 1.23

The sum of the moments about A equals zero.

Then $\Sigma M_A = 0 = 80 \times 2 + 120 \times 4 + 100 \times 7 + 90 \times 10 - R_B \times 12$

Therefore $R_B = \dfrac{160 + 480 + 700 + 900}{12}$

$= \dfrac{2\ 240}{12}$

$= 186 \cdot 7$ kN

Similarly, by considering moments about B, the other reaction R_A is found.

Then $\Sigma M_B = 0 = R_A \times 12 - 80 \times 10 - 120 \times 8 - 100 \times 5 - 90 \times 2$

Therefore $R_A = \dfrac{800 + 960 + 500 + 180}{12}$

$= \dfrac{2\ 440}{12}$

$= 203 \cdot 3$ kN

The two values so obtained can now be checked since $\Sigma (V) = 0$

$R_A + R_B =$ total downward load $= 390$ kN

Check: $186 \cdot 7 + 203 \cdot 3 = 390$ kN

It may be argued that, having obtained R_B, R_A could be found by subtracting the value of R_B from the total downward load, thus:

$$R_A = 390 - R_B$$

$$= 390 - 186 \cdot 7$$
$$= 203 \cdot 3 \text{ kN}$$

This does, of course, give the correct solution, but only if the value obtained for R_B is correct. The method involving taking moments about each reaction point (support) in turn is to be preferred, since the values obtained for R_A and R_B in this way can be checked by the application of the $\Sigma (V) = 0$ rule. Hence, if an error has been made, it will be discovered and corrected before the reaction values are used in further calculations.

1.11 Loading systems

The basic function of a structural member is to carry the applied loads and transmit the effects of these to other parts of the structure. Before any of the various *load-effects* can be considered, however, the applied loads must be rationalized into a number of ordered systems. Irregular loading cannot be dealt with, but even the most irregular loads may be reduced to a number of regular and easily definable systems. These systems have to be capable of definition in mathematical terms without departing too far from physical reality. For example, the application of a single force may be considered to occur at a point which theoretically has zero area. This, of course, is physically impossible. There must be a finite area of contact — even the point of a needle has a finite (measurable) area, albeit very small! But from a mathematical standpoint such a rationalization is not only possible and justifiable, but it can be also extremely convenient, and, providing it is not misused, errors which are more than negligible will not be incurred.

Table 1.1 shows the five most common systems encountered in general use. Any load that is applied to a relatively small area may be considered as a CONCENTRATED LOAD. Loads which are continuous and spread along the longitudinal axis of a member are treated as DISTRIBUTED LOADS, i.e. distributed along the length of the member and measured in terms of *force per unit length*. In the case of slabs and sheets, the term *distributed* load infers a distribution over the *area* of the member, i.e. measured in terms of *force per unit area*.

One of the first tasks facing the designer of a building is the rationalization of all the applied loads and reactions into a number of these loading systems. The choice of system depends largely upon the particular design calculation that is to be carried out. For instance, the reaction force at the end of a beam may be considered as a *concentrated* load when designing the beam itself, and then treated as a *uniformly distributed* load when designing the support (e.g. bracket, shoe, padstone).

1.12 Effects of loading

Having transformed the sometimes complex and sometimes irregular applied forces into definable load systems, the designer now has to

Table 1.1 Common loading systems

System	Indicated thus	Examples
Concentrated load		Reaction load from end of a beam Column load Loads at nodes in frameworks
Line load	 *w* per unit length When viewed from *A* would appear as a concentrated load.	Partition or wall
Uniformly distributed load	 *w* per unit length *w* per unit length *w* per unit length	Load due to self-mass of beam Reaction load from the edge of a slab Snow loads Wind loads
Distributed load with linear variation	 Variation from *w* per unit length to zero *w* per unit length	Loads against piling or retaining walls or the sides of liquid-retaining structures
Triangular load	 *w* per unit length	Load on lintel supporting brickwork over a door or window opening

consider how the loads will be transmitted through the structure. Loads are not transmitted as such, but as LOAD-EFFECTS. For example, when a wire is streched the load is transmitted from one end of the wire to the other by what is called *tension* in the fibres of the wire. Tension is a *load-effect*. Structural members have to be designed according to the particular load-effects that they will be required to sustain. It is therefore necessary to consider these so-called *load-effects* early in the design process.

Figure 1.24 shows a general structural member subject to some combination of load systems (not shown). It is usual to orientate the Cartesian $z-z$ axis along the longitudinal axis of the member and for the $x-x$ and $y-y$ axes to be the perpendicular cross-sectional axes respectively.

A PRIMARY LOAD-EFFECT is defined as being a force or moment which has a specific orientation with respect to these three axes. There are four *primary load-effects* and these are defined in Table 1.2 in conjunction with Fig. 1.24. *Secondary load-effects*, such as strain, deflection, etc., are derived from the four primary load-effects. In Fig. 1.24, the primary load-effects are related to the general member; in fact *six* are shown in the figure since the shearing force and bending moment can be related to either the horizontal $(x-x)$ or the vertical $(y-y)$ axes.

Any single load or combination of several loads can give rise to one or more of these primary load-effects. In most cases a member will be designed basically to sustain one load-effect, usually the one producing

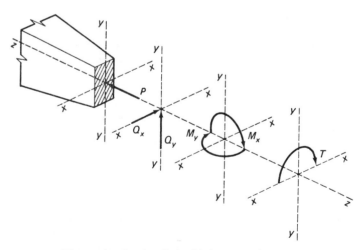

Diagram showing the relationship between primary
load-effect and the orthogonal axes of a general member

Fig. 1.24

Table 1.2 Definition of primary load-effects

Load-effect	Symbol	Definition
Direct force	P N W	Acts parallel to the longitudinal (z-z) axis
Shearing force	S Q	Acts normal to the longitudinal axis
Bending moment	M	Turning moment acting in a plane which contains the longitudinal axis
Torque	T	Turning moment acting in a plane normal to the longitudinal axis, i.e. twisting moment

the greatest effect; the others, less important maybe, must still be taken into account. The name given to the *type* of member will generally reflect its major load-effect function, as well as being related to its spatial form. Some examples are given in Table 1.3.

In the column headed 'spatial form', the shape of the member is indicated by the ratio of the dimensions in the x, y and z directions. Where two of these dimensions are approximately equal, they are given together. For example, the x and y dimensions of a beam are approximately equal compared with the z dimension (length) which may be from 10 to 50 times greater.

The structural or load-effect function of a member influences not only its size and shape, but also to some extent the choice of material from which it is made. For instance, concrete has little or no strength in tension; it could therefore hardly be used by itself to form a *tie* whose main structural function would be to resist tensile load-effects.

1.13 Load types and sources

All identifiable loads of significant magnitude and effect must be considered separately and in such reasonable combinations that they may be predicted to produce critical conditions. The loading effects on single elements of structure and on the structure as a whole must be considered. In addition to estimating the magnitude of loads their permanence or variability must be considered. The most common types and sources of loading are described below.

Table 1.3 Definitions of some common structural members

Type of member	Primary load-effect	Deformation mode	Spatial form	Most common orientation*
Beam	Bending moment	Flexural	$\dfrac{z}{x,y} = 10-50$	z horizontal
Strut	Direct force — compressive	Shortening or buckling	$\dfrac{z}{x,y} = 5-60$	Not specific
Column or stanchion	Direct force — compressive	Shortening or buckling	$\dfrac{z}{x,y} = 5-40$	z vertical
Tie	Direct force — tensile	Lengthening	$\dfrac{z}{x,y} = 10-100$	Not specific
Tendon or rod or bar	Direct force — tensile	Lengthening	$\dfrac{z}{x,y} = > 100$	z parallel to z of associate member
Slab	Bending moment — in either one or two directions	Flexural	$\dfrac{z,x}{y} = 10-150$	zx horizontal
Sheet	Direct force — tensile in two directions	Increase in length and breadth	$\dfrac{z,x}{y} = > 150$	Not specific
Rivet or bolt	Shearing force — sometimes tension	Shear — sometimes lengthening	Not relevant	According to the connection in which it is used

*The orientation of a member is not always an indication of its function, for example, vertical slabs and beams are quite common. The information given in this column is only meant to act as a guide to the *most common* orientation with which the member is associated.

Dead loads

Dead loads are permanent and constant throughout the life of the building and are related to the mass of the structure, its fabric and permanent fixtures, e.g. self-mass of members, foundations, finishes, claddings, etc.

Live loads

Live loads may vary intermittently or regularly and are related to the use of the structure and to environmental conditions, e.g. loads due to wind, snow, traffic, cranes, machinery, temperature changes, etc. Some live loads may vary from a maximum down to zero, e.g. snow loads; in other cases a change in sense or in direction may occur, e.g. wind loads. In designing a structure it is necessary to consider how the most critical combination of dead and live loads may arise. Not all of the live loads will necessarily reach their maximum effective values at the same time. In some cases, the live loading will produce an effect which is opposite in sense to that of the dead load. A typical example of such a reversal may be found in flat or low-pitched roofs where the dead loads act downward, while the wind loads produce an upward-acting load. The designer often finds that some members have to be designed as TIES for the *dead load only* condition, and also as STRUTS for the *dead load + live load* condition.

Imposed loads

Imposed loads are live loads related to the use and function of the building and take account of such things as the number of people that may use a room, the potential storage of materials, furnishings and temporary fittings, etc. For example, greater imposed loads have to be allowed for storage accommodation, such as in libraries and warehouses, than for offices and classrooms. The potential weight of unseated people in assembly rooms and dance halls is greater than that in a cinema or theatre where people will be seated. When a change of use is proposed for a room or a building, the old and new imposed load conditions must be compared and checked to avoid overloading. Some typical values of imposed loads are given in Table 1.4.

Snow loads

The magnitude and duration of snow loading is strongly governed by latitude, altitude and local weather conditions. The pitch of roof surfaces is also greatly significant: the steeper the slope the more snow will slide off. It is noticeable that in countries with high snow-fall weather the roof

Table 1.4 Typical imposed floor loads

Function use	Uniform distributed load (kN/m^2)
Bedrooms	1·5
Domestic living rooms	1·5
Classrooms	3·0
Assembly halls, gymnasia	5·0
General offices	2·5
Filing and storage rooms	5·0
Library reading rooms	2·5
Book or paper storage	6−10

Table 1.5 Examples of snow loading on pitched roofs

Location examples	Site snow load (kN/m^2)			
	Roof pitch			
	$\leq 30°$	$40°$	$50°$	$\geq 60°$
Aberystwyth, Plymouth	0·24	0·05	0·08	0
Bristol, London, Oban	0·32	0·21	0·11	0
Belfast, Carlisle,				
Manchester, Norwich	0·40	0·27	0·13	0
Hereford, Wrexham, York	0·48	0·32	0·16	0
Inverness, Sheffield	0·56	0·37	0·19	0
Newcastle-upon-Tyne, Wick	0·64	0·43	0·21	0
Aviemore, Lairg	0·80	0·53	0·27	0

Based upon data given in BS 6399: Part 3: 1988 *Loading for buildings: Code of practice for roof loads*

pitches are steeper. Formulae and charts are given for the calculation of snow loading in the UK in BS 6399: Part 3: 1988 *Loading for buildings: Code of practice for roof loads*. Some typical values are given in Table 1.5.

Wind loads

The intensity of wind pressure depends not only on wind velocity, but also on gusting effects, the degree of exposure or shelter and the height of the building. Exposure is greater on coasts and high ground; shelter is afforded by neighbouring buildings, trees, hillsides, etc. Positive (inward) pressure will normally be induced on windward faces and walls, with suction on leeward faces. Flat or low-pitched ($<26°$) roof surfaces will be subject to suction (uplift), with maximum intensity occurring near leeward corners, ridges, hips, etc. Lightweight roofs are particularly susceptible to gusting strong winds.

Settlement loads

If settlement due to ground compression or subsidence is greater under some parts of a building than others, the resulting distortion will produce additional stresses in the structure. If the structure is flexible or can articulate via joints, these stresses will be small. In stiff rigid structures differential settlement stresses may be considerable, and in brittle building fabrics, such as masonry, brickwork and unreinforced concrete, cracking and even collapse may ensue. It is usual to site separation joints between parts of a building where adjacent loading magnitudes are significantly different.

Temperature variation loads

All materials expand or contract with changes in temperature. A continuously exposed concrete member may undergo seasonal changes in length of 4 mm for every 10 m of length; a 10 m length of steel similarly might expand or contract 5 mm. In brickwork and masonry, expansion joints should be provided at 10 m intervals. Rigidly held steel members may be subject to substantial increase (or decreases) in stress as ambient temperatures fluctuate; an important consideration in the design, for example, of heat-generation or cold-storage facilities.

Shrinkage loads

Some building materials expand and contract with changes in moisture content or when chemical reactions are taking place (such as the setting of concrete and the crystallization of salts). In timber, moisture movements up to 5 per cent may occur tangential to the grain, i.e. 50 mm in 1 m. In concrete and mortar, drying *shrinkage* of 2−8 mm in 10 m may occur. Expansion joints are usually required around brick in-fill panels in concrete frames, otherwise panels may buckle due to the combination of concrete shrinkage and brickwork expansion.

Dynamic loads

Loads which vary periodically or systematically, including impact, oscillating and aerodynamic loads, can produce greatly increased levels of stress and distortion. In the treatment of dynamic loading in design it is usual to compute the *increase in static loading* caused by the dynamic system, since structural design processes essentially use static-load data.

1.14 Using computers to aid learning

Personal computers (PCs), as well as their main-frame counterparts, are now commonplace in universities, polytechnics and colleges. Students are usually placed in front of a computer and given their first 'driving lessons' in the first few weeks of their first term. Thereafter, they will be expected to make regular and frequent use of the machines that are available. As well as there being a wide variety of types of hardware, i.e. main-frames, PCs, data loggers, process controllers, materials testing machines, etc., a wide array of software packages are available. Most use will probably be made of PCs running packages for teaching programmes, data processing, word processing and drawing. In some courses, a limited amount of work may be done in writing high-level language (e.g. Basic, Pascal, Fortran, C, etc.) programmes. More advanced students will learn to use more sophisticated packages, e.g., in laboratory testing or process control routines.

At all levels computers are an invaluable aid to teaching and learning. Strict discipline in data input and operation (and especially when writing programmes) is essential, instilling similar traits in the user. Students are able to work at their own pace, with considerable opportunity for rerunning work and self-testing. The teacher can be selective with regard to what the learner is handling at any given time. Complex and complicated calculations can be carried out very quickly. Repetitious routines, which may obscure important principles and learning goals, can be hidden. Many more exercises and examples can be worked through in the time available. Clearly, computers are powerful learning aids.

Many of the topics in building engineering subjects areas (e.g. strength of materials, structural mechanics, structural analysis, structural design, soil mechanics, hydraulics, surveying, materials science, services design, etc.) are ideally suited to computer-aided learning. In the context of this book, *three* main types of software package are available.

1. *Topic-specific learning programmes.* These are designed to teach or give practice within a specific topic area through student/machine interaction. Ideas and principles are displayed and the student prompted for responses. In some, testing and grading routines are included.

2. *Professional analysis and design packages.* These are usually comprehensive packages of interactive programmes aimed at professional analysts and designers (e.g. analysis of rigid frameworks; design of steelwork, reinforced concrete, etc.). The controlled and guided use of such packages can be most instructional, even for first-year students.

3. *Spreadsheets.* Many spreadsheet packages are obtainable for many different types of machine, the most widely used being *Lotus* 1—2—3 (R).

The principles are the same for each, but the level of sophistication varies considerably. Students should be introduced to spreadsheets at the very start of their course. The great advantage of using a spreadsheet is the seemingly limitless capacity afforded for calculation and data handling. If a routine can be followed by hand, it can usually be done also using a spreadsheet, but done very much more quickly.

A number of programme packages mentioned above have been developed at Bristol Polytechnic in collaboration with QSE Limited. Lecturers and others with an interest may obtain catalogues and further information from: Dr D. M. Brohn, QSE Limited, Cambrian House, 51 Broad Street, Chipping Sodbury, Bristol BS17 6AD: Tel: (0454) 319104.

In this text, illustrations are included of the application of spreadsheet methods to the solution of numerical examples. A combined book and floppy disc publication is in preparation by the author: further details may be obtained upon request. It is assumed that readers will understand spreadsheet methods sufficiently to follow the examples given and possibly to devise their own routines on their own computers.

Spreadsheet examples on forces and reactions

Figure 1.25 shows the vdu screen print of a spreadsheet analysis of Example 1.2 (refer to Fig. 1.15 for the data). The magnitude of each

```
RESCON1     Calculation of resultant of concurrent coplanar forces

- - - - - - - - - - - - - - - - - - - - - - - - - - - - - - - - - - - - - - - - - - - -

Force No  Magnitude  Theta x    H          V
          (kN)       (deg)      (kN)       (kN)

- - - - - - - - - - - - - - - - - - - - - - - - - - - - - - - - - - - - - - - - - - - -

        1      50        90         0         50
        2      60       180       -60          0
        3      40        30   34.64102         20
        4      80       315   56.56854    -56.5685
        5                            0          0
        6                            0          0
        7                            0          0
        8                            0          0
        9                            0          0

- - - - - - - - - - - - - - - - - - - - - - - - - - - - - - - - - - - - - - - - - - - -

Resultant      33.98     23.29     31.21      13.43

- - - - - - - - - - - - - - - - - - - - - - - - - - - - - - - - - - - - - - - - - - - -

Positive -  tension     acw       right     up
```

Fig. 1.25

```
RESPAR1     Calculation of resultant of parallel coplanar forces
            EXAMPLE 1.3

- - - - - - - - - - - - - - - - - - - - - - - - - - - - - - - - - - - - - - - - - - - -

Force No  Magnitude  lever arm  Moment
          (kN)       (m)        (kNm)

- - - - - - - - - - - - - - - - - - - - - - - - - - - - - - - - - - - - - - - - - - - -

        1      80        0          0
        2     120        2        240
        3     100        5        500
        4      90        8        720
        5                           0
        6                           0
        7                           0
        8                           0
        9                           0

- - - - - - - - - - - - - - - - - - - - - - - - - - - - - - - - - - - - - - - - - - - -

Resultant     390.00     3.74    1460.00

- - - - - - - - - - - - - - - - - - - - - - - - - - - - - - - - - - - - - - - - - - - -

Positive -   down       cw         cw
```

Fig. 1.26

force is entered with the positive direction being away from the point, i.e. tension. The angle (θ_x) is measured anticlockwise (acw) from the conventionally orientated Cartesian axis, i.e. positive x to the right. The other data is recalculated as each force or angle is entered. Students should now 'invent' their own data and enter it, carefully observing the results.

Figures 1.26 and 1.27 show the screen print of spreadsheet analyses of Examples 1.3 and 1.4 respectively. These illustrate how different, but similar, types of calculation are easily accommodated with only minor modifications to the spreadsheet. Fig. 1.28 shows a printout of the analysis of Exercise 1.40, a further extension of the same basic layout.

```
REACTN1   Calculation of beam reactions: vertical forces
          Horizontal simply supported beam      EXAMPLE 1.4
-----------------------------------------------------------
NOTE: One reaction is assumed to be free to move horizontally
-----------------------------------------------------------
Reaction location.          A =      0  m
distance from LH end:       B =     12  m
-----------------------------------------------------------
Force    Magnitude  Distance      Mb
         (kN)       from LH end  (kNm)
-----------------------------------------
   1        80          2         160
   2       120          4         480
   3       100          7         700    Ra =  203.33  kN
   4        90         10         900
   5                               0     Rb =  186.67  kN
   6                               0
   7                               0
   8                               0
   9                               0
-----------------------------------------
          390                    2240
-----------------------------------------
```

Fig. 1.27

```
REACTN2 Calculation of two reactions to coplanar forces
        EXERCISE 1.40
--------------------------------------------------------------------
Reaction   Coordinates              Type (Insert 1)
           x(horiz)  y(vert)   Pin  Free horFree vert
--------------------------------------------------------------------
   A          0         2       0      0       1
   B          0         0       1      0       0
--------------------------------------------------------------------
Force  Magnitude  Angle  Distance from B     Fh        Fv       Mb
       (kN)       (deg)  x(m)    y(m)        (kN)      (kN)     (kNm)
--------------------------------------------------------------------
  1      10        225    0      4.4    -7.07107  -7.07107  -31.1127
  2      10        180    0      3.2     -10       0         -32
  3                                0         0        0
  4                                0         0        0
  5                                0         0        0
  6                                0         0        0
  7                                0         0        0
  8                                0         0        0
  9                                0         0        0
--------------------------------------------------------------------
                                      -17.07    -7.07    -63.11
--------------------------------------------------------------------
  Ha =   31.56    Va =   .00
  Hb =  -14.49    Vb =  7.07     Rb =  16.12  Theta x = 153.98
--------------------------------------------------------------------
Positive  right            up            tension            acw
```

Fig. 1.28

26

Note that, in these examples, a range printout has been used, omitting the column and row identifiers.

WORKED EXAMPLES USING GRAPHICAL METHODS

EXAMPLE 1.5 The 8 kN load shown in Fig. 1.29(a) is supported by two wires AC and BC. Determine the tension force in each wire.

Let the forces in AC and BC be F_{AC} and F_{BC} respectively. The three forces concurrent at C are then as shown in diagram (*b*). The force diagram is shown in (*c*) and from this the required values may be scaled off.

$$F_{AC} = 4 \cdot 0 \text{ kN and } F_{BC} = 6 \cdot 9 \text{ kN (Answer)}$$

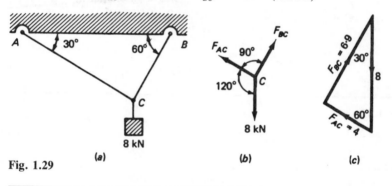

Fig. 1.29

(*a*)   (*b*)   (*c*)

EXAMPLE 1.6 Determine the magnitude of the force P required to maintain equilibrium in the system shown in Fig. 1.30(a).

Let the tension in BA equal T N.
Then the forces concurrent at A are as shown in diagram (*b*) and the force diagram is as shown in (*c*).

Scaling off from the force diagram: $P = 5 \cdot 4$ N (Answer)

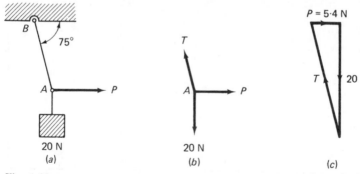

(*a*)   (*b*)   (*c*)

Fig. 1.30

EXAMPLE 1.7 Determine the total turning moment about point O of the two-force system shown in Fig. 1.31.

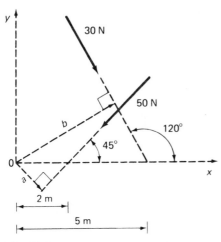

Fig. 1.31

The total moment about O (M_o) will be the algebraic sum of the moments about O of the two forces.

From a scale drawing, the lever arms a and b are determined:

$$a = 1\cdot40 \text{ m}, \ b = 4\cdot33 \text{ m}$$

Then
$$M_o = 50\,a + 30\,b$$
$$= 70 + 130$$
$$= 200 \text{ N m (Answer)}$$

WORKED EXAMPLES INVOLVING DISTRIBUTED LOADS

EXAMPLE 1.8 Determine the resultant of the coplanar system of uniformly distributed loads shown in Fig.1.32(a).

This loading system may be considered to consist of two uniformly distributed loads as shown in Fig. 1.32(b).

Rule for the resultant of a uniformly distributed load
THE RESULTANT OF A UNIFORMLY DISTRIBUTED LOAD IS EQUAL TO THE LOAD PER UNIT LENGTH × THE LENGTH OF THE LOAD, AND MAY BE CONSIDERED AS A CONCENTRATED LOAD, THE LINE OF ACTION OF WHICH PASSES THROUGH THE CENTRE OF GRAVITY OF THE DISTRIBUTED LOAD.

For convenience and to avoid using the word resultant in too many places, the resultant of a distributed load may be referred to as the EQUIVALENT CONCENTRATED LOAD (*ECL*).

The system in this example may thus be reduced to two *equivalent concentrated*

28

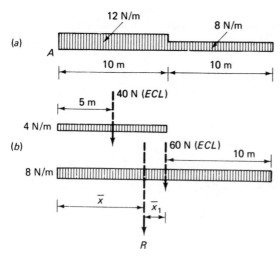

Fig. 1.32

loads of 40 N and 160 N; the resultant of these two will be the resultant of the complete system.

$$R = 40 + 160 = 200 \text{ N (Answer)}$$

To find the position of R, moments are taken about point A:

$$R \times \bar{x} = 40 \times 5 + 160 \times 10$$
$$= 200 + 1\,600$$
$$= 1\,800 \text{ Nm}$$

Therefore
$$\bar{x} = \frac{1\,800}{R}$$

$$= \frac{1\,800}{200}$$

$$= 9 \text{ m (from } A\text{) (Answer)}$$

A slightly shorter calculation results if moments are taken about one of the *ECLs*, for example the 40 N *ECL*:

$$R \times \bar{x} = 40 \times 0 + 160 \times 5$$
$$= 800 \text{ Nm}$$

Therefore
$$\bar{x} = \frac{800}{R}$$

$$= \frac{800}{200}$$

$$= 4 \cdot 0 \text{ m (from the 40 N } ECL\text{)}$$

Hence the position of R is $9 \cdot 0$ m from A.

EXAMPLE 1.9 Determine the resultant of the system of coplanar distributed loads shown in Fig. 1.33(a).

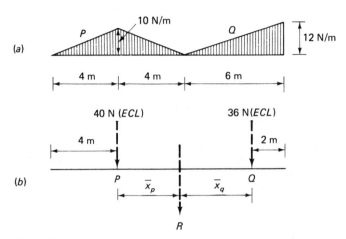

Fig. 1.33

Firstly, the *equivalent concentrated load* for each distributed load is calculated. The distributed loads are analogous to triangular prisms of equally distributed mass of which the resultant or equivalent concentrated load is given by the area of the triangle and acts through the centre of this area.

The following tabulated calculation will serve to explain the procedure:

Load	Base of triangle (m)	Height of triangle (N/m)	Area of triangle	=	Resultant or *ECL* (N)
P	8	10	$\dfrac{8 \times 10}{2}$	=	40
Q	6	12	$\dfrac{6 \times 12}{2}$	=	36

The system therefore reduces to two *ECLs* as shown in Fig. 1.34(b).

Then $$R = 40 + 36 = 76 \text{ N (Answer)}$$

To find the position of R, moments are taken about point P:

$$R \times \bar{x}_P = 36 \times 8$$

Therefore $$\bar{x}_P = \frac{36 \times 8}{76} = 3 \cdot 8 \text{ m (Answer)}$$

Or alternatively, taking moments about Q:

$$R \times \bar{x}_Q = 40 \times 8$$

Therefore $\qquad \bar{x}_Q = \dfrac{40 \times 8}{76} = 4 \cdot 2 \text{ m (Answer)}$

EXERCISES CHAPTER 1

1. Explain the basis for the quantitative measurement of force. Determine the gravity force exerted on a mass of 200 kg. ($g = 9 \cdot 81$ m/s^2.)

2. A set of solid brass weights of diameter 40 mm is to be made up consisting of the following denominations: 10 N, 5 N, 2 N and 1 N. Determine the thickness required, for each denomination. (Density of brass $= 8\,330$ kg/m^3.)

3. Determine the gravity force exerted by 800 litres of water. (Density of water $= 1 \cdot 00$ Mg/m^3; $g = 9 \cdot 81$ m/s^2.)

4–9. Obtain the resultants (magnitude and inclination to x-axis) of the coplanar force systems shown in Figures 1.34–1.39 using both the graphical method and the method of vector components.

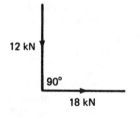

Fig. 1.34

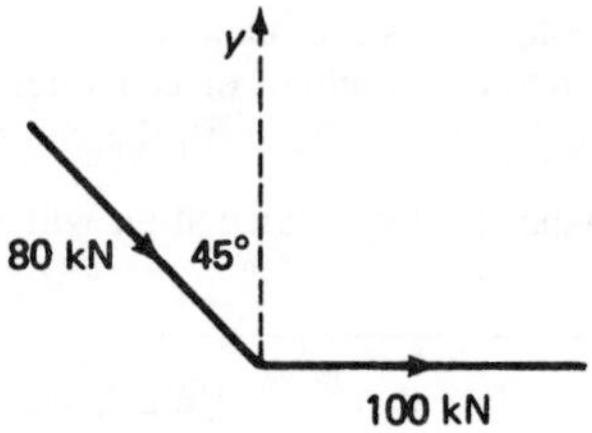

Fig. 1.35

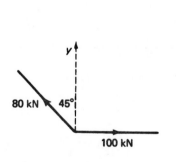

Fig. 1.36

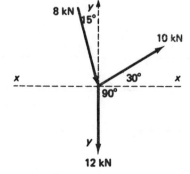

Fig. 1.37

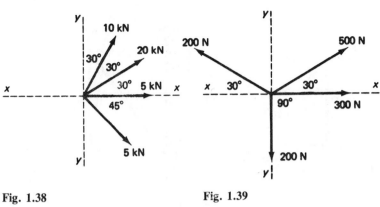

Fig. 1.38   Fig. 1.39

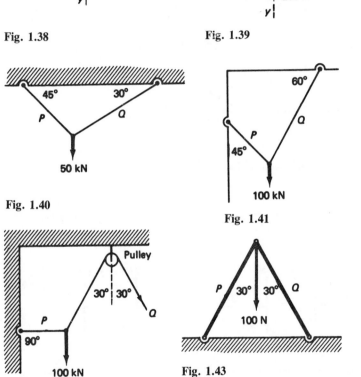

Fig. 1.40

Fig. 1.41

Fig. 1.42

Fig. 1.43

10−12. Determine the tension force acting in each of the two wires shown in Figures 1.40−1.42.

13. Determine the magnitude of the force acting in the struts P and Q (Fig. 1.43).

14−15. Determine the magnitude of the forces acting in the wires P and the struts Q shown in the Figures 1.44−1.45.

32

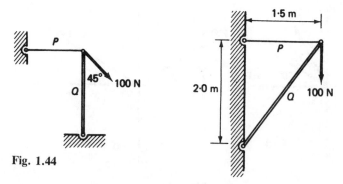

Fig. 1.44

Fig. 1.45

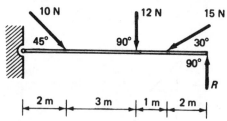

Fig. 1.46

16. Fig. 1.46 shows a bar which is held in equilibrium by a system of four forces. Determine the magnitude of force R.

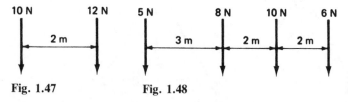

Fig. 1.47   Fig. 1.48

17–18. Determine the magnitude and position of the resultant force for each of the parallel coplanar force systems shown in Figures 1.47–1.48.

19–24. Figures 1.49–1.54 show a series of coplanar systems involving distributed loads. In each case determine the resultant force or couple.

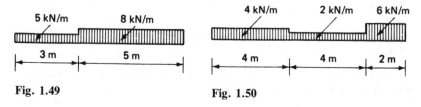

Fig. 1.49   Fig. 1.50

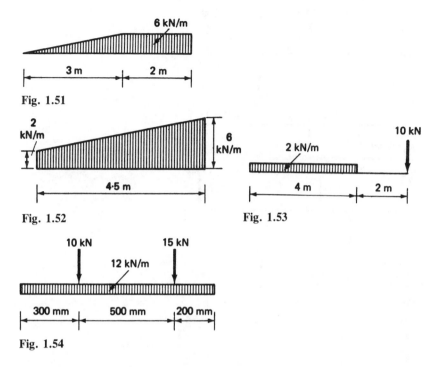

Fig. 1.51

Fig. 1.52

Fig. 1.53

Fig. 1.54

25–35. Figures 1.55–1.65 show a series of beams upon which the loads and reactions act in the same plane. Determine the magnitude of the reactions R_A and R_B in each case.

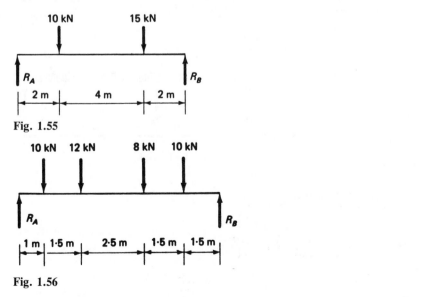

Fig. 1.55

Fig. 1.56

34

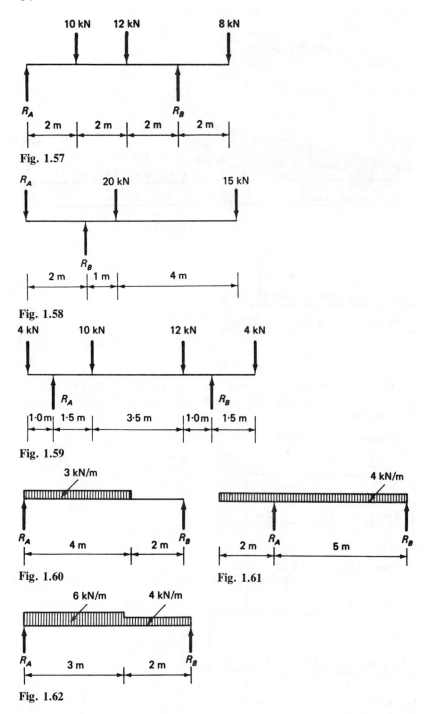

Fig. 1.57

Fig. 1.58

Fig. 1.59

Fig. 1.60

Fig. 1.61

Fig. 1.62

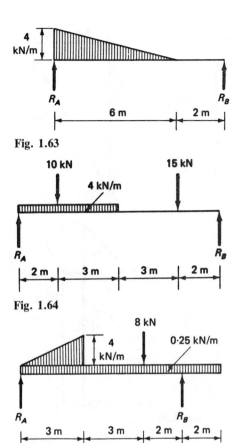

Fig. 1.63

Fig. 1.64

Fig. 1.65

36–42. Figures 1.66–1.72 show a series of structures, each of which is supported at two points A and B. It may be assumed that the members, loads and reactions are all in the same plane. Determine the reactions R_A and R_B in each case. Where the direction of the reaction is not given, this must be deduced and stated in the answer.

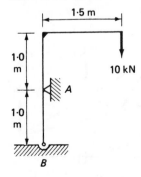

Fig. 1.66

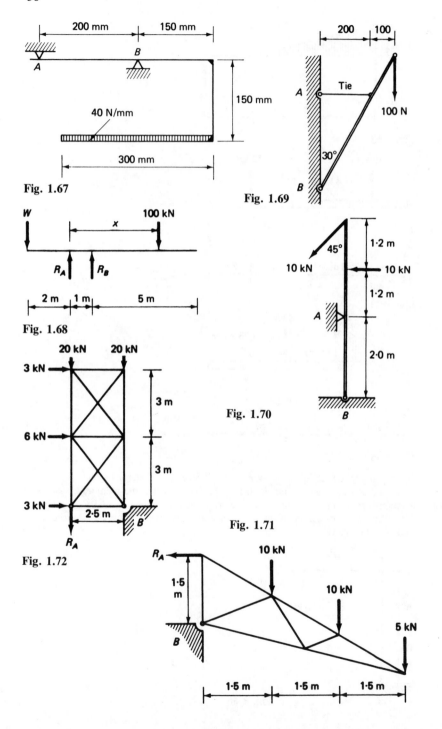

Fig. 1.67

Fig. 1.69

Fig. 1.68

Fig. 1.70

Fig. 1.71

Fig. 1.72

2 Forces in structural frameworks

Symbols

F = direct force in a member
H = horizontal force, component or reaction
j = number of joints in a framework
M = moment of force
m = number of members in a framework
R = reaction force
s = number of external support forces
V = vertical force, component or reaction
W = applied load
θ = angle

2.1 Introduction

Structural frameworks range from the very simple to the exceedingly complex; any system of three or more straight or curved members capable of transmitting structural loads may be described as a structural framework. For the purposes of analysis some simplification is necessary so as to avoid, wherever possible, lengthy and complicated calculations. This is achieved by placing frameworks into categories, where they obey mathematical rules when appropriate assumptions are made. The two main categories are as follows:

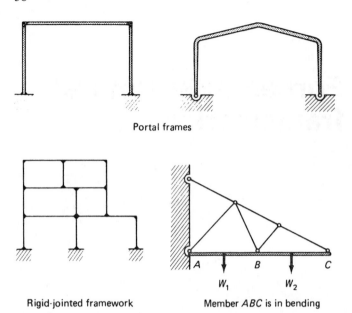

Portal frames

Rigid-jointed framework   Member *ABC* is in bending

Fig. 2.1

1. *Statically determinate frameworks*, in which the magnitude and direction of the forces acting in the members may be determined by the direct application of the *laws of static equilibrium* (Section 1.7).
2. *Statically indeterminate frameworks*, in which the number of unknown quantities exceeds the number of quantities determinable using the laws of static equilibrium.

The consideration of *statically indeterminate frameworks*, such as portal frames, rigid-jointed frames and frameworks with members in bending and/or in torsion, is beyond the scope of this work. It is nevertheless important that the student should be able to recognize to which category a particular framework belongs. Some examples of category 2 are given in Fig. 2.1.

2.2 Pin-jointed plane frameworks

In order that a framework be considered *statically determinate*, a number of conditions have to be satisfied. It is also convenient to separate frameworks where the members and applied forces lie in one plane from those that are three-dimensional; this latter group are generally referred to as *space frames*.

PIN-JOINTED PLANE FRAMEWORKS are therefore defined by the following assumptions:

1. The component members are all straight, inextensible and weightless.

2. The component members are connected at their extremities to other members by means of pin-joints, which function as perfect hinges (i.e. they are free to rotate and cannot, therefore, transmit a force-moment).
3. All members, applied forces and reactions lie in the same plane.
4. All the forces (including the reactions) are applied at nodes (i.e. joints between members).
5. The framework as a whole is capable of resisting geometrical distortion under any system of forces applied at the nodes.

A framework which could satisfy each and all of these conditions would constitute the ideal and this, of course, amounts to a practical impossibility. It is both possible and practical, however, to make these assumptions for a large number of frameworks, including those referred to as roof trusses.

One of the main difficulties encountered is in deciding whether or not a joint should be considered a *pin-joint*. Joints between members usually consist of connecting plates, (gussets, brackets, etc.) and such fastenings as bolts, rivets and welds, as well as parts of the members.

In designing the joints themselves, the size and position of all of the various elements has to be considered, but for the purpose of analysing the framework as a whole the joint conditions need to be simplified. It is usual to recognize a joint as being one of two distinct types: either it is a PIN-JOINT (or HINGED JOINT), or it is a RIGID JOINT, as shown in Fig. 2.2.

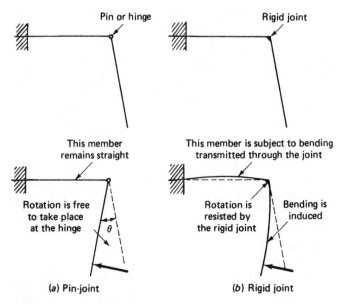

(a) Pin-joint   **(b) Rigid joint**

Fig. 2.2

2.3 Statically determinate plane frameworks rule

A framework which contains exactly the correct numbers of bars required
to keep it stable and statically determinate may be termed a *perfect
frame*. Frameworks having less than this required number will be
unstable and therefore *substatic*, whilst those having more than a
sufficient number will be *hyperstatic* and contain *redundant members*, i.e.
members which could be removed without producing instability.

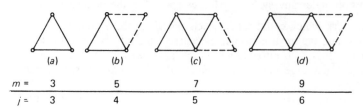

$m =$	3	5	7	9
$j =$	3	4	5	6

Fig. 2.3

In Fig. 2.3 the simplest possible framework that would satisfy the
required conditions is shown in sketch (*a*) and has three members and
three joints. The required conditions are maintained if, when this basic
framework is extended, two members are added at a time, as shown in
sketches (*b*), (*c*) and (*d*). The relationship between the number of
members and the number of joints in a *perfect frame* is then as follows:

$$m = 2j - 3 \qquad [2.1]$$

where:
$$m = \text{number of members}$$
$$j = \text{number of joints}$$

Some examples are given in Fig. 2.4.

At nodes pinned to external rigid supports the induced reaction forces
must also be accounted in the overall equilibrium. In a plane framework,
two equations of static equilibrium ($\Sigma H = 0$ and $\Sigma V = 0$) may be
obtained for each joint. Thus, a total of $2j$ equations will be available for
the solution of m member forces and s support reactions; i.e. for a
perfect frame:

$$m + s = 2j \qquad [2.2]$$

In calculating a value for s, the degrees of freedom at each joint must be
considered. For example, a joint pinned to a horizontal roller: no
horizontal reaction force can be developed, i.e. $H = 0$, and thus $s = 1$.
At a pin-joint support, $s = 2$.

$$\text{horizontal roller: } H = 0, \quad V > 0 \quad \therefore s = 1$$
$$\text{vertical roller: } \quad H > 0, \quad V = 0 \quad \therefore s = 1$$
$$\text{pin-joint support: } H > 0, \quad V > 0 \quad \therefore s = 2$$

Although rigid-joint supports are not being dealt with here it is useful to
note that, since rotation is prevented, a support moment reaction also

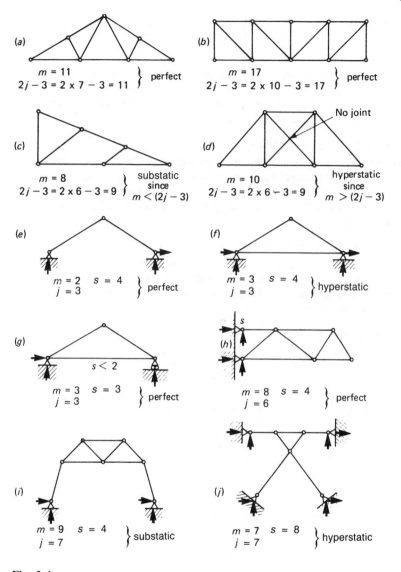

Fig. 2.4

occurs (Fig. 1.12). In this case, $H > 0$, $V > 0$ and $M > 0$, which means $s = 3$ for a rigid-joint support.

2.4 Methods of analysis

The solution of the forces in statically determinate plane frames may be obtained by using the three equations of static equilibrium.

From Section 1.7: $\Sigma\ (H) = 0$

$\quad\quad\quad\quad\quad\quad\ \Sigma\ (V) = 0$

$\quad\quad\quad\quad\quad\quad\ \Sigma\ (M) = 0$

A number of methods of applying these equations exist, of which three will be considered here. A brief description of each method is given below followed by a series of worked examples by means of which each method will be explained in detail.

The method of resolution

The forces acting concurrently at an individual pin-joint are considered, providing that not more than two of these are unknown. The trigonometric solution of the force polygon yields the magnitude of the unknown force or forces. Each pin-joint in a framework has to be considered in turn.

The force diagram method

This is a graphical method where the force polygons representing the forces concurrent at each joint are combined to form a composite force diagram. The forces acting in the individual members are scaled off the force diagram.

The method of sections

A theoretical section or cut is made through several members, including those whose forces are to be found, to divide the framework into two separate parts. The forces in the members which are cut through are treated as external forces (of which not more than three should be unknown in most cases) which, together with the applied loads and reactions, maintain one of the two parts of the framework in equilibrium. The unknown forces are determined by using the laws of static equilibrium, principally $\Sigma\ (M) = 0$.

WORKED EXAMPLES USING THE METHOD OF RESOLUTION AND THE FORCE DIAGRAM METHOD

EXAMPLE 2.1 Determine the magnitude and type of force acting in each of the members in the framework shown in Fig. 2.5.

First the reactions V_{AC} and V_{BC} must be found.

$\quad\quad (M)_{AC} = 0 = V_{BC} \times 2 + 200 \times 0\cdot5 \quad\quad \therefore\ V_{BC} = 50$ kN

$\quad\quad (M)_{BC} = 0 = V_{AC} \times 2 - 200 \times 1\cdot5 \quad\quad \therefore\ V_{AC} = 150$ kN

SOLUTION BY THE METHOD OF RESOLUTION

Consider the forces at joint CA1 and draw the polygon of forces (Fig. 2.6). Using the side-length ratios for a 60°/30° triangle, the unknown forces are determined:

$$F_{C1} = 150\ \times\ \tan 60° = 86\cdot6\ \text{kN, tension}$$

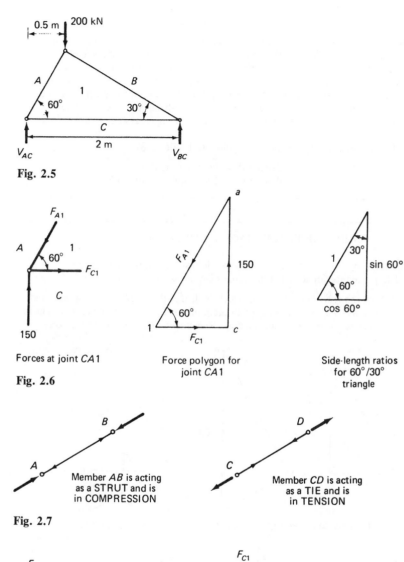

0.5 m 200 kN

A

B

60°

30°

1

C

2 m

V_{AC} V_{BC}

Fig. 2.5

F_{A1}

A

1

60°

F_{C1}

C

150

Forces at joint $CA1$

Fig. 2.6

a

F_{A1}

150

60°

1

F_{C1} c

Force polygon for
joint $CA1$

30°

1   sin 60°

60°

cos 60°

Side-length ratios
for 60°/30°
triangle

B

A

Member AB is acting
as a STRUT and is
in COMPRESSION

D

C

Member CD is acting
as a TIE and is
in TENSION

Fig. 2.7

F_{B1}

B

1

F_{C1}

C

50

Forces at joint $BC1$

F_{C1}

1   c

30°   (cos 30°)

(1)   (sin 30°)   50

F_{B1}

b

Force polygon for joint $BC1$

Fig. 2.8

$$F_{Al} = 150 \times \sin 60° = 173 \cdot 2 \text{ kN, compression}$$

The TYPE of force, i.e. whether the member is in tension or compression, is best determined by considering the consequences of removing the member from the framework. For example, if member Al is removed, the topmost joint will fall downwards; it may be concluded therefore that member Al is acting as a *strut* and is in *compression*. Similarly, member $C1$ can be seen to be acting as a tie preventing the movement to the right of the right-hand joint. A convenient way of indicating whether a member is a strut or a tie, using a system of arrows, is shown in Fig. 2.7.

Now consider the forces acting at joint $BC1$ (Fig. 2.8)

$$F_{C1} = 50/\tan 30° = 86 \cdot 6 \text{ kN, tension}$$
$$F_{B1} = 50/\sin 30° = 100 \cdot 0 \text{ kN, compression}$$

SOLUTION BY THE FORCE DIAGRAM METHOD

The force polygons for each of the three joints are combined (Fig. 2.9).
The procedure for drawing the force diagram is as follows:

1. Draw the *load line* (a, b, c), which is in fact the *external force* (loads and reactions) *polygon*, to an appropriate scale. It is usually found convenient to commence at the left-hand support. Remember that the line on the force

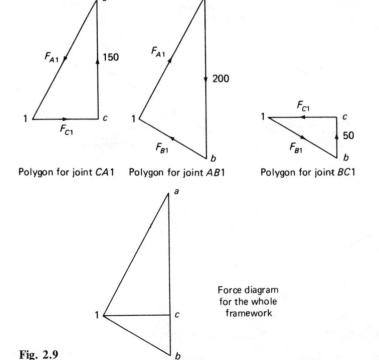

Polygon for joint $CA1$ Polygon for joint $AB1$ Polygon for joint $BC1$

Force diagram for the whole framework

Fig. 2.9

diagram is drawn parallel to the force it represents and is given a scale length equal to the magnitude of the force. If Bow's notation, i.e. labelling spaces between forces with letters or numbers is used the nodes on the force diagram can all be labelled with letters or numbers.

2. The internal spaces in the framework are represented on the force diagram by nodes labelled with numbers. Point 1 is therefore located by drawing lines parallel to members $A1$, $B1$ and $C1$ from points a, b and c respectively on the load line.

3. The magnitude of the forces can now be scaled directly off the force diagram.

$$F_{A1} = 173 \text{ kN, compression}$$
$$F_{B1} = 100 \text{ kN, compression}$$
$$F_{C1} = 87 \text{ kN, tension}$$

4. Whether a member is a strut or a tie may be determined as before, or by tracing out the individual joint polygons on the force diagram and noting the direction of the particular force at that joint.

EXAMPLE 2.2   Determine the magnitude and type of force acting in each of the members in the framework shown in Fig. 2.10.

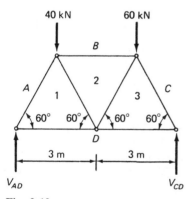

Fig. 2.10

SOLUTION BY THE METHOD OF RESOLUTION

$$\Sigma (M)_{AD} = 0 = - V_{CD} \times 6 + 40 \times 1 \cdot 5 + 60 \times 4 \cdot 5$$

$$\therefore V_{CD} = \frac{60 + 270}{6} = 55 \text{ kN}$$

$$\Sigma (M)_{CD} = 0 = V_{AD} \times 6 - 40 \times 4 \cdot 5 - 60 \times 1 \cdot 5$$

$$\therefore V_{AD} = \frac{180 + 90}{6} = 45 \text{ kN}$$

The polygons of forces for each of the joints are drawn (Fig. 2.11).

46

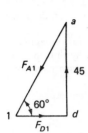

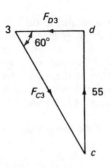

Polygon for joint $DA1$ Polygon for joint $CD3$

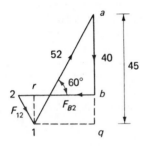

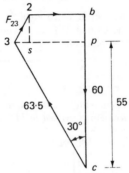

Polygon for joint $AB21$ Polygon for joint $BC32$

Fig. 2.11

Then at joint $DA1$: $F_{D1} = 45/\tan 60° = 26\cdot0$ kN, tension
                         $F_{A1} = 45/\sin 60° = 52\cdot0$ kN, compression

At joint $CD3$: $F_{D3} = 55/\tan 60° = 31\cdot8$ kN, tension
                      $F_{C3} = 55/\sin 60° = 63\cdot5$ kN, compression

At joint $AB21$:   since $a1 = 52$
                     then $aq = 52 \sin 60° = 45$
                     and $bq = r1 = 5$
                          $F_{12} = 5/\sin 60° = 5\cdot8$ kN, tension
                     also $q1 = 52 \cos 60° = 26$
                          $F_{B2} = 26 + 5/\tan 60° = 28\cdot9$ kN,
                                                 compression

At joint $BC32$:   since $c3 = 63\cdot5$
               then $cp = 63\cdot5 \sin 60° = 55$
                   and $pb = s2 = 5$
                       $F_{23} = 5/\sin 60° = 5\cdot8$ kN, compression
                  also $p3 = 63\cdot5 \cos 60° = 31\cdot8$
                       $F_{B2} = 31\cdot8 - 5/\tan 60° = 28\cdot9$ kN,
                                       compression (check!)

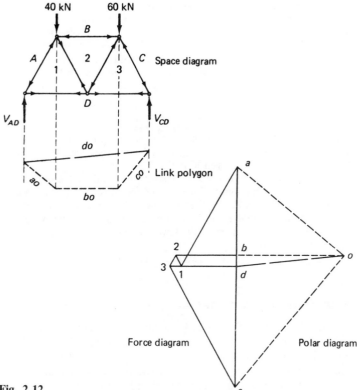

Fig. 2.12

SOLUTION BY THE FORCE DIAGRAM METHOD

To make this a fully graphical solution the reactions will be determined using the link polygon (Fig. 2.12). The procedure is as follows:

1. Draw the space diagram to an appropriate scale.
2. Draw the load line abc, so that $ab = 40$ kN and $bc = 60$ kN.
3. Select a suitable pole point to the right of the load line and join the points on the load line to the pole — ao, bo and co.
4. Draw the link polygon, commencing on the line of the reaction V_{AD} with link ao. Links ao, bo and co are parallel to the corresponding lines in the polar diagram.
5. Transfer the closing link do to the polar diagram, where it intersects the load line giving point d.
6. The reactions may now be scaled off the load line:

$$da = V_{AD} = 45 \text{ kN}$$
$$cd = V_{CD} = 45 \text{ kN}$$

7. The force diagram is now drawn, commencing with the location of point 1 by drawing $a1$ parallel to member $A1$ from a and $d1$ parallel to member $D1$ from d.

Similarly: 2 is located by 12 from 1 and b2 from b
3 is located by 23 from 2, d3 from d and c3 from c.

8. Scale off the forces in the members and enter them in the table as shown.

9. Determine the type of force in each member, either by considering the consequences of removing the member, or by tracing out the individual joint polygons. As an example of the latter method, consider joint AB21. Working around the joint in a clockwise direction, the force arrows at the joint will be:

$$a \rightarrow b \rightarrow 2 \rightarrow 1 \rightarrow a$$

Mark the direction of the force arrow on the corresponding member (near the joint under consideration) on the space diagram as shown. The type of force is now decided using the convention given in Fig. 2.7.

A single member force may be checked by considering the direction of the force arrow at either one of the joints. For example, the force arrow at the top of member 23 points upwards, since, when tracing out this force at this joint, the direction followed on the force diagram is from point 3 to point 2 (clockwise around the joint). For the force arrow at the lower end of this member, the direction would be from 2 to 3.

2.5 Force diagrams for special cases

The basic principles of the force diagram method have been set out in the solutions given to Examples 2.1 and 2.2 and they remain valid for all types of statically determinate plane frameworks. The student should first learn to apply both the method of resolution and the force diagram method to a variety of simple and straightforward frameworks. There are, however, a number of variations on the simple basic framework which require special consideration. The next series of worked examples deals with a number of these special cases as listed below.

1. Horizontal or inclined loading — roller and pinned reactions.
2. Wind loading — basic pressure, effect of slope.
3. Cantilever frameworks.
4. Solution of large frameworks by member substitution.

EXAMPLE 2.3   Determine the magnitude and type of force acting in each of the members in the framework shown in Fig. 2.13. At joint EF the support consists of a roller which is free to move horizontally.

The main consideration is that the horizontal load component(s) must be taken by the horizontal reaction component(s). In this particular case, since there is a roller at EF, the total horizontal component must be taken at AG.

The reactions are calculated as follows.
The horizontal reaction, $H = 10$ kN (acting to the left)
Vertical height, $h = 4 \tan 30° = 2·31$ m

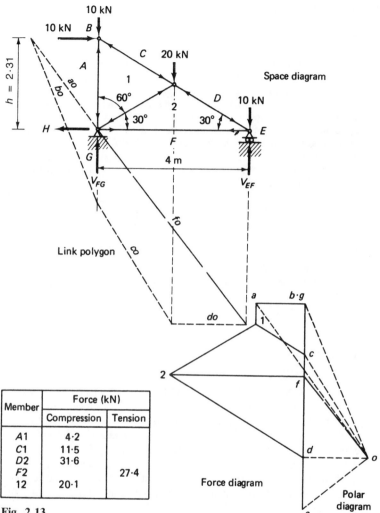

Member	Force (kN)	
	Compression	Tension
A1	4·2	
C1	11·5	
D2	31·6	
F2		27·4
12	20·1	

Fig. 2.13

For the vertical reactions:

$$\Sigma \, (M)_{AGF} \;\; = 0 = -V_{EF} \times 4 + 10 \times 2·31 + 20 \times 2 + 10 \times 4$$

$$\therefore \; V_{EF} = \frac{23·1 + 40 + 40}{4} = 25·8 \text{ kN}$$

$$\Sigma \, (M)_{EF} \;\;\; = 0 = V_{FG} \times 4 + 10 \times 2·31 - 20 \times 2 - 10 \times 4$$

$$\therefore \; V_{FG} = \frac{-23·1 + 40 + 40}{4} = 14·2 \text{ kN}$$

SOLUTION BY THE FORCE DIAGRAM METHOD

Procedure:
1. Draw the space diagram to an appropriate scale.
2. Draw the load line, commencing at *a* with a horizontal line *ab* to represent the 10 kN horizontal force, followed by *bc, cd* and *de*.
3. Draw the polar diagram to the right of the load line.
4. Draw the link polygon, commencing at the left-hand reaction with link *ao* drawn so that it interesects the line of the horizontal force *AB*. The remaining links, *bo, co* and *do*, are then added, and finally the closing link is drawn. Note that link *eo* is not used, since the reaction V_{EF} and the vertical 10 kN load lie on the same line.
5. Transfer the closing link *fo* to the polar diagram where it intersects the load line in *f*. The magnitude of the reactions can now be scaled off:

$$
\begin{aligned}
ga &= H &&= 10 \text{ kN} \\
fg &= V_{FG} &&= 14{\cdot}2 \text{ kN} \\
ef &= V_{EF} &&= 25{\cdot}8 \text{ kN}
\end{aligned}
$$

6. Draw the force diagram, locating the internal points as follows:

   1 is located by *c*1 from *c* and *a*1 from *a*
   2 is located by *d*2 from *d*, *f*1 from *f* and 12 from 1

7. Scale off the magnitude of the forces and determine the types as before.

2.6 Wind loading

When a structure is subject to a moving current of air, i.e. wind, a complex system of pressure and suction zones develop in and around the structure. The pressure within these zones is dependent on the wind velocity and is therefore constantly varying; a change in the speed or direction of the wind will produce corresponding changes in the induced pressure systems. A number of factors affect the basic pressure brought to bear on a building due to action of wind, such as the degree of exposure of the site, the roughness of the surrounding terrain, the degree of shelter afforded by adjacent structures, the incidence of gusting and the height of the building.

The basic wind pressure against a flat surface is taken as acting perpendicular to that surface, and may act as a positive pressure or as a suction depending on the orientation and slope of the surface. As a general rule the following conditions will apply:

Windward surfaces $>$ 35° to horizontal — pressure
Windward surfaces $<$ 35° to horizontal — suction
Leeward surfaces — suction

As with any inclined loading, the horizontal components of load must be contained by either one or both reactions. It should also be noted that, in the case of low-pitched roofs, wind pressure results in suction on both the leeward and windward slopes. A situation is therefore possible whereby the forces induced in the members due to the dead loading could be considerably reduced or even reversed when the wind is blowing. This wind reversal must be catered for in the design of the structure, which must also be adequately *held down* at the supports.

The following example illustrates the procedure required for the analysis of wind loading on a roof truss.

EXAMPLE 2.4   Determine the magnitude and type of force acting in each of the members in the roof truss shown in Fig. 2.14. The right-hand support consists of a roller which is free to move horizontally.

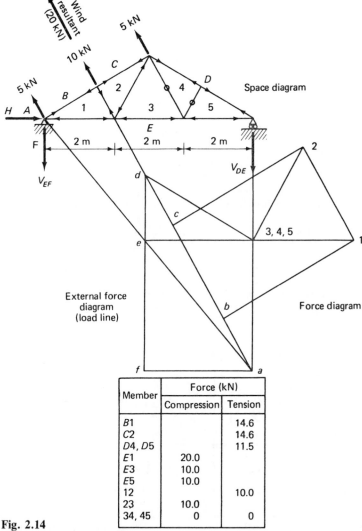

| Member | Force (kN) | |
	Compression	Tension
B1		14.6
C2		14.6
D4, D5		11.5
E1	20.0	
E3	10.0	
E5	10.0	
12		10.0
23	10.0	
34, 45	0	0

Fig. 2.14

Note that the loads are acting upward (wind suction) and therefore the reactions will be acting downward. Since the right-hand support is a roller, the horizontal component (H) must be taken entirely by the left-hand support. The reactions may be determined using a polar diagram and link diagram and link polygon as before. However, because of the symmetry of the applied loads, a somewhat simpler solution is possible. The procedure is as follows:

1. Draw the space diagram to an appropriate scale.
2. Project the line of the wind resultant until it intersects the line of the vertical right-hand reaction. For external equilibrium, the line of the left-hand reaction resultant (resultant of H and V_{EF}) must also pass through this intersection as well as the reaction point EFA. Draw in this reaction line.
3. Scale off the applied loads along the line of the wind resultant from a:

$$ab = 5 \text{ kN}$$
$$bc = 10 \text{ kN}$$
$$cd = 5 \text{ kN}$$

4. Draw the vertical line def and the horizontal line af. The diagram $abcdef$ is the external force polygon or load line, from which the reactions may be found.

$$fa = H = 10 \text{ kN}$$
$$ef = V_{EF} = 11 \cdot 5 \text{ kN}$$
$$de = V_{DE} = 5 \cdot 8 \text{ kN}$$

5. Draw the force diagram:
   1 is located by $b1$ from b and $e1$ from e
   2 is located by $c2$ from c and 12 from 1
   3 is located by $d3$ from d and 23 from 2
   4 } are located by $d4/5$ from d and $e5$ from e and are coincident with
   5 } node 3
6. Scale off the force magnitudes and determine the types as before.

EXAMPLE 2.5   Determine the magnitude and type of force acting in each member of the cantilever framework shown in Fig. 2.15.

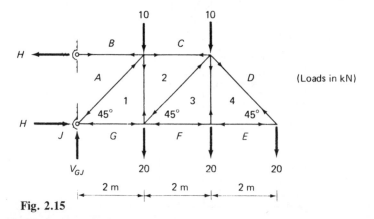

(Loads in kN)

Fig. 2.15

The framework is statically determinate, since the direction of the upper reaction can only be along the line of member AB and is therefore horizontal. Two solutions will be considered, (i) using the method of resolution, and (ii) using the force diagram method.

SOLUTION BY THE METHOD OF RESOLUTION

First determine the reactions

$$\Sigma\, (M)_A = 0 = -H \times 2 + 30 \times 2 + 30 \times 4 + 20 \times 6$$

$$\therefore H = \frac{60 + 120 + 120}{2} = 150 \text{ kN}$$

There is no vertical component at AB,

$$\therefore V_{GJ} = 10 + 10 + 20 + 20 + 20 = 80 \text{ kN}$$

It is best to start the method of resolution at the free end, i.e. at joint $DE4$ (see Fig. 2.16). Remember that a joint cannot be analysed where three or more unknowns occur.

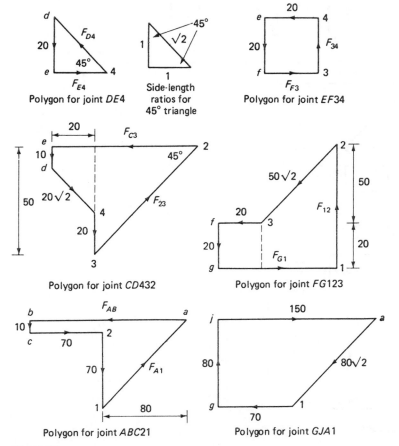

Polygon for joint $DE4$

Side-length ratios for $45°$ triangle

Polygon for joint $EF34$

Polygon for joint $CD432$

Polygon for joint $FG123$

Polygon for joint $ABC21$

Polygon for joint $GJA1$

Fig. 2.16

54

At joint *DE*4: $F_{E4} = 20 \cdot 0$ kN, compression
$F_{D4} = 20 \times \sqrt{2} = 28 \cdot 3$ kN, tension

At joint *EF*34: $F_{F3} = 20 \cdot 0$ kN, compression
$F_{34} = 20 \cdot 0$ kN, tension

At joint *CD*432: $F_{C2} = 20 + 50 = 70 \cdot 0$ kN, tension
$F_{23} = 50 \times \sqrt{2} = 70 \cdot 7$ kN, compression

At joint *FG*123: $F_{G1} = 50 + 20 = 70 \cdot 0$ kN, tension
$F_{12} = 50 + 20 = 70 \cdot 0$ kN, tension

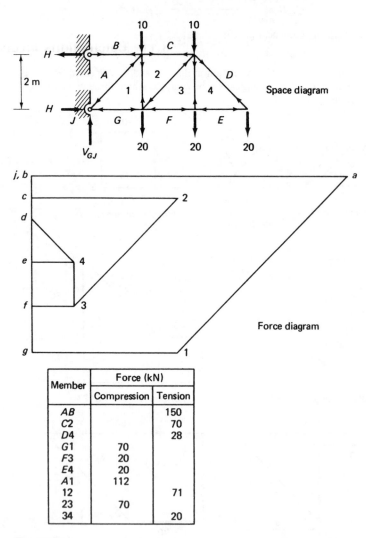

Member	Force (kN)	
	Compression	Tension
AB		150
C2		70
D4		28
G1	70	
F3	20	
E4	20	
A1	112	
12		71
23	70	
34		20

Fig. 2.17

At joint $ABC21$: F_{A1} = $(70 + 10) \sqrt{2}$ = 113·1 kN, compression
$\qquad\qquad\qquad$ F_{AB} = $70 + 80$ = 150 kN, tension ($=$ reaction H — check!)

At joint $GJA1$: $\quad$ Check values against those already obtained.

SOLUTION BY THE FORCE DIAGRAM METHOD

In the case of frameworks supported at *adjacent* nodes it is possible to draw the force diagram without first constructing the complete external force polygon. Having first drawn the load line for the applied loads, the force diagram is constructed commencing at the free end. Upon the completion of the force diagram it will be seen that the external force polygon is itself also complete and so the reactions may be determined. Most cantilever frameworks, such as this one and including many tower structures, may be analysed in this way.

Procedure
1. Draw the space diagram to an appropriate scale (Fig. 2.17).
2. Draw the load line for the applied loads.
3. Construct the force diagram, commencing at joint $DE4$:

$\qquad$ 4 is located by $d4$ from d and $e4$ from e
$\qquad$ 3 is located by $f3$ from f and 43 from 4
$\qquad$ 2 is located by $c2$ from c and 32 from 3
$\qquad$ 1 is located by $g1$ from g and 21 from 2
$\qquad$ a is located by ba from b and $1a$ from 1

4. Scale off the magnitude and estabish the type of each force.
5. Scale off the magnitude of the reactions:

$$gj = V_{GJ} = 80 \text{ kN}$$
$$ba = H = 150 \text{ kN}$$

EXAMPLE 2.6 An example involving member substitution and the use of a half-diagram. Determine the magnitude and type of force acting in each of the members of the roof truss shown in Fig. 2.18.

For frameworks of more than 7 or 8 panels the method of resolution becomes too lengthy and tedious, and the force diagram is generally chosen as the quickest method. Where the geometry of the framework and that of the applied loading are both symmetrical about a common centreline, only one-half of the complete force diagram will be required, since the second half will be the mirror-image of the first.

A further point concerning large trusses is that quite often it is found that some points on the force diagram representing internal spaces in the framework, such as point 4 in this example, cannot be located directly from the external force polygon. In such cases a hypothetical substitution member has to be introduced which replaces two of the actual members in order to facilitate the completion of the force diagram. This substitution method is illustrated in the following procedure.

1. Draw the space diagram to an appropriate scale.

56

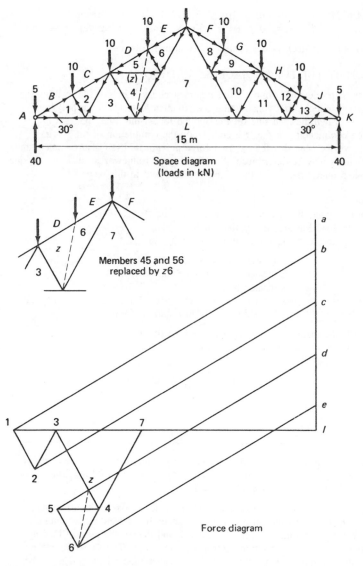

Space diagram
(loads in kN)

Members 45 and 56
replaced by z6

Force diagram

Fig. 2.18

2. Draw the external force polygon (load line) for the left-hand half of the truss. (The other half would have served equally as well.)
3. Commence the force diagram construction:

> 1 is located by $b1$ from b and $l1$ from l
> 2 is located by $c2$ from c and 12 from 1
> 3 is located by $l3$ from l and 23 from 2

4. The next point required is 4. However, it will be seen that only line 34 can, at this stage, be drawn, since points 5 and 7 have not yet been located. This difficulty may be overcome by hypothetically replacing members 45 and 56 by the member $z6$, as shown in the small sketch. This substitution can be shown to be valid, since the force in member $E6$ remains the same in both cases.
5. Continue the construction of the force diagram:

z is located by dz from d and $3z$ from 3
6 is located by $e6$ from e and $z6$ from z
7 is located by $l7$ from l and 67 from 6

6. Now replace the substitute member with the original members 45 and 56 and complete the force diagram for the half-truss:

5 is located by $d5$ from d and 65 from 6
4 is located by 34 from 3, 64 from 6 and 74 from 7

7. Scale off the magnitude and determine the type of each force as before. Include both halves of the truss in the tabulation.

WORKED EXAMPLES USING THE METHOD OF SECTIONS

EXAMPLE 2.7 Determine the magnitude and type of force acting in members C4, 34 and G3 of the Warren Girder shown in Fig. 2.19.

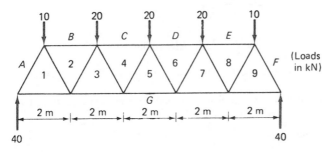

Fig. 2.19

Procedure
1. Make a hypothetical cut along section Z-Z, cutting through members $C4$, 34 and $G3$.
2. Consider the left-hand portion of the framework (Fig. 2.20). To maintain equilibrium when the right-hand portion is removed, replace the members cut through by forces having the same magnitude, direction and sense as the member-forces they replace. The sense of each force must be assumed at this stage; whether the true sense of the force is the same or is opposite to that assumed will be established later. In this case assume:

$$F_{C4} = \text{compression}$$
$$F_{34} = \text{compression}$$
$$F_{G3} = \text{tension}$$

58

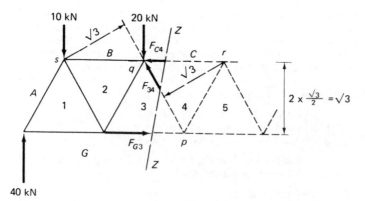

Fig. 2.20

3. The laws of static equilibrium may now be applied to this left-hand portion of the framework.

To find F_{C4}, consider the sum-moment about joint p, since the lines of action of the other two forces, F_{34} and F_{G3}, pass through this point and thus their moments about this point will be zero.

$$\Sigma\,(M)_p = 0 = -F_{C4} \times \sqrt{3} + 40 \times 4 - 10 \times 3 - 20 \times 1$$

$$\therefore F_{C4} = \frac{160 - 30 - 20}{\sqrt{3}}$$

$$= \frac{110}{\sqrt{3}} = 63 \cdot 5 \text{ kN, compression}$$

Since the value obtained is positive, the sense assumed for F_{C4} (compression) was correct.

4. To find F_{G3}, consider the sum-moment about joint q, since the lines of action of F_{C4} and F_{34} pass through this point.

$$\Sigma\,(M)_q = 0 = -F_{G3} \times \sqrt{3} + 40 \times 3 - 10 \times 2$$

$$\therefore F_{G3} = \frac{120 - 20}{\sqrt{3}} = \frac{100}{\sqrt{3}} = 57 \cdot 7 \text{ kN, tension (as assumed)}$$

5. To find F_{34}, either F_{C4} or F_{G3} must be known, since these forces are parallel and therefore only one can be eliminated by the choice of joint about which the sum-moment is considered. Assuming that F_{G3} is known, consider the sum-moment about joint r.

$$\Sigma\,(M)_r = 0 = +F_{34} \times \sqrt{3} - F_{G3} \times \sqrt{3} + 40 \times 5 - 10 \times 4 - 20 \times 2$$

$$\therefore F_{34} = \frac{-200 + 40 + 40 + 100}{\sqrt{3}} = \frac{-20}{\sqrt{3}} = -11 \cdot 6 \text{ kN, compression}$$

The negative value indicates that the actual sense of the force is opposite to that originally assumed and therefore $F_{34} = 11 \cdot 6$ kN, tension.

6. *Alternatively to find F_{34}*

$$\Sigma \,(M)_s = 0 = -F_{34} \times \sqrt{3} - F_{G3} \times \sqrt{3} + 40 \times 1 + 20 \times 2$$

$$F_{34} = \frac{40 + 40 - 100}{\sqrt{3}}$$

$$= \frac{-20}{\sqrt{3}} = -11 \cdot 6 \text{ kN, compression}$$

$$= \quad 11 \cdot 6 \text{ kN, tension}$$

EXAMPLE 2.8 Determine the magnitude and type of force acting in member X in the roof truss shown in Fig. 2.21.

Consider the portion of the truss to the left of section Z-Z (Fig. 2.22). Assume that member X is in tension and take moments about joint p.

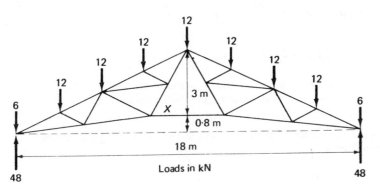

Fig. 2.21

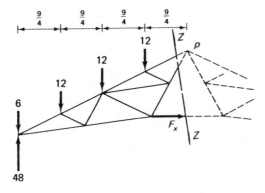

Fig. 2.22

$$\Sigma\ (M)_p = 0 = -F_x \times 3 + (48 - 6)\,9 - 12 \times 9 \times \tfrac{3}{4}$$
$$- 12 \times 9 \times \tfrac{1}{2} - 12 \times 9 \times \tfrac{1}{4}$$

$$\therefore F_x = \frac{378 - 9(9 + 6 + 3)}{3}$$

$$= \frac{216}{3} = 72 \text{ kN, tension}$$

EXAMPLE 2.9 Determine the magnitude and type of force acting in members C2, 23 and E3 of the roof truss shown in Fig. 2.14 (Example 2.4).

Procedure
1. First determine the reactions (Fig. 2.23)

$$\Sigma\ (M)_p = 0 = V_{EF} \times 6 - 20 \times 4 \times \frac{\sqrt{3}}{2}$$

$$\therefore V_{EF} = \frac{40\sqrt{3}}{6} = 11 \cdot 5 \text{ kN}$$

$H =$ the horizontal component of the resultant wind, since the roller at p offers only a vertical reaction
$$= 20 \times \tfrac{1}{2} = 10 \text{ kN}$$

2. *To find F_{C2}, consider the sum-moment at joint q, assuming that member C2 is in tension.*

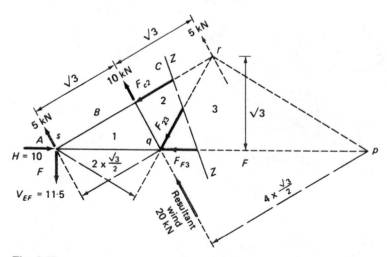

Fig. 2.23

$$\Sigma \, (M)_q = 0 = F_{C2} \times 1 - 11 \cdot 2 \times 2 + 5 \times 2 \times \frac{\sqrt{3}}{2}$$

$$\therefore F_{C2} = \frac{23 \cdot 0 - 8 \cdot 7}{1} = 14 \cdot 3 \text{ kN, tension}$$

3. *To find F_{E3}, consider the sum-moment at joint r, assuming that member $E3$ is in compression.*

$$\Sigma \, (M)_r = 0 = F_{E3} \times \sqrt{3} - 11 \cdot 5 \times 3 + 5 \times 2\sqrt{3} + 10 \times \sqrt{3} - 10 \times \sqrt{3}$$

$$F_{E3} = \frac{20 - 10}{1} = 10 \text{ kN, compression}$$

4. *To find F_{23}, consider the sum-moment at joint s, assuming that member 23 is in compression.*

$$\Sigma \, (M)_s = 0 = F_{23} \times \frac{2\sqrt{3}}{2} - 10 \times \sqrt{3}$$

$$F_{23} = 10 \text{ kN, compression}$$

EXAMPLE 2.10 Determine the magnitude and type of force acting in members C2 and 23 of the cantilever framework shown in Fig. 2.15 (Example 2.5).

It is not necessary to find the reactions. Consider, therefore, the portion of the framework to the right of section Z-Z (Fig. 2.24).

1. *To find F_{C2}, consider the sum-moment about joint p, assuming that member C2 is in tension.*

$$\Sigma \, (M)_p = 0 = -F_{C2} \times 2 + 20 \times 4 + (20 + 10)2$$
$$\therefore F_{C2} = 20 \times 2 + 20 + 10 = 70 \text{ kN, tension}$$

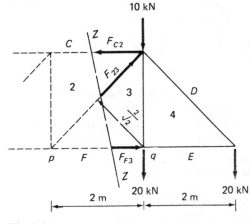

Fig. 2.24

2. *To find* F_{23}, consider the sum-moment about joint q, assuming that member 23 is in compresson.

$$\Sigma (M)_q = 0 = F_{23} \times \frac{2}{\sqrt{2}} + 20 \times 2 - 70 \times 2$$

$$\therefore F_{23} = (70 - 20) \sqrt{2} = 70 \cdot 7 \text{ kN, compression}$$

2.7 Forces in simple space frames

A space frame is a three-dimensional framework in which the applied forces, reactions and the members are located by referring to all three Cartesian axes (x, y, z). Pin-jointed space frames having the minimum number of members required for stability are statically determinate. Just as the simple basis of a *plane* frame is a *triangle*, the simple basis of a *space* frame is a *tetrahedron* having four joints and six members (Fig. 2.25). The perfect frame equation for a space frame is then as follows:

Fig. 2.25

$$m = 3j - 6 \qquad\qquad [2.3]$$

or $$m + s = 3j \qquad\qquad [2.4]$$

where: m = number of members
$\quad\quad\quad\; j$ = number of joints
$\quad\quad\quad\; s$ = number of support forces

Solutions for the forces in the members of statically determinate space frames may be obtained by the application of the laws of static equilibrium. For a completely general system of non-concurrent, non-parallel forces, SIX equations are required:

$$\left.\begin{array}{l} \Sigma (F)_x = 0 \\ \Sigma (F)_y = 0 \\ \Sigma (F)_z = 0 \end{array}\right\} \text{ Algebraic sums of the axial components of the forces in the system.}$$

$$\left.\begin{array}{l} \Sigma (M)_x = 0 \\ \Sigma (M)_y = 0 \\ \Sigma (M)_z = 0 \end{array}\right\} \text{ Algebraic sums of the moments of the forces in the system about their respective axes.}$$

However, for concurrent systems of spatial forces any *three* of these equations will be sufficient to provide a solution.

EXAMPLE 2.11 *Figure 2.26 shows a shear legs. Determine the magnitude and type of force acting in each leg and in the tie due to the application of a 50 kN load.* $AC = BC = 4m.$

$$AB = 4m$$

$$CE = \frac{4\sqrt{3}}{2} = 2\sqrt{3}m$$

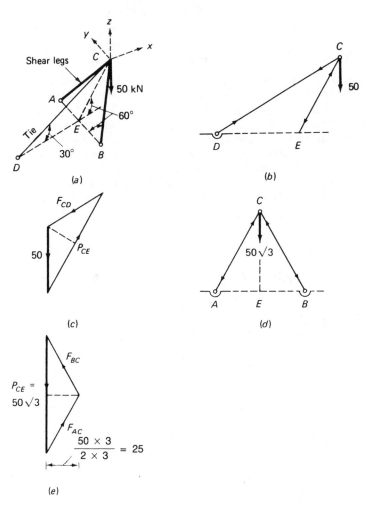

(a)

(b)

(c)

(d)

(e)

Fig. 2.26

From the elevation (b), the force polygon at joint C in the x-z plane is as shown in sketch (c).

Then $\qquad F_{CD} = 50$ kN, tension

and $\qquad P_{CE} = \dfrac{50}{2} \times \sqrt{3} \times 2 = 50\sqrt{3}$ kN, compression

Consider now the forces in the plane of the members AC and BC, sketch (d). The force polygon for joint C is shown in sketch (e).

The force P_{CE} represents the resultants of all the forces other than the member forces F_{AC} and F_{BC}.

Then $\qquad\qquad F_{AC} = F_{BC} \begin{aligned}[t] &= 25 \times 2 \\ &= 50 \text{ kN, compression} \end{aligned}$

2.8 Solutions using computers

In professional practice most framework analyses will be carried out using computers. Software packages are widely available for personal computers, but they vary greatly in their levels of sophistication. Lecturers and students should choose software that is adequate for the purpose, but not too sophisticated. It is wasteful of time and often confusing to new learners to have to deal with a lot of unnecessary data input and non-useful output. To illustrate this point several different pieces of software will be compared which give solutions to the same problem; they are all run on the same IBM clone personal computer.

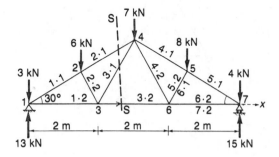

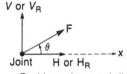

Positive values and directions

Fig. 2.27

```
┌─────────────────────────────────────────────────────────────────────────┐
│  PFRAME1  Member forces in a pin-jointed plane framework                  │
│           EXAMPLE  2.7                                                     │
└─────────────────────────────────────────────────────────────────────────┘
```

Joint:	1	2	3	4	5	6	7
LOADS							
V (up=+)	-3	-6		-7	-8		-4
H (rt=+)	0						
UNKNOWN MEMBER FORCES			(tension = +; acw = +)				
F1	-20.00	-17.00	5.20	-17.99	-21.99	-6.92	-22.00
Theta1	30	30	60	330	330	60	150
F2	17.32	-5.20	12.12	6.92	-6.93	19.04	19.05
Theta2	0	300	0	300	240	0	180
KNOWN REACTION FORCES							
V (up=+)	13						15
H (rt=+)							
KNOWN MEMBER FORCES			(tension = +; acw = +)				
F		-20	-5.2	-17	-17.99	6.92	
Theta		210	120	210	150	120	
F			17.32	5.2		12.12	
Theta			180	240		180	
F							
Theta							
F							
Theta							
F							
Theta							
F							
Theta							
F							
Sum Fh	.00	17.32	-14.72	12.12	15.58	-15.58	.00
Sum Fv	.00	10.00	-4.50	4.00	-9.00	5.99	.00

Fig. 2.28

Figure 2.27 shows a pin-jointed roof truss subject to vertical loads. Fig. 2.28 shows the printout of a spreadsheet analysis devised by the author. The method here is seen at its simplest level and consists of a semi-automatic calculation. At each joint equations of static equilibrium have been used to calculate the forces in two members whose inclinations are known. To begin with, the joints in the framework are numbered in a strict sequence so that, at each successive joint, no more than two members' forces are unknown. The analysis proceeds by entering data for each joint in turn: any horizontal or vertical loads or reaction forces and the values of any *known* (i.e. calculated at previous joints) member forces and their inclinations. The sign conventions are given in Fig. 2.27. Finally, as the inclinations of the unknown members are entered their force magnitudes (F_1 and F_2) are automatically calculated; these being used as data for subsequent joints. Checks are incorporated when the forces in some members are calculated at more than one joint.

Although the spreadsheet method described above is eminently simple and well within the scope of first-year work, it has limitations. For instance, no data on joint displacement can be obtained, and only statically determinate frameworks can be dealt with; nor can rigid joints be included.

Figures 2.29, 2.30 and 2.31 show printouts produced from more comprehensive software (which is obtainable from QSE Ltd: see Section 1.14 for further details). Students will gain valuable insights into

```
BRISTOL    POLYTECHNIC
DEPT  OF  CONSTRUCTION
     JOB TITLE...... M&S2/2          JOB No. RW90/2
     CLIENT.........              PART...........
     ANALYST........              CHECKED........
     S2-200 - SUPERTRUSS          Software by QSE Ltd
```

NODE COORDINATES

No	X (m)	Z (m)	Supports
1	0.000	0.000	FPB
2	1.500	0.866	---
3	2.000	0.000	---
4	3.000	1.732	---
5	4.500	0.866	---
6	4.000	0.000	---
7	6.000	0.000	VFB

MEMBERS

No	Node A	Node B	$A \times 10^{-3}$ (mm^2)	E (kN/mm^2)
1	1	2	1.000	205.000
2	1	3	1.000	205.000
3	2	4	1.000	205.000
4	2	3	1.000	205.000
5	3	4	1.000	205.000
6	3	6	1.000	205.000
7	4	5	1.000	205.000
8	4	6	1.000	205.000
9	5	6	1.000	205.000
10	5	7	1.000	205.000
11	6	7	1.000	205.000

LOADINGS

Node	Load type	X/Z-dir	Magnitude (kN)
1	POINT	Z	-3.000
2	POINT	Z	-6.000
4	POINT	Z	-7.000
5	POINT	Z	-8.000
7	POINT	Z	-4.000

Fig. 2.29

```
MEMBER FORCES (Local)
 Memb No    Axial Force (KN)         Stress (N/mm^2)
                                     (+ Ten, - Comp)

    1           -20.000              -20.000

    2            17.321               17.321

    3           -17.000              -17.000

    4            -5.196               -5.196

    5             5.196                5.196

    6            12.125               12.125
                   .
    7           -18.000              -18.000

    8             6.928                6.928

    9            -6.928               -6.928

   10           -22.000              -22.000

   11            19.053               19.053


REACTIONS (Global)
 Node No     X Component(kN)         Z Component (kN)

    1            0.000                13.000
    7           -0.000                15.000


EQUILIBRIUM CHECK

Total force in Z direction    = -2.800E+01 kN
Total reaction in Z direction = -2.800E+01 kN

Total force in X direction    = 0.0000E+00 kN
Total reaction in X direction = -8.440E-10 kN

Max out of balance force from nodes equilibrium check
                              = 6.6211E-10 kN
```

Fig. 2.30

structural behaviour by repeating analyses incorporating small changes in member properties or frame geometry.

Checking computer output

It is essential that checks are made that are independent of the main internal analysis process. Some of these may be included in the software

```
BRISTOL POLYTECHNIC
DEPT OF CONSTRUCTION
JOB TITLE...... M&S2/2        JOB No.......... RW90-3      REFERENCE No...
CLIENT.........              PART..........               DATE..........
ANALYST........              CHECKED.......               LOAD COMB..
SI-200 - SUPER PLANEFRAME    Software by OSE Ltd     FILENAME - A:RW90-3    Version: 11a/200989
```

MEMBER FORCES (Local)					
Memb No	Node	Node No	Axial (kN)	Shear (kN)	Moment (kNm)
1	A	1	2.913	4.616	0.130
1	B	2	-2.913	-4.616	-8.125
2	A	1	-0.215	4.546	-0.130
2	B	3	0.215	-4.546	-8.962
3	A	2	3.846	0.036	4.476
3	B	4	-3.846	-0.036	-4.539
4	A	2	0.616	-3.933	3.649
4	B	3	-0.616	3.933	0.284
5	A	3	1.859	0.398	2.664
5	B	4	-1.859	-0.398	-3.461
6	A	3	-3.898	0.237	6.013
6	B	6	3.898	-0.237	-6.487
7	A	4	3.781	0.619	4.142
7	B	5	-3.781	-0.619	-5.215
8	A	4	1.788	-0.670	3.857
8	B	6	-1.788	0.670	-2.518
9	A	5	0.946	4.519	-3.987
9	B	6	-0.946	-4.519	-0.531
10	A	5	3.262	-5.362	9.202
10	B	7	-3.262	5.362	0.086
11	A	6	-0.144	-4.725	9.536
11	B	7	0.144	4.725	-0.086

REACTIONS (Global)			
Node No	X Component(kN)	Z Component (kN)	Moment (kNm)
1	0.000	13.000	0.000
7	0.000	15.000	-0.000

EQUILIBRIUM CHECK

Total force in Z direction   = -2.000E+01 kN
Total reaction in Z direction = -2.000E+01 kN

Total force in X direction   = 0.0000E+00 kN
Total reaction in X direction = 3.5106E-10 kN

Total moment   = 9.0000E+01 kNm
Total reaction moments = 9.0000E+01 kNm

Max out of balance force from nodes equilibrium check = 8.8767E-10 kN

Max out of balance moment from nodes equilibrium check = 3.0195E-10 kNm

Fig. 2.31

package. For example, in Figs 2.30 and 2.31 the printout shows that an external equilibrium check has been made. A simple hand calculation may also provide a useful output check, as illustrated in the following example.

EXAMPLE 2.12 Use the method of sections to verify the force in member 3.2 as shown in Fig. 2.27.

The truss is divided by the hypothetical section line $S-S$.

Vertical height of joint 4 above member 3.2 = 3·0 tan 30°
$$= 1·732 \text{ m}$$

Then, for equilibrium about joint 4:
$$\Sigma M_4 = 0 = 13 \times 3 - 6 \times 1·5 - 3 \times 3 - F_{3·2} \times 1·732$$

Therefore $F_{3·2} = \dfrac{39 - 9 - 9}{1·732} = \underline{12·12 \text{ kN}\quad \text{TENSION}}$

EXERCISES   CHAPTER 2

1–16. Using graphical methods, determine the reactions and the magnitude and type of force acting in each of the members of the frameworks shown in Figures 2.32–2.47 (all loads in kN).

17–24. Determine the reactions by calculation and, using the method of resolution, determine the magnitude and type of force acting in each of the

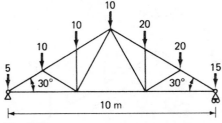
Fig. 2.32

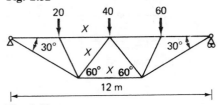
Fig. 2.33

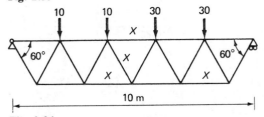
Fig. 2.34

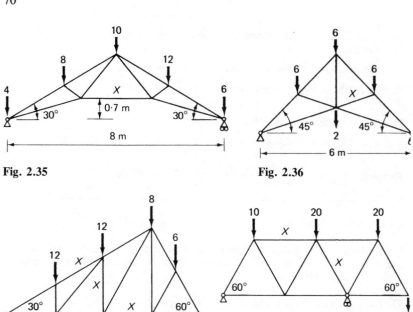

Fig. 2.35

Fig. 2.36

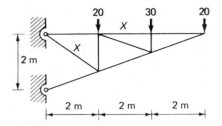

Fig. 2.37

Fig. 2.38

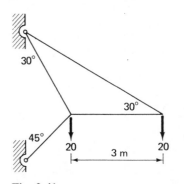

Fig. 2.39

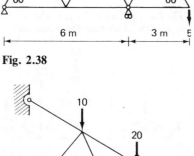

Fig. 2.40

Fig. 2.41

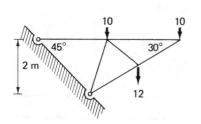

Fig. 2.42

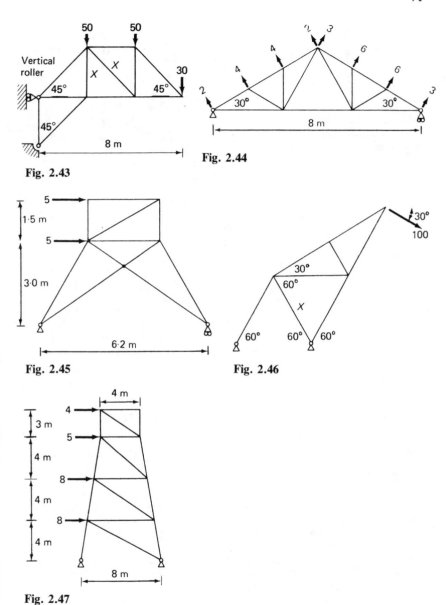

Fig. 2.43

Fig. 2.44

Fig. 2.45

Fig. 2.46

Fig. 2.47

members of the frameworks in Figures 2.32, 2.33, 2.35, 2.36, 2.41, 2.43, 2.44 and 2.46 (all loads in kN).

25–33. Determine the reactions by calculation and, using the method of sections, determine the magnitude and type of force acting in each of the members marked X in the frameworks shown in Figures 2.33–2.39, 2.43 and 2.46.

72

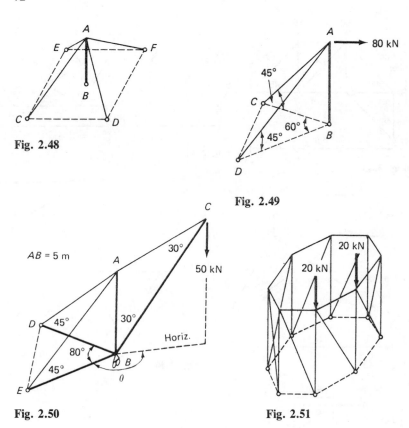

Fig. 2.48

Fig. 2.49

$AB = 5$ m

Fig. 2.50

Fig. 2.51

34. In Fig. 2.48 the vertical strut, which is 8 m long, is held by four guy wires; the force in each wire is 5 kN. Determine the force acting in the strut. $CD = DF = FE = EC = 6$ m.

35. In Fig. 2.49 the vertical strut AB, which is 6 m long, is held at A by two guy wires. Determine the forces acting in the strut and in the wires when an 80 kN horizontal force is applied at A, the line of which bisects angle $C\hat{B}D$.

36. Fig. 2.51 shows a derrick crane subject to a vertical load of 50 kN. Support B rests on a horizontal roller. Determine the magnitude and type of force acting in each member, and also the horizontal and vertical reaction components at the supports B, D and E, when the horizontal angle θ is: (a) 140°, (b) 100°, (c) 70°.

37. In Fig. 2.51 a pin-jointed framework, octagonal in plan, consists of horizontal members 2 m long and vertical members 4 m long, together with diagonal members. The feet of the framework also lie in a horizontal plane. Determine the magnitude and type of force acting in each of the members due to the two 20 kN loads.

3 Direct stress and strain

Symbols

A	=	area
E	=	Young's modulus of elasticity
F	=	force, direct force
G	=	modulus of rigidity or shear modulus
h	=	height
K	=	bulk modulus
L	=	length
m	=	modular ratio
P	=	load, static load
t	=	temperature
U	=	strain energy
V	=	volume
α	=	coefficient of linear expansion
Δ	=	finite part, e.g. ΔL, ΔA, ΔV, Δt
ϵ	=	strain
ϵ_t	=	temperature strain
μ	=	Poisson's ratio
σ	=	stress, normal stress
σ_i	=	temperature stress
$\hat{\sigma}_t$	=	maximum instantaneous stress
τ	=	tangential stress, shearing stress

3.1 Definition of stress

When a force is transmitted through a solid body the body tends to undergo a change in shape. This tendency to deform is resisted by the internal resilience of the body and the body is said to be in a state of STRESS. Thus, a stress may be described as a mobilized internal force which resists any tendency towards deformation. However, since it is more convenient to consider how these internal forces are distributed over surfaces (either real or hypothetical), the usual definition is to describe the *force transmitted per unit area* as the INTENSITY OF STRESS or UNIT STRESS. In modern conventional usage the word STRESS alone is taken to mean *intensity of stress*.

Thus   Stress = Force per unit area

or $\text{Stress} = \dfrac{\text{Force}}{\text{Area}}$

using symbols $\sigma = \dfrac{F}{A}$ [3.1]

In certain design calculations alternative symbols may be used to represent stress, e.g. f, p.

3.2 Stress is a load effect

In Section 1.2 (p. 2) the concept of force as a vector quantity was discussed. Stress, by definition, is the relationship between a force and a particular area. In order to define a stress quantitatively therefore, the force (which is a vector) must be defined and in addition the configuration of the area in question must be defined. The following information is required:

1. The magnitude of the force.
2. The direction of the force.
3. The plane in which the area lies (related to the x-, y-, z-axes).

   The direction of the force and the area over which it is applied are related in one of two ways. Either the force acts *normal* (perpendicular) to the plane producing a *normal stress* (σ), or it acts *tangential* (parallel) to the surface producing a *tangential stress* (τ). The more common relationships are given in Table 3.1.

3.3 Units of stress

The concept of stress as *force per unit area* becomes more obvious when the units of stress are considered. The basic unit will be the force unit divided by the area unit.

   For example, newtons per square millimetre (N/mm^2)

Table 3.1 Load−stress relationships

Load effect (vector)	Stress induced	
	normal	tangential
Direct tension	tensile	
Direct compression	compressive	
Shearing force		shearing
Bending moment	tensile or compressive	
Torsion		shearing
Contact force between separate bodies:		
compression	bearing or contact	
tension	adhesion	
sliding		frictional

meganewtons per square metre (MN/m^2)
kilonewtons per square metre (kN/m^2)
$1 \ N/mm^2 = 1 \ MN/m^2 = 1 \ 000 \ kN/m^2$

3.4 Direct axial stress

In this chapter, only *direct axial stress* is studied, except of course in the preceding introductory paragraphs. DIRECT stresses are those *normal* stresses in which the force-direction lies parallel to the longitudinal axis of the member. In the case of an AXIAL stress, the force is axially applied, i.e. aligned *along* the longitudinal axis.

Direct axial stresses may be either *tensile* or *compressive*, depending on the sense of the force — see Table 3.1.

CALCULATION OF SIMPLE STRESSES

EXAMPLE 3.1 Determine the tensile stress induced in the rod shown in Fig. 3.1 due to an axial load of 60 kN.

The rod may be considered as a series of cross-sectional planes $(x-x)$ across which the force is transmitted.

Area of cross-sectional plane $= \dfrac{\pi}{4} \times 20^2$

$= 100\pi \ mm^2$

Now, tensile stress $= \dfrac{\text{tension load}}{\text{area of cross-section}}$

76

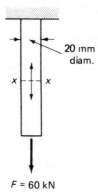

F = 60 kN

Fig. 3.1

$$\sigma_t = \frac{60 \times 1000}{100 \pi}$$

$$= 191 \text{ N/mm}^2 \text{ (Answer)}$$

EXAMPLE 3.2 Determine both the compressive stress in the shaft and the bearing stress at the base of the column shown in Fig. 3.2 due to an axial load of 10 MN.

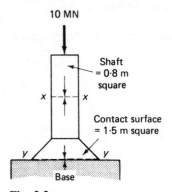

Fig. 3.2

In the shaft, consider the cross-sectional area plane $x-x$

Area of cross-section $= 0 \cdot 8 \times 0 \cdot 8$
$= 0 \cdot 64 \text{ m}^2$

Therefore compressive stress, $\sigma_c = \dfrac{10}{0\cdot64}$

$$= 15\cdot6 \text{ MN/m}^2 \text{ (Answer)}$$

At the base, consider the contact plane $y-y$

Area of contact plane $= 1\cdot5 \times 1\cdot5$
$$= 2\cdot25 \text{ m}^2$$

Therefore bearing stress, $\sigma_b = \dfrac{10}{2\cdot25}$

$$= 4\cdot4 \text{ MN/m}^2 \text{ (Answer)}$$

EXAMPLE 3.3 Determine the shearing stress induced in the rivet shown in Fig. 3.3 due to a shearing load of 40 kN.

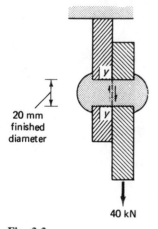

20 mm
finished
diameter

40 kN

Fig. 3.3

Consider the area $y-y$, over which the shear stress acts tangentially

Area of shear plane $= \dfrac{\pi}{4} \times 20^2$

$$= 100\pi \text{ mm}^2$$

Therefore shearing stress $= \dfrac{40 \times 1\,000}{100\pi}$

$$127\cdot3 \text{ N/mm}^2 \text{ (Answer)}$$

3.5 Definition of strain

As stated previously, when a force is transmitted through a solid body the body tends to be deformed. The measure of this change in shape is called STRAIN. When a body is placed in a state of stress it undergoes strain according to the configuration of the stress applied. Thus, direct stresses cause changes in length, torsional or shearing stresses cause twisting and bearing stresses cause indentation in the bearing surface.

The conventional way of expressing strain is to relate changes in a particular dimension to the original value of that dimension, thus:

$$\text{Tensile strain} = \frac{\text{increase in length}}{\text{original length}} \quad \text{(see Fig. 3.4)}$$

$$\epsilon_t = \frac{\Delta L}{L} \tag{3.2}$$

$$\text{Compressive strain} = \frac{\text{decrease in length}}{\text{original length}} \quad \text{(see Fig. 3.5)}$$

$$\epsilon_c = \frac{\Delta L}{L} \tag{3.3}$$

$$\text{Areal strain} = \frac{\text{change in area}}{\text{original area}}$$

$$\epsilon_A = \frac{\Delta A}{A} \tag{3.4}$$

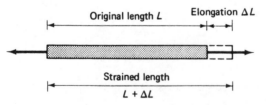

Fig. 3.4

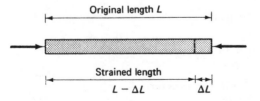

Fig. 3.5

$$\begin{array}{l} \text{Volumetric or} \\ \text{bulk strain} \end{array} = \frac{\text{change in volume}}{\text{original volume}}$$

$$\epsilon_V = \frac{\Delta V}{V} \hspace{3cm} [3.5]$$

Shearing strain = angular displacement caused by
shearing stress (see Chapter 8)

Torsional strain = angular displacement caused by
torsional shearing stress (see Chapter 9)

3.6 Units of strain

By definition, strain is the ratio of the change in a dimension to the
original dimension value, i.e. length divided by length, area divided by
area, volume divided by volume, etc. Consequently strain is a
dimensionless quantity and has no units.

3.7 Elasticity

All material bodies will deform when placed in a state of stress, and as
the stress is increased the deformation also increases. It is noticeable,
however, that in some cases, when the loads causing the deformation are
removed, the body returns to its original size and shape. A material or a
body having this property is said to be ELASTIC. It is also noticeable
that if the stress is steadily increased, a point is sooner or later reached
when, after the removal of the load, not all of the induced strain is
recovered. This limiting value of stress is called the ELASTIC LIMIT.

3.8 Hooke's law and the modulus of elasticity

The relationship between the induced strain and the stress causing it is
found to be constant in elastic materials. According to Hooke's law:
'STRAIN IS PROPORTIONAL TO THE STRESS CAUSING IT,
PROVIDING THAT THE LIMIT OF PROPORTIONALITY HAS NOT
BEEN EXCEEDED'.

Stress $\propto$ Strain

or $\dfrac{\text{Stress}}{\text{Strain}} = \text{constant}$

Thus, if a graph is produced of stress against strain as the load is
gradually applied, then the first portion of the graph will be a straight
line (Fig. 3.6). The slope of this straight line is the constant of
proportionality known as the MODULUS OF ELASTICITY. Three such
moduli may be defined according to the stress configuration induced:

$$\frac{\text{Direct stress}}{\text{Direct strain}} = \text{YOUNG'S MODULUS or THE MODULUS OF ELASTICITY } (E)$$

$$\frac{\text{Shearing stress}}{\text{Shearing strain}} = \text{THE MODULUS OF RIGIDITY } (G)$$

$$\frac{\text{Bulk stress}}{\text{Volumetric strain}} = \text{THE BULK MODULUS } (K)$$

In this chapter only Young's modulus will be needed. For reference purposes, some typical values of E and G are given in Table 3.2.

Table 3.2 Approximate mechanical properties of materials

Material	E (kN/mm^2)	G (kN/mm^2)	Ultimate stress (N/mm^2)	Poisson ratio ν	Density (Mg/m^3)	Yield stress (N/mm^2)
Steel						
High strength	200−210	75−80	500−620	0·30	7·85	300−450
Mild Steel	200−210	75−80	370−520	0·30	7·85	245−275
Stainless	205−215	80−85	500−950	0·33	7·85	250−450
Titanium alloy	120−140	45−55	600−800	0·30	4·50	350−450
Magnesium alloy	40−70	15−30	250−400	0·33	1·80	150−180
Wrought iron	170−200	65−80	300−360	0·28	7·85	180−200
Cast iron	180−210	70−80	80−200	0·27	7·20	
Aluminium (pure)	65−85	25−40	70−140		2·65	
Aluminium alloy	65−80	25−40	100−400		2·70	
Copper	90−120	35−45	200−350	0·33	8·9	120−160
Brass	75−100	30−40	200−500	0·33	8·4	100−260
Bronze	80−120	30−45	250−500	0·33	8·8	100−200
Polyamide (Nylon)	2−8		50−100		1·14	50−80
Polystyrene	2−6		40−80		1·05	40−70
Polyethylene	0·1−2		10−30		0·92	6−10
Timber (parallel to grain)						
English hardwood	7−12		15−23		0·72	
Imported hardwood	5−14		12−30		0·7	
Greenheart	13−17		37−42		1·06	
Spruce, cedar	4−7		3−8		0·45	
Pine, fir, hemlock	4−10		5−17		0·55	
Concrete		U_{28}*				
Unreinforced	12−45		15−70		2·2−2·4	
Lightweight	12−30		15−35		0·8−1·8	
Aerated	1·5−3·0		2−15		0·4−1·5	

*Compressive strength after 28 days

3.9 Behaviour of mild steel in tension

A convenient way of demonstrating elastic behaviour is to plot a graph of the results of a simple tensile test carried out on a thin mild steel rod or wire. The rod or wire may be hung vertically and a series of forces applied at the lower end. Two gauge points are marked on the rod and the distance between them measured after each force increment has been added. The test is continued until the rod breaks.

The stress and the strain values at each incremental stage are calculated and plotted as shown in Fig. 3.6. The following characteristics may now be observed from the graph:

Elastic range of stress: the straight-line portion of the graph between the origin and the elastic limit; i.e. the portion over which Hooke's law is obeyed.

Plastic range of stress: the curved portion of the graph between the elastic limit and the failure point; a permanent set (change in length) will be observed if the load is removed in this region.

Elastic limit: the stress up to which complete recovery of the strain takes place upon removal of the load.

Limit of proportionality: the stress up to which Hooke's law is obeyed, i.e. the graph is a straight line, usually occurring fractionally below the elastic limit.

Upper yield point: the stress at which the first sudden increase in strain occurs.

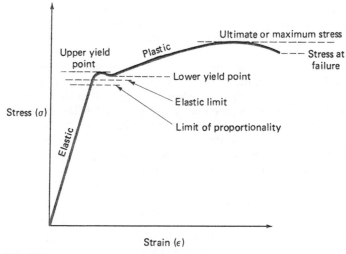

Fig. 3.6

Lower yield point: the lowest value of stress during the first sudden yielding; usually this can only be located by using a controlled strain method of testing.

Failing stress: the stress at which the rod breaks; allowance has to be made for the reduction in cross-section (known as 'necking') which occurs due to plastic yielding immediately prior to failure.

Ultimate or maximum stress: the maximum value of stress attained before the final rapid yielding leading to failure.

3.10 Limiting values of stress. Factor of safety and load factors

In the design of members and components that are to be placed in a state of stress, it is necessary to consider their safety against failure. There are two stress values at which failure may be said to occur, one is the *ultimate or maximum stress* which is reached just prior to a complete break, and the other is the *yield point stress* at which the first irreversible deformation takes place. The permissible or allowable working stress may be related to either:

$$\text{Permissible stress} = \frac{\text{ultimate stress}}{\text{factor of safety}}$$

$$\text{Permissible stress} = \frac{\text{yield stress}}{\text{yield factor}}$$

The ultimate stress relationship is more applicable to brittle materials, whilst that involving the yield stress would apply in the case of ductile materials. The factors of safety used vary in value between $1 \cdot 5$ and 5, depending on the material in question and on the reliability of the critical values both at the time of manufacture and during the life of the member or component.

With many non-ferrous metals, such as copper and aluminium, the yield point is difficult to define. In such cases an alternative to the yield stress is required and in a number of British Standards a limiting value termed the PROOF STRESS is specified. The PROOF STRESS is defined as the *stress causing a permanent strain of $0 \cdot 1 \%$*, or, from a more practical point of view, the *stress at which the stress/strain graph departs by $0 \cdot 1 \%$ strain from the line of proportionality (Fig. 3.7)*. The permissible stress may then be defined as follows:

$$\text{Permissible stress} = \frac{\text{proof stress}}{\text{proof factor}}$$

In modern design practice involving ductile materials, such as steel, reinforced concrete and timber, it is more common to apply a *load factor* as opposed to a factor of safety. If the limiting or ultimate design value

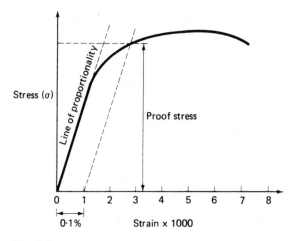

Fig. 3.7

of stress is taken as the yield stress, then the load required to induce this value may be defined as the *ultimate load*. The ratio of the ultimate load to the working (actual) load is termed the *load factor*.

Ultimate load = Working load × load factor

As is the case with factors of safety, load factors are related to the type of load and the reliability of load information. For example, in the design of structural steelwork in accordance with BS 5950: Part 1: 1985 *The structural use of steelwork in building*, the following are some of the load factors recommended:

Dead load	1.4
Dead load acting with wind load and imposed load combined	1.2
Imposed	1.6
Wind load	1.4
Forces due to temperature effects	1.2
Vertical and horizontal crane loads	1.6

The aim in building design is to establish that a number of *limit states* will not occur during the lifetime of a building. A *limit state* is defined as a state at which the building, a component, or a member, would become unfit for the intended purpose. Two main categories are determined in building design:

Ultimate limit state: at which instability or collapse is likely to occur, e.g. yielding, buckling, becoming a mechanism, overturning, sliding, brittle fracturing.

Serviceability limit state: at which function and use are impaired to an undesirable degree, e.g. deflection, twisting, settling of supports, vibration, corrosion.

Thus, if a structural member is to be subject, for example, to a dead load of W_d and an imposed load of W_i, the *ultimate load* will be:

$$W_U = 1 \cdot 4 \ W_d + 1 \cdot 6 \ W_i$$

and the *serviceability load* will be:

$$W_s = W_d + W_i$$

3.11 Calculations involving direct stress and strain

Elastic deformations resulting from direct stress are evaluated by the application of Hooke's law.

$$\frac{\text{Direct stress}}{\text{Direct strain}} = \text{Young's modulus of elasticity}$$

$$\frac{\sigma}{\epsilon} = E \qquad\qquad [3.6]$$

Now from equation [3.1]:

$$\sigma = \frac{F}{A}$$

and from equation [3.2] or [3.3]

$$\epsilon = \frac{\Delta L}{L}$$

Then substituting

$$\frac{F}{A} \times \frac{L}{\Delta L} = E$$

and rewriting for ΔL

$$\Delta L = \frac{FL}{AE} \qquad\qquad [3.7]$$

where: ΔL = change in length
L = original length
F = applied axial force
A = area of cross-section
E = Young's modulus of elasticity

EXAMPLE 3.4 *Determine the increase in length of a steel tie-rod 3 m long and 30 mm diameter when subjected to a tensile load of 120 kN. E = 205 kN/mm²*

Area of cross-section, $\quad A = \dfrac{\pi}{4} \times 300^2$

$$= 225\pi \text{ mm}^2$$

Using equation [3.7], $\quad \Delta L = \dfrac{FL}{AE}$

$$= \dfrac{120 \times 3 \times 1\ 000}{225\pi \times 205}$$

$$= 2 \cdot 48 \text{ mm (Answer)}$$

EXAMPLE 3.5 *A timber specimen of 50 mm square cross-section was initially 200 mm long parallel to the grain. After an axial compressive load of 40 kN was applied the specimen shortened by 0·29 mm. Determine the longitudinal elastic modulus for the timber.*

$$\text{Stress, } \sigma = \dfrac{F}{A} = \dfrac{40}{50 \times 50}$$

$$\text{Strain, } \epsilon = \dfrac{\Delta L}{L} = \dfrac{0 \cdot 29}{200}$$

and the modulus of elasticity, $\quad E = \dfrac{\sigma}{\epsilon} = \dfrac{40 \times 200}{50 \times 50 \times 0 \cdot 29}$

$$= 11 \cdot 03 \text{ kN/mm}^2$$

This value is the longitudinal modulus, since the load was applied parallel to the grain. The transverse modulus, i.e. perpendicular to the grain, will be a somewhat lower value. Materials which have a modulus of elasticity which is equal in all directions are said to be ISOTROPIC and those, such as timber, where E varies from one axis to another are ANISOTROPIC.

EXAMPLE 3.6 *A mass concrete pier of rectangular cross-section 600 mm × 800 mm and 2·20 m long carries an axial compressive load of 2·5 MN. Determine (a) the stress in the concrete at the base of the pier, (b) the amount of shortening that will occur in the pier. Density of concrete = 2 500 kg/m³; Young's modulus = 13 kN/mm²*

(a) First the 'inclusive' load must be calculated:
    Applied load $\qquad\qquad\qquad\qquad\qquad\qquad\qquad$ = 2 500 kN
    Load due to the mass of the concrete

$$= 0 \cdot 6 \times 0 \cdot 8 \times 2 \cdot 2 \times 2\,500 \times \frac{9 \cdot 81}{1\,000} \qquad = \qquad 26 \text{ kN}$$

Total inclusive load at base of pier $\qquad = 2\,526$ kN

Then the stress at the base of the pier,

$$= \frac{F}{A}$$

$$= \frac{2\,526 \times 1\,000}{600 \times 800}$$

$$= 5 \cdot 26 \text{ N/mm}^2 \text{ (Answer)}$$

(b) The change in length is determined from the strain, which is made up of the strain due to the applied load plus the strain due to the self-mass load of the concrete.

$$\text{Strain due to applied load} = \frac{\sigma}{E} = \frac{2\,500}{600 \times 800 \times 13}$$

$$= 0 \cdot 000401$$

Average strain due to self-mass load of concrete

$$= \frac{\text{average stress}}{E}$$

$$= \frac{26}{2 \times 600 \times 800 \times 13} = 0 \cdot 000002$$

Therefore the total strain $\qquad = 0 \cdot 000401 + 0 \cdot 000002$
Hence the total change in length $= 0 \cdot 000403 \times 2\,200 \qquad = 0 \cdot 887$ mm
(Answer)

EXAMPLE 3.7 The following data were recorded during a tensile test on a mild steel test piece of circular cross-section (Fig. 3.8).
Initial diameter of cross-section = 25 mm, gauge length = 250 mm, diameter of cross-section at fracture after failure = 18·6 mm.

Load/extension readings

Load (kN) :	20	40	60	80	100	120	140	150	160	170
Extension (mm) :	0·05	0·10	0·16	0·21	0·26	0·31	0·36	0·38	0·41	0·44

172	174	176	175	178	180	190	200	210	220	230	240
0·47	0·50	0·55	0·62	0·72	0·76	0·90	1·07	1·25	1·46	1·70	1·99

250	257	260	261	259	256	250	242	229
2·51	3·12	3·75	4·50	5·00	5·40	5·60	5·80	5·85

87

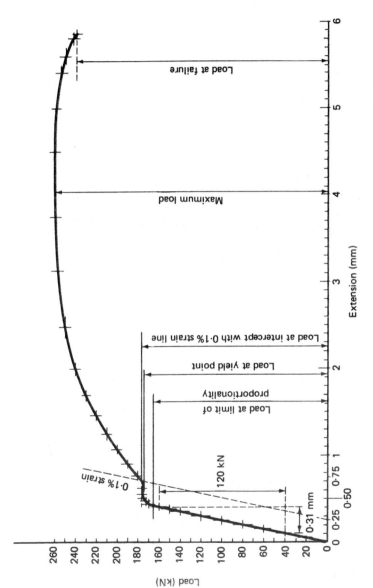

Load at failure

Maximum load

Load at intercept with 0·1% strain line

Load at yield point

Load at limit of
proportionality

0·1% strain

120 kN

0·31 mm

Extension (mm)

Load (kN)

Fig. 3.8

Plot the load/extension graph and from it find:
(a) *the modulus of elasticity of the material*
(b) *the limit of proportionality*
(c) *the yield stress*
(d) *the proof stress*
(e) *the percentage elongation*
(f) *the ultimate or maximum stress*

(a) The modulus of elasticity is determined from the slope of the straight portion of the graph.
A change in load of 120 kN produces a change in length of 0·31 mm

$$\text{Area of cross-section} = \frac{\pi}{4} \times 25^2 = 491 \text{ mm}^2$$

$$\text{Therefore modulus of elasticity, } E = \frac{120 \times 250}{491 \times 0·31}$$

$$= 197 \text{ kN/mm}^2 \tag{a}$$

(b) From the graph, the load at the limit of proportionality, i.e. the end of the straight-line portion, is found to be 166 kN.

$$\text{Therefore limit of proportionality} = \frac{166 \times 1\ 000}{491}$$

$$= 338 \text{ N/mm}^2 \tag{b}$$

(c) From the graph the load at yield point = 176 kN

$$\text{Therefore the yield stress} = \frac{176 \times 1\ 000}{491}$$

$$= 359 \text{ N/mm}^2 \tag{c}$$

(d) The proof stress is the stress causing a strain of 0·1% (Section 3.10).

$$\text{Extension at } 0·1\% \text{ strain} = 250 \times \frac{0·1}{100} = 0·25 \text{ mm}$$

A line is drawn from 0·25 mm extension on the extension axis parallel to the straight-line portion of the graph.
Load at the curve intercept of 0·1% strain = 178 kN

$$\text{Therefore the proof stress} = \frac{178 \times 1\ 000}{491}$$

$$= 362 \text{ N/mm}^2 \tag{d}$$

(e) The percentage elongation = percentage increase in gauge length

$$= \frac{5·85 \times 100}{250}$$

$$= 2·34\% \tag{e}$$

(f) From the graph the maximum load = 261 kN

Therefore the ultimate or maximum stress $= \dfrac{261 \times 1\,000}{491}$

$$= 532 \text{ N/mm}^2 \qquad (f)$$

(g) From the graph, the load at failure $= 239$ kN

Reduced area at failure $= \dfrac{\pi}{4} \times 18 \cdot 6^2$

$$= 272 \text{ mm}^2$$

Therefore failing stress $= \dfrac{239 \times 1\,000}{272}$

$$= 879 \text{ N/mm}^2 \qquad (g)$$

EXAMPLE 3.8   A steel tie-rod is required to transmit a load of 125 kN. The steel will have the same mechanical properties as the specimen for which the test results are given in Example 3.7. Determine the diameter of rod required (a) if the ultimate stress factor of safety is 4, and (b) if the yield factor is 1·8.

(a) From Example 3.7 the ultimate stress $= 532$ N/mm^2

Therefore the permissible stress $= \dfrac{532}{4} = 133$ N/mm^2

Then the area required $= \dfrac{125 \times 1\,000}{133}$

$$= 940 \text{ mm}^2$$

and the diameter required $= \sqrt{\left(940 \times \dfrac{4}{\pi}\right)}$

$$= 34 \cdot 6 \text{ mm}$$

(b) From Example 3.7 the yield stress $= 359$ N/mm^2

Therefore the allowable stress $= \dfrac{359}{1 \cdot 8}$

$$= 199 \text{ N/mm}^2$$

Then the area required $= \dfrac{125 \times 1\,000}{199} = 628 \text{ mm}^2$

and the diameter required $= \sqrt{\left(628 \times \dfrac{4}{\pi}\right)}$

$$= 28 \cdot 3 \text{ mm}$$

3.12 Compound bars

Members made up of two or more components of different materials and constructed in such a way that the components deform equally under axial loading are called COMPOUND BARS. The stresses in the components will be in the same proportions as their moduli of elasticity.

Consider the bar shown in Fig. 3.9 which is made up of two components, one of material A and the other of material B. The two components are secured together along their length so that they remain of equal length when the load F is varied.

Then change in length of A = change in length of B

or since their original lengths were equal,

$$\text{strain in } A = \text{strain in } B$$
$$\epsilon_A = \epsilon_B$$

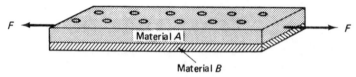

Fig. 3.9

Substituting from equation [3.6] gives:

$$\frac{E_A}{\sigma_A} = \frac{E_B}{\sigma_B} \qquad [3.8]$$

This is called the COMPATIBILITY EQUATION and specifies the ratio of stresses in A and B.

Rearranged, the equation becomes

$$\frac{\sigma_A}{\sigma_B} = \frac{E_A}{E_B} = m \text{ (the MODULAR RATIO)} \qquad [3.9]$$

The addition to the *modular ratio equation*, a further equation can be obtained from the fact that the portions of the load carried by the individual components must add up to the total load.

$$F_A + F_B = F$$
$$\sigma_A A_A + \sigma_B A_B = F \qquad [3.10]$$

This is called the EQUILIBRIUM EQUATON

Problems involving compound bars may be solved by the application of equations [3.9] and [3.10].

EXAMPLE 3.9 *A short reinforced concrete column is 450 mm square and contains four steel bars of 25 mm diameter. Determine the stresses in the steel and the concrete when the total load on the column is 1·5 MN. Young's moduli: steel = 210 kN/mm², concrete = 14 kN/mm²*

$$\text{Area of steel,} \quad A_s = \frac{\pi}{4} \times 25^2 \times 4 \ = 1\,963 \text{ mm}^2$$

$$\text{Area of concrete,} \quad A_c = 450^2 - 1\,963 \ = 200\,540 \text{ mm}^2.$$

From equation [3.9],

$$\frac{\sigma_s}{\sigma_c} = \frac{210}{14} = 15$$

$$\therefore \ \sigma_s = 15\,\sigma_c$$

and from equation [3.10],

$$\sigma_s\,A_s + \sigma_c\,A_c = \text{total load}$$
$$15\,\sigma_c \times 1963 + \sigma_c \times 200\,540 = 1\cdot5 \times 10^6$$
$$230\,000\,\sigma_c = 1\cdot5 \times 10^6$$

$$\therefore \ \sigma_c = \frac{1\cdot5}{0\cdot23}$$

$$= 6\cdot52 \text{ N/mm}^2 \text{ (Answer)}$$
$$\text{and} \quad \sigma_s = 15 \times 6\cdot52$$
$$= 97\cdot8 \text{ N/mm}^2 \text{ (Answer)}$$

EXAMPLE 3.10 *A rigid horizontal beam of length 1·5 m carries a uniform inclusive load of 4 kN. The beam is supported by three vertical wires, each initially 2·4 m long when unstressed; two brass wires of diameter 4·0 mm are attached to the ends of the beam and a steel wire of diameter 1·8 mm is attached to the mid-point. Determine the stresses and elongations occurring in the wires due to the loading described. Young's moduli: steel = 205 kN/mm²; brass = 85 kN/mm².*

$$\text{Area of steel,} \quad A_s = \frac{\pi}{4} \times 1\cdot8^2 = 2\cdot54 \text{ mm}^2$$

$$\text{Area of brass,} \quad A_b = \frac{\pi}{4} \times 4\cdot0^2 \times 2 = 25\cdot13 \text{ mm}^2$$

From equation [3.9],

$$\frac{\sigma_s}{\sigma_b} = \frac{205}{85} = 2\cdot41$$

$$\therefore \ \sigma_s = 2\cdot41\,\sigma_b$$

and from equation [3.10],

$$\sigma_s A_s + \sigma_b A_b = \text{total load}$$
$$2\cdot41\ \sigma_b \times 2\cdot54 + \sigma_b \times 25\cdot13 = 4 \times 1\ 000$$
$$31\cdot25\ \sigma_b = 4\ 000$$

$$\therefore\ \sigma_b = \frac{4\ 000}{31\cdot25}$$

$$= 128\ \text{N/mm}^2\ \text{(Answer)}$$
$$\text{and}\qquad \sigma_s = 2\cdot41 \times 128$$
$$= 308\ \text{N/mm}^2\ \text{(Answer)}$$

Now $\epsilon_s = \epsilon_b = \dfrac{\sigma}{E} = \dfrac{308}{205 \times 10^3}$

$\therefore$ elongation (in all three wires) $= \dfrac{308 \times 2\cdot40 \times 10^3}{205 \times 10^3} = 3\cdot6\ \text{mm}$ (Answer)

EXAMPLE 3.11   *A short reinforced concrete column of section 200 mm ×
220 mm is to be reinforced with 4 No steel bars and is required to carry an axial
load of 850 kN (Fig. 3.10). The stress in the concrete must not exceed 7 N/mm²*

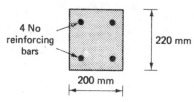

Fig. 3.10

*and the stress in the steel must not exceed 150 N/mm². Determine the diameter
required for the bars and the subsequent stresses occurring in the concrete and
the steel under the specified load. Young's moduli: concrete = 14 kN/mm², steel
= 210 kN/mm².*

From equation [3.9],

$$\frac{\sigma_s}{\sigma_c} = \frac{210}{14} = 15$$

$$\therefore\ \sigma_s = 15\ \sigma_c$$

But $\sigma_s \not> 150\ \text{N/mm}^2$
and $\sigma_c \not> 7\ \text{N/mm}^2$

If σ_c is at maximum value permitted, $\sigma_s = 7 \times 15 = 105\ \text{N/mm}^2\ (<150)$

If σ_s is at maximum value permitted, $\sigma_c = \dfrac{150}{10} = 10\ \text{N/mm}^2\ (>7)$

The limiting case is therefore with σ_c at the maximum permitted value.

From equation [3.10],

$$\sigma_s A_s + \sigma_c A_c = \text{total load}$$
$$105 A_s + 7(220 \times 200 - A_s) = 850 \times 1\,000$$

$$\therefore A_s = \frac{850\,000 - 308\,000}{98}$$

$$= \frac{542\,000}{98}$$

$$= 5\,530 \text{ mm}^2$$

Then diameter required
$$= \sqrt{\left(\frac{5530 \times 4}{4 \times \pi}\right)}$$
$$= \sqrt{1760}$$
$$= 41 \cdot 95 \text{ mm}$$

Nearest practical size
$$= 42 \text{ mm (Answer)}$$

Actual area $= \dfrac{\pi}{4} \times 42^2 \times 4 = 5542 \text{ mm}^2$

From equation [3.10],

$$15 \sigma_c \times 5542 + \sigma_c (220 \times 200 - 5542) = 850\,000$$
$$83\,130 \sigma_c + 38\,458 \sigma_c = 850\,000$$

$$\therefore \sigma_c = \frac{850\,000}{121\,588}$$

$$= 6 \cdot 99 \text{ N/mm}^2 \text{ (Answer)}$$
$$\text{and} \quad \sigma_s = 6 \cdot 99 \times 15$$
$$= 104 \cdot 8 \text{ N/mm}^2 \text{ (Answer)}$$

EXAMPLE 3.12   A threaded steel rod of minimum diameter 20 mm passes centrally through a copper tube of outside diameter 40 mm and inside diameter 25 mm. Nuts and washers are fitted and the components assembled as shown in Fig. 3.11 so that the nuts are just hand-tight with the rod and tube both still unstressed. Determine the stresses in the rod and the tube if one of the nuts is now given a quarter-turn tighter. The thread on the steel rod has a pitch of 2·50 mm. Young's moduli: copper = 105 kN/mm^2; steel = 210 kN/mm^2.

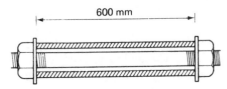

600 mm

Fig. 3.11

Area of steel rod $\qquad A_s = \dfrac{\pi}{4} \times 20^2 = \dfrac{\pi}{4} \times 400 \text{ mm}^2$

Area of copper tube $\qquad A_c = \dfrac{\pi}{4}(40^2 - 25^2) = \dfrac{\pi}{4} \times 975 \text{ mm}^2$

When the nut is tightened, the steel rod is placed in tension and the copper tube in compression.

Then, for equilibrium,

the tension in the steel rod = the compression in the copper tube
$$\sigma_s A_s = \sigma_c A_c$$

$$\sigma_s \times \frac{\pi}{4} \times 400 = \sigma_s \times \frac{\pi}{4} \times 975$$

$$\therefore \sigma_s = \frac{975}{400}\sigma_c = 2\cdot44\,\sigma_c$$

The nut is tightened a $\frac{1}{4}$ turn, therefore the movement along the thread $= 2\cdot50/4$ mm

Let Δ_c = contraction of copper tube
and Δ_s = extension of steel rod

Then $\qquad\qquad\qquad \dfrac{2\cdot50}{4} = \Delta_s + \Delta_c$

$$= \frac{\sigma_s L}{E_s} + \frac{\sigma_c L}{E_c}$$

$$\frac{2\cdot50}{4 \times 600} = \frac{\sigma_s}{210} + \frac{\sigma_c}{105}$$

$$\frac{210 \times 2\cdot50 \times 10^3}{4 \times 600} = \sigma_s + 2\sigma_c$$

$$218\cdot75 = 2\cdot44\,\sigma_c + 2\,\sigma_c = \sigma_s + 2\sigma_s/2\cdot44$$

$$\therefore \sigma_c = \frac{218\cdot75}{4\cdot44} = 49\cdot3 \text{ N/mm}^2 \text{ (Answer)}$$

and $\qquad \sigma_s = \dfrac{218\cdot75}{1\cdot820} = 120\cdot2 \text{ N/mm}^2 \text{ (Answer)}$

3.13 Temperature stress

All solid bodies undergo a linear expansion or contraction when the temperature is either raised or lowered. The change in length of an unrestrained bar which is subject to a change in temperature is given by the following expression:

$$\Delta L = \alpha L \,\Delta t \qquad\qquad\qquad\qquad [3.11]$$

where: ΔL = the change in length of the bar
α = the coefficient of linear expansion of the material
L = the original length of the bar
Δt = the change in temperature

Then the new length of the bar $= L + \Delta L$
$$= L(1 + \alpha \Delta t)$$

Now supposing the ends of the bar are restrained, thus preventing a change in length; this will be the same as compressing a bar of length $L(1 + \alpha \Delta t)$ to a new length of L (Fig. 3.12).

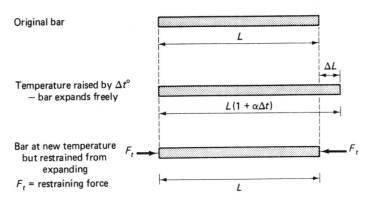

Fig. 3.12

The compressive strain induced in the bar will be,

$$\epsilon_t = \frac{\Delta L}{L + \Delta L} = \frac{\alpha L \, \Delta t}{L(1 + \alpha \Delta t)}$$

and since $\alpha \Delta t$ will be very small compared with unity, it may be neglected in the denominator, giving:

$$\epsilon_t = \alpha \Delta_t \qquad [3.12]$$

The temperature stress set up in the bar will therefore be

$$\sigma_t = \epsilon_t E = A \Delta t \, E \qquad [3.13]$$

This equation is valid providing that σ_t does not exceed the yield point of the material. It should be noted particularly that the temperature stress is *independent of both the length and cross-section size of the bar*.

EXAMPLE 3.13 Determine the variation in stress in fully restrained steel members resulting from changes in ambient temperature.
Coefficient of linear expansion for steel, $\alpha = 12 \times 10^{-6}$ per $°C$; Young's modulus of elasticity for steel, $E = 205 \text{ kN/mm}^2$.

96

Temperarure stress, $\sigma_t = \alpha \Delta t\, E$

$$= \frac{12 \times 1 \times 205 \times 10^3}{10^6}$$

$$= 2\cdot 52 \text{ N/mm}^2 \text{ per } °C \text{ (Answer)}$$

3.14 Strain energy

When a load is gradually applied to a bar and the bar undergoes a change
in length, the load moves as the bar deforms and therefore work is done
by the load on the bar. Since the effect of this work is to *strain* the bar,
the measure of the work done is called *strain energy* (U). The strain
energy stored in the bar whilst it is in a state of strain is also called the
RESILIENCE of the bar.

For a gradually applied tensile load, the work done in straining the bar
is as shown in Fig. 3.13.

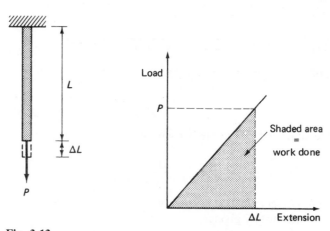

Fig. 3.13

$$\begin{aligned}
\text{Work done} &= \text{strain energy } (U)\\
&= \text{average load} \times \text{extension}\\
&= \tfrac{1}{2} P \times \Delta L
\end{aligned}$$

Hence $\qquad\qquad U = \tfrac{1}{2} P\, \Delta L \qquad\qquad$ [3.14]

Substituting for $\quad \Delta L\left(= \dfrac{PL}{AE} \right)$

$$U = \frac{P^2 L}{2AE} \qquad\qquad [3.15]$$

Substituting for $P(= \sigma A)$,

$$U = \frac{\sigma^2 AL}{2E} \qquad [3.16]$$

Now AL = the volume of the bar
Hence equation [3.16] may be written

$$U = \frac{\sigma^2}{2E} \times \text{volume of bar}$$

The term $\sigma^2/2E$ is called the *modulus of resilience*, or the *resilience per unit volume*. The greatest amount of strain energy that can be stored in a bar without it being permanently deformed is called the *proof resilience*, and corresponds to the value of U when the stress in the bar is equal to the yield stress or the proof stress, whichever is appropriate.

Beyond the elastic limit only a small quantity of the work done is stored as resilience. Most of the remainder is either used up in producing permanent deformation in the bar or is dissipated in the form of heat.

EXAMPLE 3.14 *A solid steel rod of length 1 m has a diameter of 25 mm except for a centre portion 300 mm long where the diameter has been turned down to 15 mm. Determine the strain energy stored in the rod when a tensile load of 50 kN is gradually applied. Determine also the diameter required for a brass rod of the same length and of constant diameter, which will have the same proof resilience as the steel rod. Young's moduli: steel = 210 kN/mm²; brass = 85 kN/mm². Yield stresses: steel = 220 N/mm²; brass = 140 N/mm².*

$$A_{s1} = \frac{\pi}{4} \times 25^2 = 491 \text{ mm}^2$$

$$A_{s2} = \frac{\pi}{4} \times 15^2 = 177 \text{ mm}^2$$

Strain energy, $U = \dfrac{P^2 L}{2AE} = \dfrac{P^2}{2E}\left[\dfrac{L_1}{A_{s1}} + \dfrac{L_2}{A_{s2}}\right]$

$$= \frac{50^2 \times 10^6}{2 \times 210 \times 10^3}\left[\frac{300}{177} + \frac{700}{491}\right]$$

$$= \frac{2\,500 \times 10^3}{420}\left[\frac{300}{177} + \frac{700}{491}\right]$$

$$= 18\,575 \text{ Nmm (Answer)}$$

Proof resilience of steel rod, $U_y = \dfrac{220^2\,(0\cdot 3 \times 177 \div 0\cdot 7 \times 491)}{2 \times 210 \times 10^3}$

$$= \frac{220^2 \times 397}{2 \times 210 \times 10^3}$$

Proof resilience of brass rod, $U_y = \dfrac{140^2 \times 1 \times A_b}{2 \times 85 \times 10^3}$

$$\therefore A_b \text{ (reqd)} = \frac{220^2 \times 397 \times 85}{140^2 \times 210} = 397 \text{ mm}^2$$

and diameter required $= \sqrt{\dfrac{397 \times 4}{\pi}}$

$$= 22 \cdot 5 \text{ mm (Answer)}$$

3.15 Suddenly applied loads

If a load is *suddenly* (as opposed to *gradually*) applied to an elastic member, the member will tend to behave as a spring showing damped oscillations on either side of the equilibrium position. The equilibrium position corresponds to the average strain which is equal to the static strain caused by the same load gradually applied. The minimum instantaneous strain is zero, so therefore the maximum instantaneous strain must be twice the average strain (Fig. 3.14).

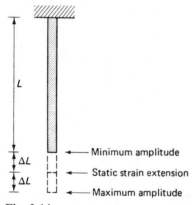

Fig. 3.14

The *maximum instantaneous stress* is the stress occurring in the bar at the maximum instantaneous strain and is *twice the static stress set up by the gradual application of the same load.*

$$\text{Maximum instantaneous stress, } \hat{\sigma}_i = \frac{2P}{A} \qquad [3.17]$$

3.16 Falling loads

When a load *falls* (as opposed to being *suddenly* applied — Section 3.15) onto a bar, the work done on the bar includes the kinetic energy of the

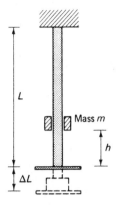

Fig. 3.15

load at the moment of impact. As a result, the maximum instantaneous stress may be several times the value of the equivalent static stress.

Consider the vertical bar shown in Fig. 3.15. The mass m falls through height h onto the collar at the lower end of the bar causing a maximum extension in the bar of ΔL.

Let P = the static force which, when gradually applied, will cause the same extension ΔL.

Then the maximum instantaneous strain energy in the bar is

$$U_{max} = \tfrac{1}{2} P \, \Delta L = \frac{P^2 L}{2AE}$$

Neglecting the losses on impact, the total work done by the falling load is

$$U = mg(h + \Delta L) = mg\left(h + \frac{PL}{AE}\right)$$

Since the work done by the load = the strain energy stored

$$\frac{P^2 L}{2AE} = mgh + \frac{mgPL}{AE}$$

Rewriting, $P^2 - 2mgP - \dfrac{2mghAE}{L} = 0$

This is a quadratic equation which can now be solved, the positive root yielding an expression for P.

$$P = mg + \sqrt{\left\{(mg)^2 + \frac{2mghAE}{L}\right\}}$$

or
$$P = mg \left[1 + \sqrt{\left(1 + \frac{2hAE}{mgL}\right)} \right] \qquad [3.18]$$

From this expression the maximum instantaneous stress and the maximum extension can be found:

$$\hat{\sigma}_i = \frac{P}{A}$$

$$\Delta L = \frac{PL}{AE}$$

It will be seen that the case for a *suddenly applied load* is obtained by putting $h = 0$

Then $P = 2mg$

i.e. the same result as equation [3.17].

EXAMPLE 3.15   In an arrangement such as that shown in Fig. 3.15, a mass of 120 kg falls 50 mm onto a collar at the lower end of a vertical steel bar of diameter 25 mm and length 3 m. Determine the maximum instantaneous stress occurring in the bar and also the maximum extension that takes place. E = 210 kN/mm².

$$\textit{Area of cross-section of bar} = \frac{\pi}{4} \times 25^2 = 491 \text{ mm}^2$$

From equation [3.18]:

The equivalent static load, $P = 120 \times 9 \cdot 81$

$$\left[1 + \sqrt{\left(1 + \frac{2 \times 50 \times 491 \times 210 \times 10^3}{120 \times 9 \cdot 81 \times 3 \times 10^3} \right)} \right]$$

$$= 120 \times 9 \cdot 81 \, [1 + \sqrt{(1 + 2\,920)}]$$
$$= 120 \times 9 \cdot 81 \times 55 \cdot 04$$
$$= 64\,797 \text{ N}$$

Then the maximum instantaneous stress

$$\hat{\sigma}_i = \frac{64\,797}{491}$$

$$= 132 \text{ N/mm}^2 \text{ (Answer)}$$

Extension of the bar under the equivalent static load equals the maximum instantaneous extension:

$$\Delta L = \frac{PL}{AE}$$

$$= \frac{64\ 797 \times 3 \times 10^3}{491 \times 210 \times 10^3}$$

$$= 1 \cdot 89 \text{ mm (Answer)}$$

3.17 Creep under sustained loading

It has been found by experiment that in most metals a slow, progressive deformation takes place under conditions of constant sustained stress. This phenomenon is called CREEP; it occurs in most structural materials to some degree and is attributed to changes in the internal crystallographic structure of the material.

If a tensile test is carried out in the usual way, over a period of several minutes at a constant room temperature, a definite ultimate stress value can be measured. A similar test piece of the same material, however, may be made to fail at a much lower stress if the test is prolonged over a period of days or weeks.

In the case of soft metals (such as lead and zinc) and timber, a noticeable amount of creep will take place in a few hours at room temperature. At normal temperatures, concrete shows a measurable tendency to creep over periods of several years; this fact must be taken into consideration when designing prestressed concrete members.

For most steels, however, there appears to be a limiting stress value, corresponding to a particular temperature, below which the amount of creep that is likely to take place is negligible. The higher the ambient temperature the lower this *limiting creep stress* will be. For work at higher temperatures special alloy steels containing small amounts of chromium, molybdenum and vanadium have been developed which have very low creep rates.

3.18 Fluctuating stress. Fatigue

It has been found that, if a material has been subjected to fluctuating stresses at fairly high frequencies, there is a tendency for failure to occur at a stress well below the ultimate stress value as obtained in a static test. This phenomenon is called FATIGUE and is more likely to occur under conditions of either *repeated* tensile stress, or *reversed* stress. In the case of *repeated* tensile stress, the stress varies from some maximum tensile value to some lower tensile value, or maybe zero. The *reversed* stress condition is said to occur when the mean stress is zero and the fluctuations range between a maximum compressive value and a maximum tensile value. Fatigue is not usually associated with repeated compressive stress.

Results of laboratory tests show that, for steels and steel alloys, there is a limiting stress value below which fatigue failure will not take place, regardless of the number of fluctuation cycles. This limiting value is called the FATIGUE LIMIT of the material (Fig. 3.16).

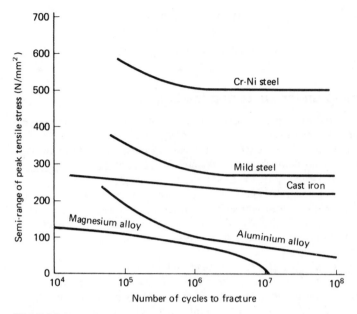

Fig. 3.16

The majority of non-ferrous metals do not show a fatigue limit, but the stress/No-cycles curve continues to slope at a shallow angle. It is usual in these cases to quote the ENDURANCE LIMIT, this being the stress range over which a specific large number of cycles may be expected to occur (usually 50×10^6 cycles) before fatigue failure is likely.

3.19 Solutions using computers

Most of the exercises in this chapter are relatively easy to solve by hand calculation. Students engaged in laboratory work, however, should make use of graph plotting, curve fitting and statistical packages. The study of engineering behaviour is well served by making small changes to key properties or dimensions and then observing the consequences in calculated results.

Figure 3.17 shows a spreadsheet printout of elasticity calculations. In this spreadsheet starting data may be entered in a variety of ways and sequences. This is illustrated using Examples 3.1, 3.4, 3.6 and 3.7. When dealing with compound members (Section 3.12), calculations will normally start with load data and finish with stress, or vice versa. The spreadsheet shown in Fig. 3.18 enables either mode to be followed after basic dimensional data has been entered; data from Examples 3.9 and 3.11 is illustrated. For more information and sources of software see Section 1.14.

```
SSE1              Stress,  strain  and  elasticity  calculations
-------------------------------------------------------------------
   Calc  No:       3.1        3.4        3.6a       3.6b        3.7a
-------------------------------------------------------------------
  AREA DIMENSIONS    (mm)
  Breadth                      0.6       0.6
  Depth                        0.8       0.8
  Diameter          20         30                               25
-------------------------------------------------------------------
    Area           314.16    706.86      0.48       0.48      490.87
-------------------------------------------------------------------
  LOAD (kN)  60000  120000   2.526      2.513    120000
  Stress           190.99    169.77      5.26                5.24   244.46
-------------------------------------------------------------------
  LENGTH             3000                 2.2       250
  MODULUS E         205000              13000    197146.7
  Strain           0.000828            0.000402   0.00124
  Elongation       2.484369                       0.00885       0.31
-------------------------------------------------------------------
```

Fig. 3.17

```
SRCCOL1         Short reinforced concrete column (elastic method)
                Compound bar
===================================================================
Modulus A         210        210
Modulus B          14         14
Area A           5542       1963
Area B          38458     200537
Total Area      44000     202500
===================================================================
SAFE LOAD: stresses and moduli known in materials A and B
-------------------------------------------------------------------
Stress A          105
Stress B            7
Safe load       85116
===================================================================
STRESSES: Load and moduli known for materials A and B
-------------------------------------------------------------------
Applied load             1500000
Stress A                    6.52
Stress B                   97.83
===================================================================
```

Fig. 3.18

EXERCISES   CHAPTER 3

1. A steel tie-bar of 100 mm × 9 mm cross-section is subject to an axial pull of 180 kN. Calculate the stress in the bar. By how much must the thickness of the bar be increased if the stress must not be allowed to exceeed 135 N/mm^2?

2. Calculate the safe axial load in tension for a steel bar of cross-section 75 mm × 12 mm, if the allowable maximum stress is 155 N/mm^2.

3. A steel rod of diameter 25 mm is subject to an axial tensile load of 80 kN. Calculate the stress in the rod.

4. Calculate the required diameter for a steel rod that is to transmit an axial pull of 50 kN, if the allowable maximum stress is 155 N/mm^2.

5. A hollow steel tube of external diameter 100 mm and wall thickness 12 mm is subject to an axial tension of 550 kN. Calculate the stress induced by this load. What diameter of solid rod of the same steel would be of equal strength?

6. Calculate the safe load for a timber tie of cross-section 75 mm × 50 mm if the allowable maximum stress is 8·5 N/mm^2. Calculate also the diameter of steel bar that would be of equal strength if the allowable maximum stress for the steel is 155 N/mm^2.

7. The total compression transmitted at the base of a brick pier is 122 kN. Calculate the size of pier required if the allowable maximum stress for the brickwork is 1·10 MN/m^2.

8. A steel slab base plate for a steel stanchion is required to transmit a load of 1·21 MN to a concrete footing. Calculate the size of square slab required if the allowable maximum direct compression on the concrete is 5·3 MN/m^2.

9. Calculate the size of square concrete footing required to transmit a load of 870 kN to the foundation soil if the safe bearing capacity of the soil is 0·30 MN/m^2.

10. A steel tie-bar of gross cross-section 75 mm × 9 mm has a 20 mm diameter hole drilled in it at each end to receive bolts. Calculate the safe load for the member if the allowable tensile stress is 155 N/mm^2.

11. A steel tie-rod of diameter 28 mm has a screw thread at each end which has a root diameter of 23·9 mm. Calculate the safe maximum load if the allowable stress is 215 N/mm^2. What will be the stress in the main part of the rod at this maximum load?

12. A steel tie of rectangular section is to transmit an axial load of 145 kN. The member is connected at each end by two lines of rivets in 20 mm diameter holes (Fig. 3.19). Calculate the breadth required if the member is to be 9 mm thick and the allowable stress is 155 N/mm^2.

13. Calculate the number of 19 mm diameter rivets driven into 20 mm diameter holes that are required to transmit a shearing force of 260 kN. The rivets will be in single shear as shown in Fig. 3.3 (p. 77). The permissible shearing stress for the rivets is 110 N/mm^2.

14. A timber cube of side dimension 50 mm failed in a compression test at a load of 85 kN. Determine the allowable stress (based on this test) for the timber in direct compression if the factor of safety is to be five.

15. Concrete cubes tested to failure in compression showed an average strength of

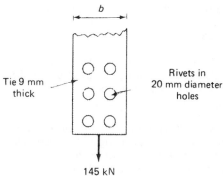

Fig. 3.19

$22 \cdot 7$ N/mm^2 at an age of 28 days after mixing. Calculate the factor of safety if the allowable direct compressive stress is specified at $5 \cdot 5$ N/mm^2.

16. A steel specimen of diameter 6 mm sustained a maximum load of $22 \cdot 9$ kN in a tensile test. Adopting a factor of safety of $3 \cdot 5$, calculate the breadth of flat bar having a thickness of 12 mm (in the same steel) that would safely transmit an axial tensile load of $33 \cdot 5$ kN.

17. A tie-bar of diameter 32 mm is 2 m long. Calculate the stress in the bar and the extension that takes place when an axial tension of 100 kN is applied. Young's modulus = 210 kN/mm^2.

18. A steel rod of diameter 25 mm is subject to an axial pull of 150 kN. Calculate the increase in length that takes place over a gauge length of 350 mm. Young's modulus = 210 kN/mm^2.

19. A steel wire of diameter $4 \cdot 76$ mm and length 3 m has a modulus of elasticity of 210 kN/mm^2. Calculate the load required to cause the wire to stretch 12 mm.

20. A steel wire of diameter $2 \cdot 0$ mm and length 8 m stretches $10 \cdot 3$ mm when subject to a load of 890 N. Calculate Young's modulus of elasticity for the steel.

21. A timber specimen of 100 mm square cross-section was observed to shorten by $0 \cdot 052$ mm over a gauge length of 200 mm when it was subject to an axial compression of $21 \cdot 2$ kN. Calculate Young's modulus for the timber.

22. During a tensile test on a specimen of steel the following data was recorded:

Diameter of specimen = $6 \cdot 6$ mm   Gauge length = $50 \cdot 0$ mm
Elongation at a load of $6 \cdot 0$ kN = $0 \cdot 043$ mm
Load at yield point = $9 \cdot 7$ kN   Load at failure = $14 \cdot 2$ kN
Reduction of area at failure = 45 per cent

Calculate: (a) Young's modulus of elasticity for the steel,
          (b) the yield stress
          (c) the ultimate stress.

106

23. A steel cable hangs freely down a shaft. If the density of the cable is 7 800 kg/m^3 and Young's modulus is 210 kN/m^2, calculate the elongation that takes place over an original length of 220 m.

24. A brass tube of external diameter 50 mm and 6 mm wall thickness is 1·8 m long and is compressed between end plates under a load of 27 kN. Calculate the reduction in length that takes place if Young's modulus for the brass is 80 kN/mm^2.

25. Figure 3.20 shows how two rods A and B support a rigid bar. The diameter of rod A is 20 mm. If the rods are of equal length and the rigid bar is horizontal both before and after the application of the 80 kN load, calculate the diameter of rod B. Also determine the stress in each rod. Young's moduli: Rod A = 200 kN/mm^2; rod B = 85 kN/mm^2.

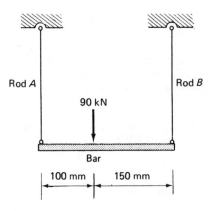

Fig. 3.20

26. A short concrete column of 350 mm square section is reinforced with four 20 mm diameter bars and carries an axial load of 1 000 kN. Calculate the stresses in the steel and concrete.
Young's moduli: Steel = 210 kN/mm^2; concrete = 14 kN/mm ^{2}.

27. A concrete column 4 m high and 460 mm × 380 mm in section is reinforced with six 25 mm diameter bars. Calculate the safe axial load for the column if the permissible stresses are 7 N/mm^2 and 140 N/mm^2 for the concrete and steel respectively. Calculate also the amount of shortening that will take place under this load.
Young's moduli: Steel = 210 kN/mm^2; concrete = 14 kN/mm^2.

28. A copper rod of 25 mm diameter is enclosed by and securely attached at each end to a steel tube of 40 mm external diameter and 35 mm internal diameter. Calculate the stresses in the copper and the steel when the combination is subject to an axial tension of 40 kN.
Young's moduli: Steel = 200 kN/mm^2; copper = 95 kN/mm^2.

29. A compound tie consists of a square timber core 75 mm × 75 mm with a 75 mm × 12 mm steel plate bolted to opposite sides. Calculate the safe axial load for the member if the permissible stresses in the timber and steel are $6 \cdot 3$ N/mm^2 and 140 N/mm^2 respectively.
Young's moduli: Timber = 8 kN/mm^2; steel = 200 kN/mm^2.

30. A steel bolt of diameter 25 mm passes through a steel tube of external diameter 65 mm, internal diameter 50 mm and length 406 mm. The bolt is tightened onto the tube through rigid end blocks until the tensile force in the bolt is 40 kN. At this point the distance between the head of the bolt and the nut is 508 mm. If an additional tension of 30 kN is now applied to the end-blocks, calculate the forces now acting in the bolt and the tube.

31. A steel rod of diameter 25 mm and length $0 \cdot 5$ m is placed concentrically inside a brass tube having an internal diameter of 28 mm and a thickness of 5 mm. The length of the tube exceeds that of the rod by $0 \cdot 15$ mm. An axial compressive load is applied through rigid plates at the ends of the tube.
Calculate: (a) the load at which the rod and tube first become of the same length,
(b) the stresses in the steel and the brass when a load of 50 kN is applied.
Young's moduli: Steel = 208 kN/mm^2; brass = 100 kN/mm^2.

32. Calculate the increase in stress that will occur in a restrained steel strut subject to a rise in temperature from 12 to 40 °C.
Coefficient of linear expansion of steel = $12 \cdot 0 \times 10^{-6}$ per °C.
Young's modulus for steel = 205 kN/mm^2.

33. A straight copper rod of uniform section is held in clamps which are 400 mm apart and exert a pull on the rod giving rise to a tensile stress of 35 N/mm^2. If the temperature of the rod is then raised by 50 °C and the distance between the clamps remains unchanged, calculate the stress in the rod.
Coefficient of linear expansion of copper = $17 \cdot 0 \times 10^{-6}$ per °C.
Young's modulus for copper = 110 kN/mm^2.

34. Determine the minimum gap to be left between successive lengths of crane running rails each of 10 m if the increase in stress in the steel is not to exceed 20 N/mm^2 for a temperature increase of 50 °C.
Coefficient of linear expansion of steel = $12 \cdot 0 \times 10^{-6}$ per °C.
Young's modulus for steel = 205 kN/mm^2.

35. A steel beam is 25 m long between supports. If each support can accommodate $2 \cdot 5$ mm of expansion movement, determine the increase in compressive stress that will take place due to a rise in temperature of 40 °C.
Coefficient of linear expansion of steel = $12 \cdot 0 \times 10^{-6}$ per °C.
Young's modulus for steel = 205 kN/mm^2.

36. Figure 3.21 shows a compound strut consisting of a brass portion AB of diameter 75 mm and a steel portion BC of diameter 40 mm. The strut is held between rigid supports at A and C. If the stresses in the strut are zero at 10 °C and the temperature is then raised to 150 °C, determine:

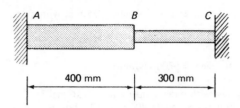

A   B   C

| 400 mm | 300 mm |

Fig. 3.21

(a) the force exerted on the supports, assuming that the length between them remains constant.

(b) The relative movement of junction B.

Coefficient of linear expansion of steel $= 11 \cdot 0 \times 10^{-6}$ per °C.

Coefficient of linear expansion of brass $= 10 \cdot 0 \times 10^{-6}$ per °C.

Young's moduli: Steel $= 210$ kN/mm²; brass $= 85$ kN/mm².

37. A steel rod of length 1 m is of diameter 75 mm over a portion 0·7 m long and of diameter 45 mm over the remainder. Calculate the strain energy in the rod when it is subject to an axial tension of 50 kN.

Young's modulus $= 210$ kN/mm².

38. A steel tie-rod of length 1·5 m and a diameter of 30 mm has a screw thread at each end for a length of 200 mm which has a root diameter of 25·2 mm. Calculate the difference in the strain energy in the rod under a load of 100 kN when the length between the nuts is reduced from 1·415 to 1·132 m.

39. A compound tie is made up of a copper rod and a brass rod connected end to end; each component has a diameter of 100 mm and is 2·5 m long. Calculate the strain energy in the tie when it is subject to a load of 250 kN. What diameter of steel rod can replace the brass rod if the strain energy is to remain constant at this load?

Young's moduli: Steel $= 210$ kN/mm²; copper $= 100$ kN/mm²; brass $= 80$ kN/mm².

40. A load of 2·2 kN falls through a height of 35 mm onto a collar at the lower end of a vertical bar which is 9 m long and of diameter 27 mm.

Calculate the maximum stress induced in the bar.

Young's modulus $= 210$ kN/mm².

41. A load of 40 kN falls onto a collar at the lower end of a vertical bar which has a length of 3·2 m and a diameter of 35 mm. Calculate the height above the collar from which the load must fall in order to produce a maximum stress in the bar of 200 N/mm².

Young's modulus $= 210$ kN/mm².

42. A vertical bar of length 2·5 m and diameter 20 mm is held rigidly at the upper end and is fitted with a collar at the lower end onto which a weight W may fall freely from a height of 350 mm above the collar. Calculate the maximum value allowable for W if the stress in the bar must not exceed 120 N/mm².

Young's modulus $= 210$ kN/mm².

4 Properties of structural sections

Symbols

A, a = area, total area, distance
B, b = breadth of section
c = distance, distance between parallel axes
D, d = diameter
d = depth of section
G = position of centroid
I = second moment of area (about a given axis)
J = polar second moment of area (about the centroid)
R, r = radius
r_{xx} = radius of gyration about the $x-x$ axis
r_{yy} = radius of gyration about the $y-y$ axis
x = distance in the x-direction
$\bar{x}$ = distance to centroid in the x-direction
y = distance in the y-direction
$\bar{y}$ = distance to centroid in the y-direction
Z = elastic section modulus (about a given axis)

4.1 Introduction

It will be necessary in later chapters to refer to a number of basic geometrical properties of the cross-sections of structural members. In order that the various expositions in structural theory, which are complex

Table 4.1 Properties of structural sections

Section	Area (A)	Distance from extreme fibre to centroid $(\bar{y})$	Second moment of area (I_{xx})	Polar second moment of area $(J = I_{xx} + I_{yy})$	Radius of gyration (r_{xx})
	bd	$\dfrac{d}{2}$	$\dfrac{bd^3}{12}$	$\dfrac{bd^3}{6}$	$\dfrac{d}{2\sqrt{3}}$
Rectangle or rhombus	bd	$\dfrac{d}{2}$	$\dfrac{bd^3}{3}$	—	$\dfrac{d}{\sqrt{3}}$
	$\dfrac{bd}{2}$	$\dfrac{2d}{3}$	$\dfrac{bd^3}{36}$	—	$\dfrac{d}{3\sqrt{2}}$
	$\dfrac{bd}{2}$	$\dfrac{2d}{3}$	$\dfrac{bd^3}{12}$	—	$\dfrac{d}{\sqrt{6}}$
	$\dfrac{bd}{2}$	$\dfrac{2d}{3}$	$\dfrac{bd^3}{4}$	—	$\dfrac{d}{\sqrt{2}}$
Triangle	$\dfrac{bd}{2}$	$\dfrac{2a+c}{3}$	$\dfrac{d}{36}(b^3 - abc)$	—	$\sqrt{\left(\dfrac{b^2 - ac}{18}\right)}$
Trapezium	$\dfrac{(B+b)d}{2}$	$\dfrac{(B+2b)d}{(B+b)3}$	$\dfrac{d^3}{36}\left[B + b + \dfrac{2Bb}{B+b}\right]$	—	—
Circle	$\dfrac{\pi D^2}{4}$	$\dfrac{D}{2}$	$\dfrac{\pi D^4}{64}$	$\dfrac{\pi D^4}{32}$	$\dfrac{D}{4}$

Table 4.1 continued

Section	Area (A)	Distance from extreme fibre to centroid (ȳ)	Second moment of area (I_{xx})	Polar second moment of area ($J = I_{xx} + I_{yy}$)	Radius of gyration (r_{xx})
Annulus	$\frac{\pi}{4}(D^2 - d^2)$	$\frac{D}{2}$	$\frac{\pi}{64}(D^4 - d^4)$	$\frac{\pi}{32}(D^4 - d^4)$	$\frac{1}{4}\sqrt{(D^2 - d^2)}$
Semicircle	$\frac{\pi D^2}{8}$	$\frac{2D}{3\pi}$	$\frac{\pi D^4}{4\,58}$	—	$\frac{D}{23 \cdot 93}$

enough in themselves, are not encumbered needlessly, this chapter will be devoted to the explanation in purely geometric terms of the principal properties of structural sections.

A brief explanation of each property, together with suggested methods for its determination, is given. In Table 4.1 formulae for the common standard cases are set out.

4.2 Area

Cross-sectional areas will generally be calculated in mm², since the dimensions of most structural members are given in mm.

The following notation is used:

A = total area of the section
ΔA = small finite portion of the area of the section
δA = an infinitely small portion of the area of the section

Then $A = \Sigma (\Delta A) = \Sigma (\delta A)$ [4.1]
or $A = \int_A dA$ [4.2]

4.3 Centre of area. Centroid

The centre of area, or centroid, of a section is the point about which the area of the section is distributed evenly. A useful analogy that helps the understanding of this idea may be found by considering the centre of gravity (or centre of mass) of a thin plate of uniform thickness and of homogeneous material. The centre of gravity of such a plate will occur at the point where (theoretically) a single point support could be placed to obtain a perfect balance. It should be noted, however, that if the plate is either non-uniform or non-homogeneous the centre of area and the centre of gravity will occur at different points instead of being coincident.

In order to locate the position of the centroid the moments of area of the section are considered about convenient axes. Moments of area are similar to moments of force:

moment of force = force magnitude × distance to line of force
moment of area = area magnitude × distance to centre of area

and the same principle of superposition applies — *the total moment of area equals the sum of the component moments of area about the same axis.*

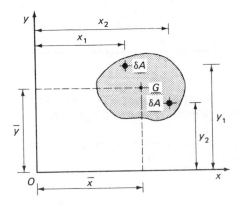

Fig. 4.1

Consider the shaded area shown in Fig. 4.1. It is required to find the position of the centroid (G) in relation to the two axes OX and OY. Let the area be made up of an infinite number of very small elements of area δA, the co-ordinates of these elements being x_1y_1, x_2y_2, etc. Also let the distances to the centroid from the respective axis be $\bar{x}$ and $\bar{y}$, as shown.

Then for equilibrium, $A\bar{x} = \Sigma\ (\delta A\ x_1 + \delta Ax_2 + \ ...)$

or

$$\bar{x} = \frac{\Sigma\ (\delta Ax)}{A}$$ [4.3]

also

$$\bar{y} = \frac{\Sigma\ (\delta Ay)}{A}$$ [4.4]

EXAMPLE 4.1   Find the position of the centroid of the triangular area shown in Fig. 4.2, related to the y-axis.

Consider an elemental strip of the area of thickness δx and situated at a distance x from the axis.

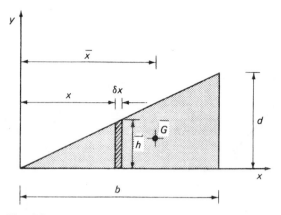

Fig. 4.2

Area of strip $= \delta A = h\delta x$

$$= d\,\frac{x}{b}\,\delta x$$

From equation [4.3], $\quad \bar{x} = \dfrac{\Sigma\,(\delta Ax)}{A}$

$$= \frac{\Sigma\left(d\,\dfrac{x}{b}\,\delta xx\right)}{\frac{1}{2}\,bd}$$

Then, in the limit, $\quad \bar{x} = \displaystyle\int_0^b \frac{2x^2\,\mathrm{d}x}{b^2}$

$$= \frac{2}{b^2}\left[\frac{x^3}{3}\right]_0^b$$

$$= \frac{2}{b^2} \times \frac{b^3}{3}$$

$$= \frac{2}{3}\,b\ \text{(Answer)}$$

EXAMPLE 4.2   Find the position of the centroid of the triangular area shown in Fig. 4.3, referred to the y-axis.

The area is first divided into geometrically finite elements, in this case two triangles (Fig. 4.4).

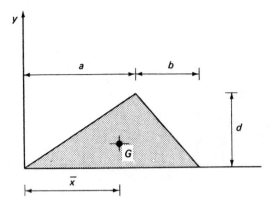

Fig. 4.3

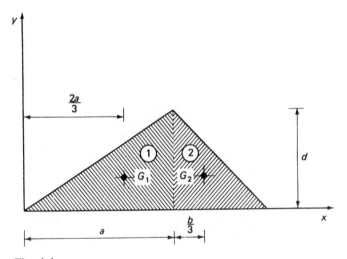

Fig. 4.4

From equation [4.3], $\bar{x} = \dfrac{\Sigma\,(\delta A\;x)}{A}$

$$= \frac{\text{moment of area 1 + moment of area 2}}{\text{total area}}$$

Now,   moment of area 1 = area 1 × distance to G_1

and   moment of area 2 = area 2 × distance to G_2

$$\therefore \bar{x} = \frac{\frac{1}{2}\,ad \times \dfrac{2a}{3} + \frac{1}{2}\,bd\left(a + \dfrac{b}{3}\right)}{\frac{1}{2}\,(a + b)d}$$

$$= \frac{\dfrac{2a^2}{3} + ab + \dfrac{b^2}{3}}{a + b}$$

$$= \frac{2a^2 + 3ab + b^2}{3\,(a + b)}$$

$$= \frac{2a + b}{3} \quad \text{(Answer)}$$

4.4 Graphical determination of centroids

It may be more convenient at times, particularly in graphical analyses, to locate the position of the centroid using a graphical procedure. Some useful examples are given in Figs. 4.5–4.7.

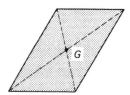

Fig. 4.5

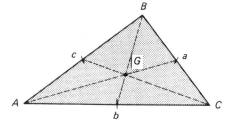

Fig. 4.6

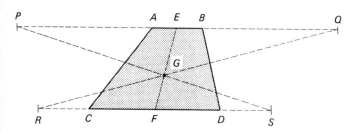

Fig. 4.7

116

Centroid of a parallelogram or rectangle	The centroid lies at the intersection of the diagonals.
Centroid of a triangle	Bisect sides *AB, BC, AC*.
	The centroid lies at the intersection of the three medians *Aa, Bb, Cc*.
Centroid of a trapezium	Bisect *AB* in *E* and *CD* in *F̄*.
	Produce *AB* to *P* and *Q* so that *AP* = *BQ* = *CD*.
	Produce *CD* to *R* and *S* so that *CR* = *DS* = *AB*.
	The centroid lies at the intersection of lines *EF, PS, QR*.

4.5 Second moment of area

The *first moment* of an area about a given axis, as stated previously, is the product of the area and the distance from its centroid to the axis. Written mathematically, the general expression is

$$\text{First moment of area} = \int_0^A y \, dA \qquad [4.5]$$

The *second moment* of an area about a given axis is the sum product of the area and the *square* of the distance from the centroid to the axis. Consider the shaded area in Fig. 4.8 and let it consist of an infinite number of very small elements of area δA. Then about the x-axis,

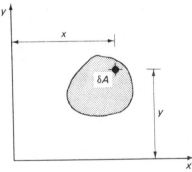

Fig. 4.8

$$\text{First moment of area of an element} = \delta A \times y$$
$$\text{Second moment of an area of an element} = \delta A \times y^2$$

Applying the principle of superposition, the *total second moment of area* about the x-axis is given by,

$$I_{xx} = \Sigma(\delta A \, y^2)$$

or

$$I_{xx} = \int_A y^2 \, dA \qquad [4.6]$$

The units of the second moment of area will be $(\text{length})^4$, since the value is the product of area and distance squared, e.g. m^4 or mm^4.

EXAMPLE 4.3 Determine the second moment of area of a rectangle about an axis passing through one of the shorter sides.

Consider the section to be made up of elemental strips of area

$$b \times \delta y \,(\text{Fig. 4.9})$$

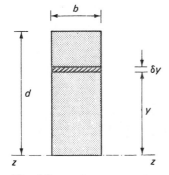

Fig. 4.9

Then second moment of area of a strip about $z - z = b \times \delta y \times y^2$

The total second moment of area of the section about $z - z$ is therefore:

$$I_{zz} = \int_0^d by^2 \, dy$$

$$= b \left[\frac{y^3}{3} \right]_0^d$$

$$= \frac{bd^3}{3} \text{ (Answer)}$$

Numerical solutions may be obtained either by applying a standard case formula, such as the one just established, or by considering the section as a series of small, but finite, elements. See Example 4.4.

EXAMPLE 4.4 Determine the second moment of area of the rectangle shown in Fig. 4.10 about the $z - z$ axis.

The first solution will be obtained by considering the section as four finite elements of area 50 mm × 50 mm.

$$I_{zz} = \Sigma(\Delta A \, y^2)$$

where, $\quad \Delta A = 50 \times 50 = 2\,500 \text{ mm}^2$

118

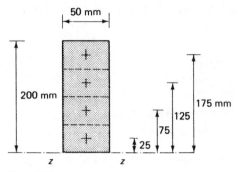

Fig. 4.10

and
$$y_1 = 25 \text{ mm}$$
$$y_2 = 75 \text{ mm}$$
$$y_3 = 125 \text{ mm}$$
$$y_4 = 175 \text{ mm}$$

Therefore, I_{zz}
$$= 2\,500\,(25^2 + 75^2 + 125^2 + 175^2)$$
$$= 2\,500\,(625 + 5\,625 + 15\,625 + 30\,625)$$
$$= 2\,500 \times 52\,500$$
$$= 131 \cdot 25 \times 10^6 \text{ mm}^4 \text{ (Answer)}$$

Now consider the same section as ten finite elements of area 50 mm $\times$ 20 mm (Fig. 4.11).

$$\Delta A = 50 \times 20 = 1\,000 \text{ mm}^2$$

and I_{zz}
$$= 1\,000\,(10^2 + 30^2 + 50^2 + 70^2 + 90^2 + 110^2 + 130^2 + 150^2 + 170^2 + 190^2)$$
$$= 1\,000 \times 133\,000$$
$$= 133 \cdot 0 \times 10^6 \text{ mm}^4 \text{ (Answer)}$$

Thus the second solution gives a greater value than the first. This is so because a greater number of finite elements was considered in the second solution and if, in a further calculation, still more strips were considered, a higher value again

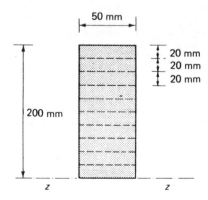

Fig. 4.11

would be obtained. The reason for this error is that, using *finite* elements, only an approximate value for the second moments of area can be obtained. The greater the number of elements considered, the nearer to the correct value the solution will be. If an *infinite* number of very small elemental strips are taken, then the calculus (equation [4.6]) yields the correct answer.

$$I_{zz} = \int y^2 \, \mathrm{d}A = \frac{bd^3}{3}$$

$$= \frac{50 \times 200^3}{3}$$

$$= 133 \cdot 3 \times 10^6 \text{ mm}^4 \text{ (Answer)}$$

4.6 Second moment of area about a centroidal axis

It is usual to consider the reference axes of structural sections as those passing through the centroid. In general, the $x - x$ axis is drawn perpendicular to the greatest lateral dimension of the section, and the $y - y$ axis is drawn perpendicular to the $x - x$ axis, intersecting it at the centroid (Fig. 4.12).

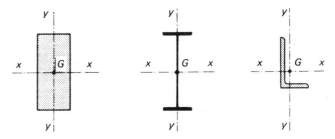

Fig. 4.12

For design purposes, it is therefore convenient to use the second moment of area of the section about one or both centroidal axes.

EXAMPLE 4.5 Determine the second moments of area of the rectangle shown in Fig. 4.13 about the $x - x$ and $y - y$ axes.

Consider the section to be made up of elemental strips of area $b \times \delta y$. Then the second moment of area of a strip about $x - x = b \times \delta y \times y^2$. Therefore the total second moment of area of the section is

$$I_{zz} = \int_{-d/2}^{d/2} by^2 \, \mathrm{d}y$$

$$= b \left[\frac{y^3}{3} \right]_{-d/2}^{d/2}$$

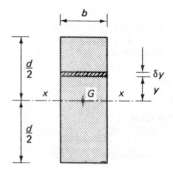

Fig. 4.13

$$= b \left[\frac{d^3}{24} - \left(-\frac{d^3}{24} \right) \right]$$

$$= \frac{bd^3}{12} \quad \text{(Answer)}$$

and, using the same procedure

$$I_{yy} = \frac{b^3 d}{12} \quad \text{(Answer)}$$

4.7 The parallel axis principle

It is often necessary to obtain the second moment of area of a standard shape section about an axis parallel to one of the centroidal axes. Consider the shaded area in Fig. 4.14 to be made up of elemental strips of area δA. The second moment of area of a strip about $z-z = \delta A$

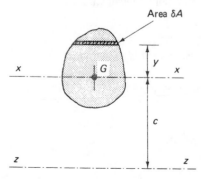

Fig. 4.14

$(y + c)^2$. Then the total second moment of area of the section about $z - z$ is

$$I_{zz} = \int (y + c)^2 \, dA$$
$$= \int y^2 \, dA + 2c \int y \, dA + c^2 \int dA$$

But $\qquad \int y^2 \, dA = I_{xx}, \int y \, dA = 0$ and $\int dA = A$

Hence $\qquad I_{zz} = I_{xx} + Ac^2$ $\qquad\qquad\qquad\qquad$ [4.7]

This expression is known as the PARALLEL AXIS PRINCIPLE.

EXAMPLE 4.6 Determine the second moment of area of the rectangle shown in Fig. 4.10 about a centroidal axis parallel to the $z - z$ axis shown. Assume $I_{zz} = bd^3/3$.

$$I_{zz} = \frac{bd^3}{3} = 133 \cdot 3 \times 10^6 \text{ mm}^4$$

$$c = 100 \text{ mm}$$

Then from equation [4.7],

$$I_{xx} = I_{zz} - Ac^2$$
$$= 133 \cdot 3 \times 10^6 - 50 \times 200 \times 100^2$$
$$= (133 \cdot 3 - 100 \cdot 0) \, 10^6$$
$$= 33 \cdot 3 \times 10^6 \text{ mm}^4 \text{ (Answer)}$$

Alternatively

$$I_{zz} = \frac{bd^3}{3}$$

$$c = \frac{d}{2}$$

Then

$$I_{xx} = I_{zz} - Ac^2$$
$$= \frac{bd^3}{3} - bd \left(\frac{d}{2}\right)^2$$
$$= \frac{bd^3}{12}$$
$$= \frac{50 - 200^3}{12}$$
$$= 33 \cdot 3 \times 10^6 \text{ mm}^4 \text{ (Answer)}$$

4.8 Compound sections

A section made up of a number of geometrically finite areas is called a *compound section*. It is usually necessary first to locate the position of

122

the centroid. The second moments of area can then be determined as required, using the parallel axis principle together with standard case expressions obtained from Table 4.1.

EXAMPLE 4.7 Determine the second moments of area, I_{xx} and I_{yy}, for the compound section shown in Fig. 4.15.

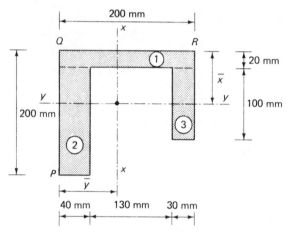

Fig. 4.15

$$\text{Area 1} = 200 \times 20 = 4\ 000\ \text{mm}^2$$
$$\text{Area 2} = 180 \times 40 = 7\ 200\ \text{mm}^2$$
$$\text{Area 3} = 100 \times 30 = \underline{3\ 000\ \text{mm}^2}$$
$$\text{Total area} = 14\ 200\ \text{mm}^2$$

To find $\bar{x}$, take moments of area about QR

$$\Sigma(\Delta Ax) = 4\ 000 \times 10 + 7\ 200 \times 100 + 3\ 000 \times 70$$
$$= 1\ 042\ 000\ \text{mm}^3$$
$$\therefore \bar{x} = \frac{\Sigma(\Delta Ax)}{A} = \frac{1\ 042\ 000}{14\ 200}$$
$$= 73\cdot 4\ \text{mm}$$

To find $\bar{y}$, take moments of area about QP.

$$\Sigma(\Delta Ay) = 4\ 000 \times 100 + 7\ 200 \times 20 + 3\ 000 \times 185$$
$$= 1\ 099\ 000\ \text{mm}^3$$
$$\therefore \bar{y}) = \frac{\Sigma(\Delta Ay)}{A} = \frac{1\ 099\ 000}{14\ 200}$$
$$= 77\cdot 4\ \text{mm}$$

Element	Area (mm²)	c_x	Ac^2 (mm⁴ × 10⁶)	I_c (mm⁴ × 10⁶)	I_{xx} (mm⁴ × 10⁶)
1	4 000	22·6	2·04	$\dfrac{20 \times 200^3}{12 \times 10^6} = 13·33$	15·37
2	7 200	57·4	23·72	$\dfrac{180 \times 40^3}{12 \times 10^6} = 0·96$	24·68
3	3 000	107·6	34·73	$\dfrac{100 \times 30^3}{12 \times 10^6} = 0·23$	34·96
				Σ	75·01

c_x = distance between centroid of element and $x-x$ axis
I_c = second moment of area of element about its centroidal axis parallel to $x-x$ (Table 4.1)

Element	Area (mm²)	c_y	Ac^2 (mm⁴ × 10⁶)	I_c (mm⁴ × 10⁶)	I_{yy} (mm⁴ × 10⁶)
1	4 000	63·4	16·08	$\dfrac{200 \times 20^3}{12 \times 10^6} = 0·13$	16·21
2	7 200	36·6	9·64	$\dfrac{40 \times 180^3}{12 \times 10^6} = 19·44$	29·08
3	3 000	3·4	0·03	$\dfrac{30 \times 100^3}{12 \times 10^6} = 2·50$	2·53
				Σ	47·82

Hence $I_{xx} = 75·01 \times 10^6$ mm⁴   (Answer)
          $I_{yy} = 47·82 \times 10^6$ mm⁴

4.9 Radius of gyration

If all of the area of a section could be concentrated into a single dimensionless point, without altering the value of the second moment of area about a given axis, then the distance between this point and the axis would have to be equal to r, where

$$Ar^2 = I = \int y^2 \, dA$$

In which case
$$r = \sqrt{\dfrac{I}{A}} \qquad [4.8]$$

and r is termed the RADIUS OF GYRATION.

The radius of gyration is used in the design of compression members, such as columns and struts. Slender compression members fail by buckling rather than crushing or shearing, and tend to do so about the axis for which the radius of gyration is a minimum value. From equation

[4.8] it will be seen that the *least radius of gyration* is related to the axis about which the *least second moment of area* occurs, in many cases this being the $y - y$ axis.

$$r_{xx} = \sqrt{\frac{I_{xx}}{A}}$$

$$r_{yy} = \sqrt{\frac{I_{yy}}{A}}$$

EXAMPLE 4.8 Determine the radii of gyration, r_{xx} and r_{yy}, for the section described in Example 4.7.

$$r_{xx} = \sqrt{\frac{I_{xx}}{A}} = \sqrt{\frac{75 \cdot 66 \times 10^6}{14\,200}}$$

$$= \sqrt{5\,328}$$

$$= 73 \cdot 0 \text{ mm (Answer)}$$

$$r_{yy} = \sqrt{\frac{I_{yy}}{A}} = \sqrt{\frac{47 \cdot 82 \times 10^6}{14\,200}}$$

$$= \sqrt{3\,368}$$

$$= \sqrt{58 \cdot 0} \text{ mm (Answer)}$$

4.10 Polar second moment of area

In the study of torsion in Chapter 9, it will be necessary to obtain the second moment of area of a cross-section about an axis *perpendicular* to the cross-section. In Fig. 4.16 the $x - x$ and $y - y$ axes are shown, as is

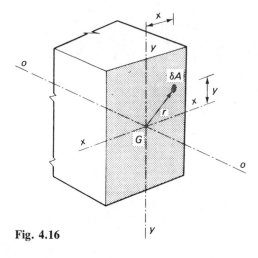

Fig. 4.16

usual, in the plane of the cross-section. The longitudinal axis is, however, perpendicular to the cross-section, and is sometimes referred to as the POLAR axis. The second moment of area about this longitudinal, or polar, axis is called the POLAR SECOND MOMENT OF AREA (J).

Consider the shaded area in Fig. 4.16 to be made up of elements of area, δA, situated at distance r from the polar axis $o-o$.

Second moment of area of an element about $o-o = \delta A \times r^2$. Then the total polar second moment of area about $o-o$ is

$$J = \int r^2 \, dA \qquad [4.9]$$

But $$r^2 = x^2 + y^2$$

Then $$J = \int (x^2 + y^2) \, dA$$

$$= \int x^2 \, dA + \int y^2 \, dA$$

Hence $$J = I_{xx} + I_{yy} \qquad [4.10]$$

i.e. the sum of any two mutually perpendicular centroid-axial second moments of area for the same section.

EXAMPLE 4.9   Determine the polar second moment of area for a circle of diameter D, and from this obtain the centroid-axial second moment of area.

Consider the shaded area in Fig. 4.17 to be made up of elemental rings of variable radius r and thickness δr.

Fig. 4.17

Then $$\delta A = 2\pi r \, \delta r$$

Second moment of area of the ring about the polar axis $= 2\pi r \, \delta r r^2$. Then the total polar second moment of area is

$$J = 2\pi \int_0^{D/2} r^3 \, dr$$

$$= 2\pi \left[\frac{r^4}{4} \right]_0^{D/2}$$

$$= 2\pi \times \frac{D^4}{4 \times 16}$$

$$= \frac{\pi D^4}{32} \text{ (Answer)}$$

Any two mutually perpendicular centroidal axes will give equal second moments of area, i.e. $I_{xx} = I_{yy}$.

Therefore

$$I_{xx} = I_{yy} = \frac{J}{2}$$

$$= \frac{\pi D^4}{64} \text{ (Answer)}$$

4.11 Elastic section modulus

In problems involving bending stresses in beams, including the design of beam sections, a property called the ELASTIC SECTION MODULUS is used to express the bending moment/stress relationship (Chapter 6). This too may be treated as a geometrical property.

Theoretically each point within the cross-section of a beam will have a *section modulus*, this being simply the ratio of the second moment of area to the distance between the point in question and the relevant axis.

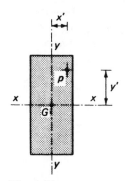

Fig. 4.18

In the section shown in Fig. 4.18 the section moduli of point p are,

$$Z_{xx} = \frac{I_{xx}}{y'} \qquad [4.11]$$

$$Z_{yy} = \frac{I_{yy}}{x'} \qquad [4.12]$$

In practice it is usually found that only the section moduli for the fibres at the most extreme distance from the axes are required.

4.12 Principal axes and principal second moments of area

In the preceding examples axes have been considered (for convenience) that are generally horizontal or vertical, or parallel to the defining sides of the section. The corresponding second moments of area, however, will only be maximum or minimum values where these are *axes of symmetry* or *principal axes*. In sections of rectangular, circular, I or H shape, the $x-x$ and $y-y$ axes are indeed axes of symmetry and are therefore principal axes. For angle, channel and other non-symmetrical sections, the $x-x$ and $y-y$ axes may *not* be principal axes.

A *principal axis* is defined as an axis about which the area-moment sums are equal to zero, i.e. $\int x \, dA = 0$ and $\int y \, dA = 0$.

Consider an element of area δA at point A in the section shown in Fig. 4.19. The co-ordinates of A about horizontal and vertical axes will be

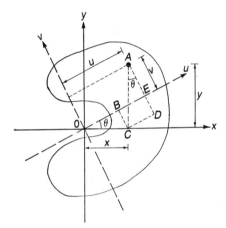

Fig. 4.19

(x,y). For the orthogonal axes u and v, at angle θ to axes x and y, the co-ordinates of A will be:

$$u = OB + CD = x \cos\theta + y \sin\theta$$
$$v = AD - DE = y \cos\theta - x \sin\theta$$

The *product moment of area* is defined as $I_{xy} = \int xy \, dA$, and with respect to axes u and v it will be:

$$
\begin{aligned}
I_{uv} = \int uv \, dA &= \int (x \cos\theta + y\sin\theta)(y \cos\theta - x\sin\theta)dA \\
&= \int [xy(\cos^2\theta - \sin^2\theta) + (y^2 - x^2)\sin\theta \, \cos\theta]dA \\
&= \cos 2\theta \, I_{xy} + \tfrac{1}{2}\sin 2\theta (I_{xx} - I_{yy}) \qquad [4.13]
\end{aligned}
$$

Now, for u and v to be principal axes, $I_{uv} = 0$, since either $\int u \, dA$ or $\int v \, dA$ must be zero.

Therefore $$\tan 2\theta = \frac{-2I_{xy}}{I_{xx} - I_{yy}} \qquad [4.14]$$

Thus, the orientation of the principal axes may be obtained.

Also, $I_{uu} = \int v^2 dA = \int (y^2 \cos^2\theta + x^2 \sin^2\theta - 2xy \sin\theta \cos\theta) dA$.
However, from eqn [4.14], $I_{xy} = -\frac{1}{2}\tan2\theta(I_{xx} - I_{yy})$.

Therefore, $I_{uu} = \cos^2\theta I_{xx} + \sin^2\theta I_{yy} + \dfrac{\sin^2 2\theta}{\cos2\theta}(I_{xx} - I_{yy})$

$$= \tfrac{1}{2}(1 + \cos2\theta)I_{xx} + \tfrac{1}{2}(1 - \cos2\theta)I_{yy} + \dfrac{\sin^2 2\theta}{\cos2\theta}(I_{xx} - I_{yy})$$

$$= \tfrac{1}{2}(I_{xx} + I_{yy}) + \tfrac{1}{2}\left(\cos2\theta + \dfrac{\sin^2 2\theta}{\cos2\theta}\right)(I_{xx} - I_{yy})$$

$$= \tfrac{1}{2}(I_{xx} + I_{yy}) + \tfrac{1}{2}(I_{xx} - I_{yy})\sec2\theta \qquad [4.15]$$

Furthermore, the sum of the orthogonal second moments of area, i.e. the polar second moment of area, is constant for all sets of axes passing through the centroid.

$$I_0 = I_{xx} + I_{yy} = I_{uu} + I_{vv} \qquad [4.16]$$

Hence, $$I_{vv} = I_{xx} + I_{yy} - I_{uu} \qquad [4.17]$$

Where the second moments of area are required for axes not passing through the centroid of the section, the *parallel axis principle* is used (see Section 4.7). To determine the product second moment of area in such cases, the area product $xydA$ must be integrated with respect to the two variables (x and y). Consider a rectangular element having side dimensions of b and d, which are parallel to the centroidal axes x and y respectively (Fig. 4.20).

The product moment of area will be:

$$I_{xy} = \int xy \, dA^{\cdot} = \int xy \, dx \, dy$$

$$= \tfrac{1}{2}[y^2] \, \frac{C_x + d/2}{C_x - d/2} \times \tfrac{1}{2}[x^2] \, \frac{C_y + b/2}{C_y - b/2}$$

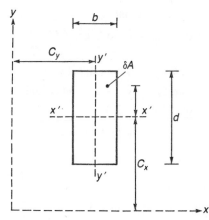

Fig. 4.20

$$= \tfrac{1}{4}(2C_x d + 2C_x d) \times \tfrac{1}{4}(2C_y b + 2C_y b)$$

$$= bd \; C_x \; C_y \qquad\qquad [4.18]$$

EXAMPLE 4.10 Determine the orientation of the principal axes (uu and vv) and hence the principal second moments of area for the steel angle section shown in Fig. 4.21.

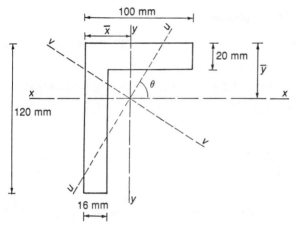

Fig. 4.21

First, the position of the centroid is determined.

$$\bar{x} = \frac{100 \times 20 \times 50 + 100 \times 16 \times 8}{100 \times 20 + 100 \times 16} = \frac{112\ 800}{3600} = 31\cdot33 \text{ mm}$$

$$\bar{y} = \frac{100 \times 20 \times 10 + 100 \times 16 \times 70}{100 \times 20 + 100 \times 16} = \frac{132\ 000}{3600} = 36\cdot67 \text{ mm}$$

The orthogonal second moments of area with respect to $x - x$ and $y - y$ are:

$$I_{xx} = \frac{100 \times 20^3}{12} + 100 \times 20(36\cdot67 - 10)^2 + \frac{16 \times 100^3}{12}$$

$$+ \; 16 \times 100(70 - 36\cdot67)^2$$

$$= 4\cdot600 \times 10^6 \text{ mm}^4$$

$$I_{yy} = \frac{20 \times 100^3}{12} + 100 \times 20(50 - 31\cdot33)^2 + \frac{100 \times 16^3}{12}$$

$$+ \; 100 \times 16(31\cdot33 - 8)^2$$

$$= 3\cdot269 \times 10^6 \text{ mm}^4$$

The product moment of area will be:

$$I_{xy} = 200 \times 20(50 - 31 \cdot 33)(36 \cdot 67 - 10)$$
$$- 16 \times 100(31 \cdot 33 - 8)(70 - 36 \cdot 67)$$
$$= 2 \cdot 240 \times 10^6 \text{ mm}^4$$

The angle between the *xy* and *uv* (principal) axes can now be calculated from eqn [4.14]

$$\text{Tan } 2\theta = \frac{-2I_{xy}}{I_{xx} - I_{yy}} = \frac{-2 \times 2 \cdot 240}{4 \cdot 600 - 3 \cdot 269} = -3 \cdot 366$$

Giving $\qquad \theta = \underline{36 \cdot 73°}$ (Answer)

So now the principal second moments of area are, from eqns [4.15] and [4.17]:

$$I_{uu} = \tfrac{1}{2}(I_{xx} + I_{yy}) + \tfrac{1}{2}(I_{xx} - I_{yy})\sec 2\theta$$
$$= \tfrac{1}{2}[4 \cdot 600 + 3 \cdot 269 + (4 \cdot 600 - 3 \cdot 269)\sec(-73 \cdot 46°)]10^6$$
$$= \underline{6 \cdot 271 \times 10^6 \text{ mm}^4} \text{ (Answer)}$$

$$I_{vv} = I_{xx} + I_{yy} - I_{uu}$$
$$= (4 \cdot 600 + 3 \cdot 269 - 6 \cdot 271) \times 10^6$$
$$= \underline{1 \cdot 598 \times 10^6 \text{ mm}^4} \text{ (Answer)}$$

Figure 4.22 shows the printout of an analysis of this example using a spreadsheet designed to deal with sections comprising rectilinear elements. With only slight

SECT1	Section Properties						
Element	1	2	3	4	5	6	TOTAL
b	100	16					
d	20	100					
x	0	0					
y	-20	-120					
Area dA	2000	1600					3600
xdA	100000	12800					112800
ydA	-20000	-112000					-132000
cx	18.66667	-23.3333					
cy	26.66667	-33.3333					
Igx	66666.67	1333333					1400000
Igy	1666667	34133.33					1700800
dA*cx^2	696888.9	871111.1					1568000
dA*cy^2	1422222	1777778					3200000
dA*cx*cy	995555.6	1244444					2240000
xbar	31.33333		Ixx	4.60	x 10^6		Ixy	2.24 x 10^6
ybar	-36.6667		Iyy	3.27	x 10^6		
Theta	-0.64098	-36.7255			Iuu	6.27	x 10^6
					Ivv	1.60	x 10^6

Fig. 4.22

modification complementary spreadsheets can be developed from this one which can handle other types of composite section. (See also Section 1.14).

EXERCISES   CHAPTER 4

1. Determine the position of the centroid of a semicircle of diameter D.

2. Determine the position of the centroid of a circular quadrant of diameter D.

3−6. Determine the position of the centroid of the shaded portion of the shapes shown in Figures 4.23−4.26.

7. Show that if a piece of thin wire is bent to form a semicircular arc its centre of gravity will be located as indicated in Fig. 4.27.

8−11. Determine the centre of gravity of the bent wire shown in Figures 4.28−4.31. The location of the centroid of a piece of wire bent in the form of a semicircle is shown in Fig. 4.27.

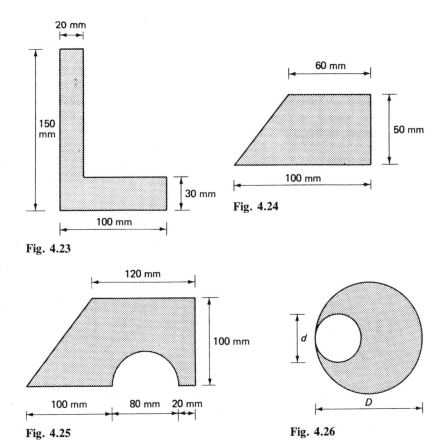

Fig. 4.23

Fig. 4.24

Fig. 4.25

Fig. 4.26

132

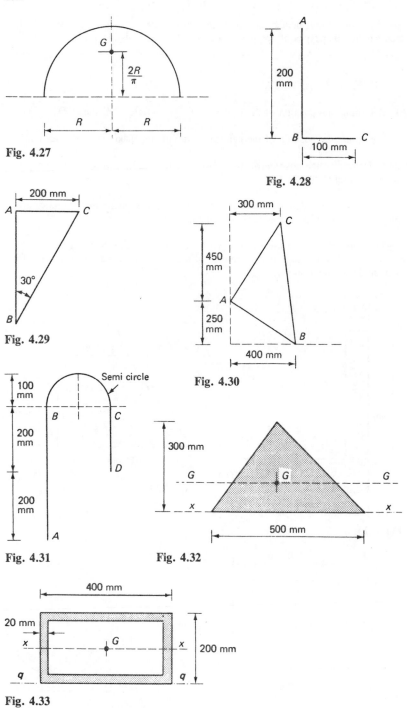

Fig. 4.27

Fig. 4.28

Fig. 4.29

Fig. 4.30

Fig. 4.31

Fig. 4.32

Fig. 4.33

12. Calculate the second moment of area of the triangle shown in Fig. 4.32, (a) about the $x - x$ axis, (b) about the axis $g - g$ which passes through the centre of area of the triangle and is parallel to axis $x - x$.

13. The box section shown in Fig. 4.33 has a constant wall thickness of 20 mm. Calculate the second moments of area of the section about the $x - x$ and $q - q$ axes.

14. Calculate the second moments of area of the section shown in Fig. 4.34 about the $x - x$ and $y - y$ axes.

15. A folded-plate box section has a shape approximating to that shown in Fig. 4.35, the thickness of the plate being 10 mm. Determine the second moments of area of the section about the centroidal axes $x - x$ and $y - y$.

16. Calculate the second moment of area of the section shown in Fig. 4.36 about the axis $x - x$ which coincides with the edge of the section and also about the axis $o - o$ which is parallel to $x - x$.

17. Calculate the second moments of area of the section shown in Fig. 4.37 about the centroidal axes $x - x$ and $y - y$.

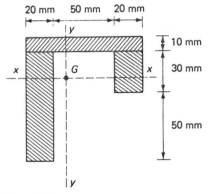

Fig. 4.34

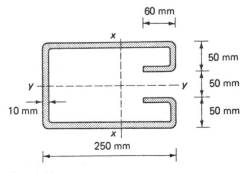

Fig. 4.35

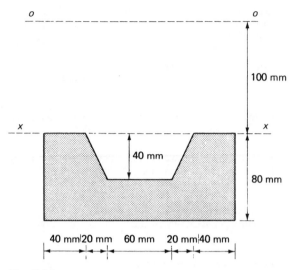

Fig. 4.36

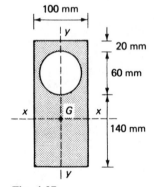

Fig. 4.37

18. Calculate the polar second moment of area of an annulus of external diameter 150 mm and internal diameter 120 mm.

19. Calculate the polar second moment of area of the section shown in Fig. 4.33.

20–34. Figures 4.38–4.52 show a series of structural sections with the positions of their centroidal axes approximately indicated. In each case calculate the second moment of area and the radius of gyration about each centroidal axis shown. Calculate also the elastic section modulus of the section for the point marked 'A', referred to the $x-x$ axis.

35. The two steel channels shown in Fig. 4.53 are arranged to form a compound section in which I_{xx} is equal to I_{yy}. Determine the value of the dimension s that will satisfy this condition.

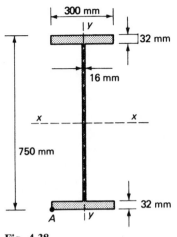

Fig. 4.38

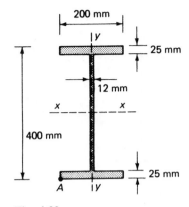

Fig. 4.39

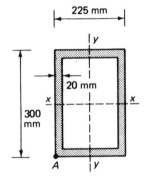

Fig. 4.40

Fig. 4.41

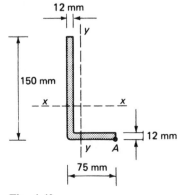

Fig. 4.42

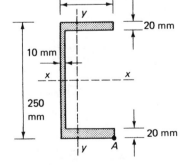

Fig. 4.43

136

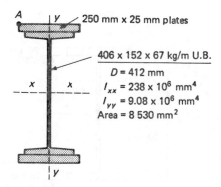

Fig. 4.44

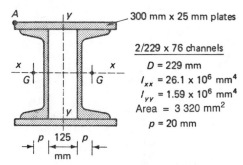

Fig. 4.45

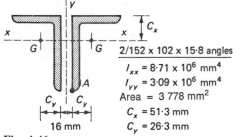

Fig. 4.46

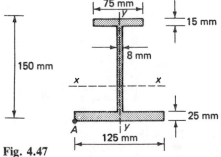

Fig. 4.47

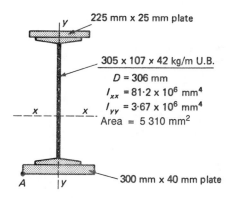

225 mm x 25 mm plate

305 x 107 x 42 kg/m U.B.

$D = 306$ mm
$I_{xx} = 81.2 \times 10^6$ mm^4
$I_{yy} = 3.67 \times 10^6$ mm^4
Area $= 5\,310$ mm^2

300 mm x 40 mm plate

A

Fig. 4.48

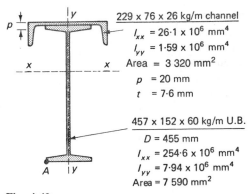

229 x 76 x 26 kg/m channel

$I_{xx} = 26.1 \times 10^6$ mm^4
$I_{yy} = 1.59 \times 10^6$ mm^4
Area $= 3\,320$ mm^2
$p = 20$ mm
$t = 7.6$ mm

457 x 152 x 60 kg/m U.B.

$D = 455$ mm
$I_{xx} = 254.6 \times 10^6$ mm^4
$I_{yy} = 7.94 \times 10^6$ mm^4
Area $= 7\,590$ mm^2

A

Fig. 4.49

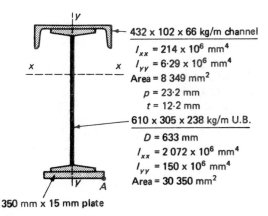

432 x 102 x 66 kg/m channel

$I_{xx} = 214 \times 10^6$ mm^4
$I_{yy} = 6.29 \times 10^6$ mm^4
Area $= 8\,349$ mm^2
$p = 23.2$ mm
$t = 12.2$ mm

610 x 305 x 238 kg/m U.B.

$D = 633$ mm
$I_{xx} = 2\,072 \times 10^6$ mm^4
$I_{yy} = 150 \times 10^6$ mm^4
Area $= 30\,350$ mm^2

A

350 mm x 15 mm plate

Fig. 4.50

138

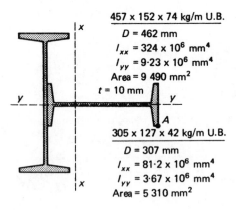

457 × 152 × 74 kg/m U.B.
$D = 462$ mm
$I_{xx} = 324 \times 10^6$ mm^4
$I_{yy} = 9.23 \times 10^6$ mm^4
Area = 9 490 mm^2
$t = 10$ mm

305 × 127 × 42 kg/m U.B.
$D = 307$ mm
$I_{xx} = 81.2 \times 10^6$ mm^4
$I_{yy} = 3.67 \times 10^6$ mm^4
Area = 5 310 mm^2

Fig. 4.51

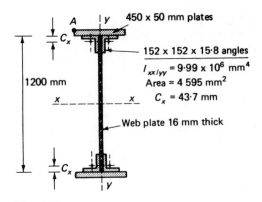

450 × 50 mm plates

152 × 152 × 15·8 angles
$I_{xx/yy} = 9.99 \times 10^6$ mm^4
Area = 4 595 mm^2
$C_x = 43.7$ mm

1200 mm

Web plate 16 mm thick

Fig. 4.52

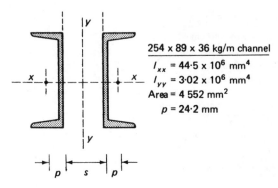

254 × 89 × 36 kg/m channel
$I_{xx} = 44.5 \times 10^6$ mm^4
$I_{yy} = 3.02 \times 10^6$ mm^4
Area = 4 552 mm^2
$p = 24.2$ mm

Fig. 4.53

5 Shearing force and bending moment

Symbols

a	=	distance
b	=	distance
FBM	=	'free' bending moment
h	=	height
i	=	slope of a tangent to a curve
L	=	length, span
l	=	length, part-span
M	=	bending moment
M_{max}	=	maximum bending moment
Q	=	shearing force
R	=	reaction force
UDL	=	uniformly distributed load
W	=	concentrated load, total uniform load
w	=	uniform load per unit length
x	=	distance in the x-direction
y	=	distance in the y-direction
z	=	distance, distance to point of zero shearing force

5.1 Beams in bending

In Chapter 1, a BEAM is defined as a structural member subject to lateral loading in which the developed resistance to deformation is of a

140

flexural character (Table 1.3, p. 19). The primary load-effect that a beam is designed to sustain is that of bending moment, but, in addition, the effects of shearing force must be considered. See Table 1.2 for definitions.

In design, once the load systems and member configurations have been defined, the next step is to determine the distribution of the critical load-effects throughout the member. This stage in the design procedure is of considerable importance, since it is from these distributions that the stresses and deformations induced in the member are evaluated. Bending moment and shearing force are closely related to each other, as well as to the applied loading — in fact all three can be shown as mathematical functions of each other (Chapter 6). It is usual, therefore, to deal with the determination of bending moments and shearing forces at the same time.

5.2 The nature of shearing force and bending moment

When a beam is laterally loaded it must transmit to each of its supports some proportion of the applied load. This is achieved by the development of resistance to shearing forces within the fabric of the beam. The beam must also remain nearly straight and therefore has to develop resistance to bending.

A CANTILEVER is a beam which is supported only at one end. Consider the cantilever *AB* shown in Fig. 5.1(*a*). For equilibrium, the

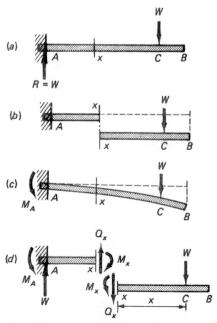

Fig. 5.1

reaction force at A must be vertical and equal to the load W. The cantilever must therefore transmit the effect of load W to the support at A by developing resistance (on vertical cross-sectional planes between the load and the support) to the load-effect called SHEARING FORCE. Failure to transmit the shearing force at any given section, say section $x - x$, will cause the beam to shear as in Fig. 5.1(b). The bending effect of the load will cause the beam to deform as in Fig. 5.1(c).

The shearing force and the bending moment transmitted across the section $x - x$ may be considered as the force and moment respectively that would be necessary to maintain equilibrium if a cut were to be made severing the beam at $x - x$, as shown in Fig. 5.1(d).

Then the shearing force between A and $C = Q_x = W$ and the bending moment between A and $C = M_x = Wx$.

Note: Both the shearing force and the bending moment will be zero between C and B.

5.3 Definitions for calculation purposes

Since shearing force and bending moment are calculable effects, it is necessary to define them quantitatively. The following definitions are those conventionally used.

At any transverse section in a loaded beam:

the **Shearing force**   IS THE ALGEBRAIC SUM OF ALL THE
FORCES ACTING ON ONE (EITHER) SIDE
OF THE SECTION

the **Bending moment**   IS THE ALGEBRAIC SUM OF THE
MOMENTS ABOUT THE SECTION OF ALL
THE FORCES ACTING ON ONE (EITHER)
SIDE OF THE SECTION

In the beam shown in Fig. 5.2, the shearing force and bending moment at section X can be obtained as follows:

Working to the left of X

$$Q_x = R_A - W_1 - W_2 - W_3 \qquad [5.1]$$

$$M_x = R_A a - W_1 a_1 - W_2 a_2 - W_3 a_3 \qquad [5.2]$$

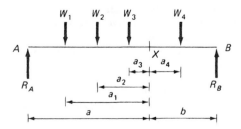

Fig. 5.2

Working to the right of X (which will give equal values)

$$Q_x = -R_B + W_4 \qquad\qquad [5.3]$$

$$M_x = R_B b - W_4 a_4 \qquad\qquad [5.4]$$

5.4 Sign convention and units

Both shearing force and bending moment are vector quantities requiring a convention of signs in order that values of opposite sense may be separated. Mathematical signs are chosen since it is in calculation problems that it becomes necessary to use such a convention. The units will be basically those of *force* for shearing force and those of *force* × *distance* for bending moment. Details of the sign convention and units are given in Table 5.1.

Table 5.1 Sign convention and units for shearing force and bending moment

Load effect	Symbol	Sign Convention		Units
		Positive (+)	Negative (−)	
SHEARING FORCE	Q	↑ ↓ 'up on the left'	↓ ↑ 'down on the left'	N kN
BENDING MOMENT	M	'SAGGING' (top fibres in *compression*)	'HOGGING' (top fibres in *tension*)	N m kN m N mm

5.5 Shearing force and bending moment diagrams

Representative diagrams of the distribution of shearing force and bending moment are often required at several stages in the design process. These diagrams are obtained by plotting graphs with the beams as the base and the values of the particular effect as ordinates. It is usual to construct these diagrams in sets of three, representing the distribution of loads, shearing forces and bending moments respectively.

The following rules should be observed so that a conventional system of graphical representation is adopted:

1. Draw the *SFD* (shearing force diagram) directly below the loading diagram and draw the *BMD* (bending moment diagram) directly below the *SFD*.

2. If the diagrams are drawn to scale, give the scale used for each, adopting standard units wherever possible (Table 5.1).
3. For the *SFD*, use the sign convention given in Table 5.1 plotting positive values above the line and negative values below the line. (*Note*: This rule is easily adopted by starting at the left-hand end of the beam and plotting from there in the direction of the forces.)
4. For the *BMD*, use the sign convention given in Table 5.1 *plotting positive values on the tension side of the beam*. (This diagram also gives a useful indication of the deflected shape of the beam.)
5. Mark the diagram, whether it is to scale or not, with all the important values, such as those at node points and at supports and also peak values. Also mark all positive and negative zones.
6. Remember that the diagrams and graphs are that the ordinates perpendicular to and measured from the base line give the value of the particular load-effect at that section of the beam.

5.6 Standard cases

A number of particular loading cases occur frequently and it is useful to evaluate standard expressions for these, thus obviating the need to go back to first principles on each occasion that they arise. There are four common simple cases:

1. A cantilever with a single concentrated load (Fig. 5.3)
2. A cantilever with a uniformly distributed load (Fig. 5.4)
3. A simply supported beam with a single concentrated load (Fig. 5.5)
4. A simply supported beam with a uniformly distributed load (Fig. 5.6)

At section X when	Shearing force Q_x	Bending moment M_x
$x < a$	W	$-W(a - x)$
$x \geqslant a$	0	0
$x = 0$	W	$-Wa$

$$M_{max} = M_A$$

Special case

when $a = L$ $\quad M_x = -W(L - x)$
$\quad\quad\quad\quad\quad\quad M_A = -WL$

Fig. 5.3

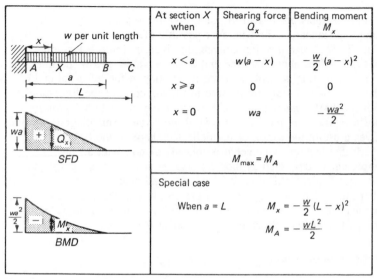

Fig. 5.4

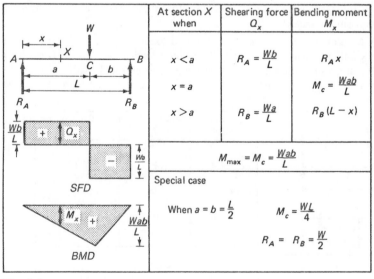

Fig. 5.5

5.7 Construction of parabolas

In cases 2 and 4 described in the preceding section, it will be seen that the equation for the bending moment diagram (graph) contains a second-order term:

for the cantilever,

$$M_x = -\frac{wx^2}{2}$$

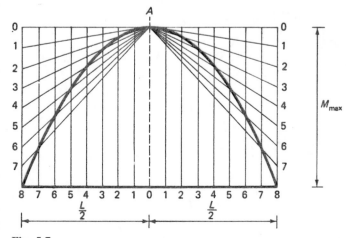

At section X when	Shearing force Q_x	Bending moment M_x
$x = 0$	$R_A = \dfrac{wL}{2}$	0
$x = L$	$R_B = -\dfrac{wL}{2}$	0
$x < \dfrac{L}{2}$	$\dfrac{wL}{2} - wx$	$\left.\rule{0pt}{22pt}\right\}\ \dfrac{wLx}{2} - \dfrac{wx^2}{2}$
$x > \dfrac{L}{2}$	$-\dfrac{wL}{2} + wx$	
$M_{max} = M_c = \dfrac{wL^2}{8}$		

Fig. 5.6

and for the beam,
$$M_x = \frac{wLx}{2} - \frac{wx^2}{2}$$

Mathematically these are called second-order equations, the graphs of which are parabolic curves. It is a significant fact that, *for any portion* of a beam carrying a uniformly distributed load, that portion of the bending moment diagram will be in the form of a parabolic curve. Since the parabola is a geometric figure, a simple procedure exists for its construction. The parabola shown in Fig. 5.7 is constructed as follows:

Fig. 5.7

1. Construct the enclosing rectangle in which:

$$\text{base} = \text{span of beam}$$
$$\text{height} = \text{maximum bending moment } (M_{\text{max}})$$

2. Divide each half-span into a number of equal parts, points 0 to 8, and erect an ordinate line at each point.
Any number of points may be chosen, but usually 4, 5, 6 or 8 are found to be convenient.

3. Divide each side of the rectangle into the *same number* of equal parts as each half-span — as shown in the diagram.

4. Draw a series of radial lines from the apex A to intersect the sides of the rectangle in points 1 to 8.

5. The intersection of a radial with an ordinate line of the same position value is a point on the parabolic curve. The parabola is therefore drawn through these intersections.

It is sometimes necessary to construct a parabola on a sloping base line, although the base line of the bending moment diagram itself may still be horizontal. The procedure for this construction (Fig. 5.8) is identical with that given for the standard parabola, with the following modifications:

1. The parabola is enclosed by a parallelogram in which:

$l =$ the horizontal length of the parabolic portion of the *BMD* (i.e. the length of the *UDL*)

$h =$ the apex height of the parabola
$=$ the 'free bending moment' value (see Example 5.9)

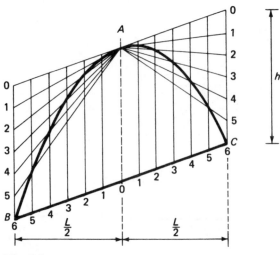

Fig. 5.8

2. Neither of the points B and C necessarily lies on the base of the *BMD* and therefore their positions must be established before the parabola can be constructed.

5.8 Properties of curves and diagrams

The following definitions will be useful when considering the properties of curves — see also Figs 5.9, 5.10 and 5.11.

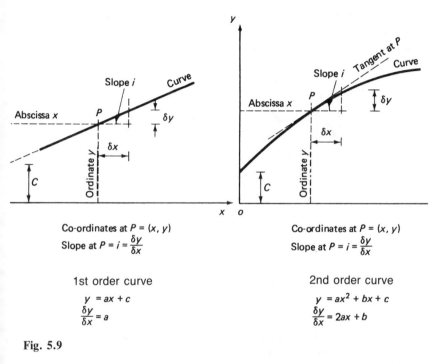

Co-ordinates at $P = (x, y)$
Slope at $P = i = \dfrac{\delta y}{\delta x}$

Co-ordinates at $P = (x, y)$
Slope at $P = i = \dfrac{\delta y}{\delta x}$

1st order curve

$$y = ax + c$$
$$\frac{\delta y}{\delta x} = a$$

2nd order curve

$$y = ax^2 + bx + c$$
$$\frac{\delta y}{\delta x} = 2ax + b$$

Fig. 5.9

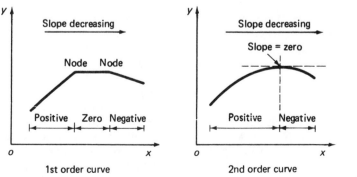

1st order curve

2nd order curve

Fig. 5.10

148

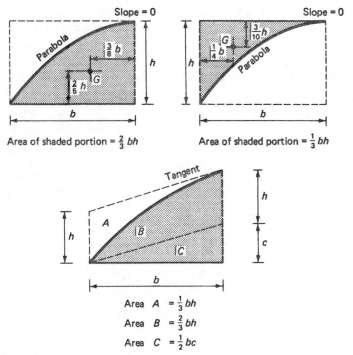

Fig. 5.11

Diagram	a line related to the x and y axes and consisting of one or more curves
Curve	any line which is precisely defined by a mathematical equation, including a straight line
Abscissa	a line drawn parallel to the x-axis between a point on the curve and the y axis (length of abscissa $= x$)
Ordinate	a line drawn parallel to the y-axis between a point on the curve and the x-axis (length of ordinate $= y$)
Co-ordinates	The lengths respectively of the abscissa and ordinate to a point on the curve and used to locate this point (x,y)
Slope	the slope of the tangent to the curve at a particular point

$$i = \frac{dy}{dx}$$

Node	a point on a diagram at which the equation for the curve changes (i.e. two different curves meet); it should be noted that the slope always *decreases* at a node on a bending moment diagram between supports.
Contraflexure	a point of contraflexure or inflexion is a point where the bending moment value (ordinate) is zero and represents

the point at which the mode of bending changes from sagging to hogging or vice versa.

5.9 The mathematical relationship between load, shearing force and bending moment

In a beam *AB* (Fig. 5.12), which is subject to some system of lateral loading, consider an elemental length δx situated at distance x from a support. Since δx is very small, the load on this length may be assumed to be uniformly distributed. Then the load on the element $= w\delta x$.

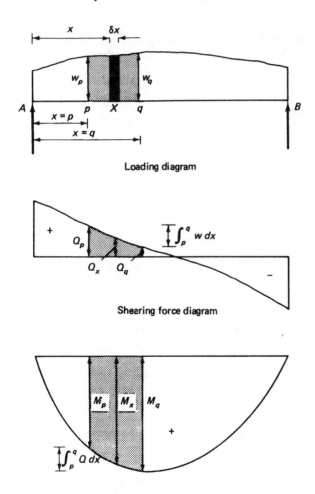

Loading diagram

Shearing force diagram

Bending moment diagram

Fig. 5.12

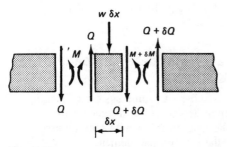

Fig. 5.13

The arrangement of shearing force and bending moment on the element is shown in Fig. 5.13. Resolving the vertical forces on the element:

$$Q - w\delta x - (Q + \delta Q) = 0$$

Therefore
$$\delta Q = -w\delta x$$

or
$$\frac{\delta Q}{\delta x} = -w \qquad [5.5]$$

If this expression is integrated between $x = p$ and $x = q$,

then
$$\int_p^q dQ = Q_q - Q_p = -\int_p^q w\, dx$$

or
$$Q_p - Q_q = \int_p^q w\, dx \qquad [5.6]$$

Thus THE CHANGE IN SHEARING FORCE BETWEEN ANY TWO POINTS ON THE BEAM IS EQUAL TO THE AREA OF THE LOADING DIAGRAM BETWEEN THOSE TWO POINTS.

Resolving the moment on the element for equilibrium:

$$Q\delta x + M - (M + \delta M) - w\delta x\, \frac{\delta x}{2} = 0$$

The term $w\delta x\, \dfrac{\delta x}{2}$ will be very small and may be neglected

Then
$$\delta M = Q\delta x$$

or
$$\frac{\delta M}{\delta x} = Q \qquad [5.7]$$

If this expression is integrated between $x = p$ and $x = q$,

Then
$$\int_p^q dM = M_q - M_p = \int_p^q Q\, dx \qquad [5.8]$$

Thus THE CHANGE IN BENDING MOMENT BETWEEN ANY TWO

POINTS ON THE BEAM IS EQUAL TO THE AREA OF THE
SHEARING FORCE DIAGRAM BETWEEN THOSE TWO POINTS.

The relationships may be summarized as follows:

In terms of differential coefficients

$$-w = \frac{dQ}{dx} \qquad [5.9]$$

$$Q = \frac{dM}{dx} \qquad [5.10]$$

In terms of integrals

$$Q = -\int w\, dx \qquad [5.11]$$

$$M = \int Q\, dx \qquad [5.12]$$

$$= -\int\int w\, dx\, dx \qquad [5.13]$$

Two important and extremely useful conclusions may be drawn from
these relationships:

1. Since the maximum value of a function $y = f(x)$ occurs when
   $dy/dx = 0$, the maximum value of the bending moment occurs when
   $dM/dx = 0$, i.e. when the shearing force $Q = 0$.
2. In order to obtain the expression for the bending moment the
   expression for the loading condition is integrated *twice* with respect
   to x.

Worked examples

In the series of worked examples which follow the requirements of the
question are basically the same in each case:

(a) Calculate the reactions
(b) Draw the shearing force diagram
(c) Draw the bending moment diagram
(d) Determine the position and magnitude of the maximum bending
    moment

Where additional answers are required these will be indicated.

EXAMPLE 5.1   Also find Q_k and M_k (Fig. 5.14)

Reactions	R_A	$= 20 + 30 + 10 = 60$ kN
Shearing forces	Q_D	$= 0$
	Q_C	$= 10$ kN

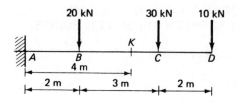

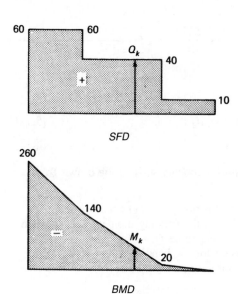

SFD

BMD

Fig. 5.14

$$
\begin{aligned}
Q_B &= 10 + 30 = 40 \text{ kN} \\
Q_A &= 10 + 30 + 20 = 60 \text{ kN}
\end{aligned}
$$

Bending moments

$$
\begin{aligned}
M_D &= 0 \\
M_C &= -10 \times 2 = -20 \text{ kNm} \\
M_B &= -10 \times 5 - 30 \times 3 \\
&= -50 - 90 = -140 \text{ kNm} \\
M_A &= 10 \times 7 - 30 \times 5 - 20 \times 2 \\
&= -70 - 150 - 40 = -260 \text{ kNm} \\
M_{\max} &= M_A
\end{aligned}
$$

At section K

$$
\begin{aligned}
Q_k &= 10 + 30 = 40 \text{ kN} \\
M_k &= -10 \times 3 - 30 \times 1 \\
&= -30 - 30 \\
&= -60 \text{ kNm}
\end{aligned}
$$

EXAMPLE 5.2   Also find Q_k and M_k (Fig. 5.15)

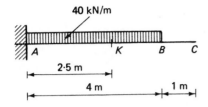

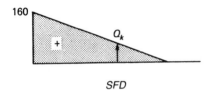

SFD

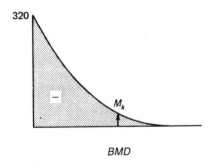

BMD

Fig. 5.15

Reactions	R_A	$= 40 \times 4 = 160$ kN
Shearing forces	Q_B	$= Q_C = 0$
	Q_A	$= 160$ kN
Bending moments	M_B	$= M_C = 0$
	M_A	$= -40 \times 4 \times \frac{4}{2}$
		$= -320$ kNm
	M_{max}	$= M_A$
At section K	Q_k	$= 40 \times 1 \cdot 5 = 60$ kN
	M_k	$= -40 \times \dfrac{1 \cdot 5^2}{2} = -45$ kNm

154

EXAMPLE 5.3   *Also find Q_k and M_k (Fig. 5.16)*

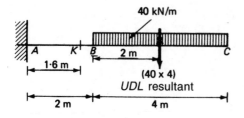

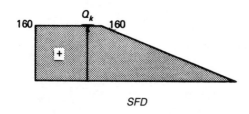

SFD

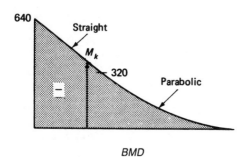

BMD

Fig. 5.16

Reactions	R_A	$= 40 \times 4 = 160$ kN
Shearing forces	Q_C	$= 0$
	Q_A	$= Q_B = 160$ kN
Bending moments	M_C	$= 0$
	M_B	$= -40 \times 4 \times \frac{4}{2} = -320$ kNm
	M_A	$= -40 \times 4 \times 4 = -640$ kNm
	M_{max}	$= M_A$
At section K	Q_k	$= 160$ kN
	M_k	$= -40 \times 4 (2 \cdot 0 + 0 \cdot 4)$
		$= -384$ kNm

EXAMPLE 5.4   (Fig. 5.17)

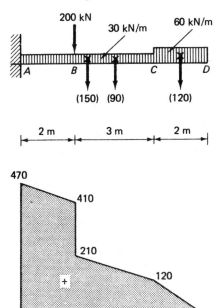

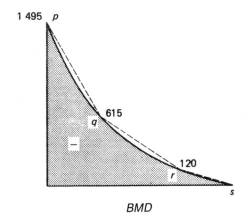

SFD

BMD

Fig. 5.17

Reactions	R_A	$= 200 + 30 \times 5 + 60 \times 2$
		$= 200 + 150 + 120 = 470$ kN
Shearing forces	Q_D	$= 0$
	Q_C	$= 60 \times 2 = 120$ kN

$$Q_B \quad = 120 + 30 \times 3 = 210 \text{ kN}$$
$$Q_A \quad = 470 \text{ kN}$$

UDL resultants $\quad (A - C) = 30 \times 5 = 150 \text{ kN}$
$$(B - C) = 30 \times 3 = 90 \text{ kN}$$
$$(C - D) = 60 \times 2 = 120 \text{ kN}$$

Bending moments (using the *UDL* resultants)

$$M_D \quad = 0$$
$$M_C \quad = -120 \times 1 = -120 \text{ kNm}$$
$$M_B \quad = -120 \times 4 - 90 \times 1 \cdot 5$$
$$\quad = -480 - 135 = -615 \text{ kNm}$$
$$M_A \quad = -120 \times 6 - 150 \times 2 \cdot 5 - 200 \times 2$$
$$\quad = -720 - 375 - 400 = -1495 \text{ kNm}$$
$$M_{max} \quad = M_A$$

Two intermediate nodes occur on the *BMD* at q and r. If the diagram is to be constructed accurately to scale, separate parabolas will be required on bases pq, qr and rs respectively. The heights of the parallelograms enclosing the parabolas are obtained from the expression $wl^2/8$, where l =length of the base of the parabola parallel to the x-axis (Section 5.7).

$$h(pq) = \frac{30 \times 2^2}{8} = 15 \text{ kNm}$$

$$h(qr) = \frac{30 \times 3^2}{8} = 37 \cdot 5 \text{ kNm}$$

As the figures suggest, these values are often too small to allow them to be scaled with any accuracy.

$$h(rs) = \frac{60 \times 2^2}{8} = 30 \text{ kNm}$$

EXAMPLE 5.5   Also find Q_k and M_k (Fig. 5.18)

Reactions $\quad R_A \quad = \dfrac{50 \times 3}{5} = 30 \text{ kN}$

$$R_B \quad = \dfrac{50 \times 2}{5} = 20 \text{ kN}$$

Bending moments $\quad M_A \quad = M_B = 0$
$$M_C \quad = 30 \times 2 = 60 \text{ kNm}$$
$$\text{or } M_C \quad = 20 \times 3 = 60 \text{ kNm}$$

or, alternatively, using the expression given in Fig. 5.5:

$$M_C \quad = \frac{Wab}{L}$$

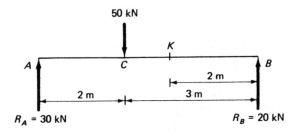

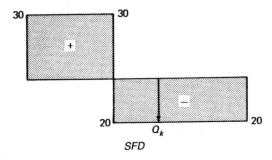

SFD

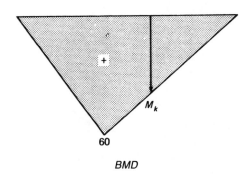

BMD

Fig. 5.18

$$= \frac{50 \times 2 \times 3}{5} = 60 \text{ kN}$$

At section K $\qquad Q_k \quad = -20 \text{ kN}$

$\qquad M_k \quad = 20 \times 2 = 40 \text{ kNm}$

EXAMPLE 5.6 Also find Q_k and M_k (Fig. 5.19)

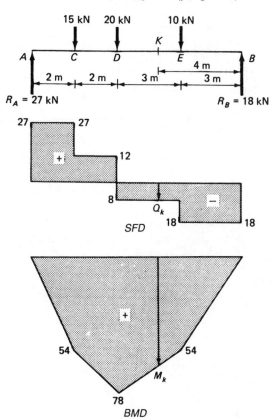

Fig. 5.19

Reactions

$$\Sigma(M)_B = 0 = 10R_A - 15 \times 8 - 20 \times 6 + 10 \times 3$$

$$\therefore R_A = \frac{120 + 120 + 30}{10} = 27 \text{ kN}$$

$$\Sigma(M)_A = 0 = -10R_B + 15 \times 2 + 20 \times 4 + 10 \times 7$$

$$\therefore R_B = \frac{30 + 80 + 70}{10} = 18 \text{ kN}$$

Check: $R_A + R_B = 45$ kN = total load

Bending moments

$$M_A = M_B = 0$$
$$M_C = 27 \times 2 = 54 \text{ kNm}$$
$$M_D = 27 \times 4 - 15 \times 2$$
$$= 108 - 30 = 78 \text{ kNm}$$

$$\text{or } M_D = 18 \times 6 - 10 \times 3$$
$$= 108 - 30 = 78 \text{ kNm}$$
$$M_E = 18 \times 3 = 54 \text{ kNm}$$
$$M_{max} = M_D$$

(*Note* that D is the point of zero shearing force.)

At section K
$$Q_k = -8 \text{ kN}$$
$$M_k = 18 \times 4 - 10 \times 1$$
$$= 72 - 10 = 62 \text{ kNm}$$

EXAMPLE 5.7 Find also Q_p and M_p and determine the position of the point of contraflexure. (Fig. 5.20).

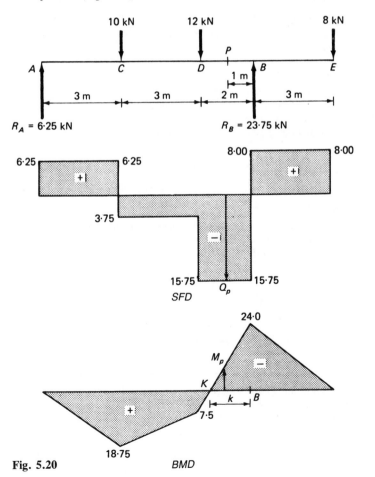

Fig. 5.20   BMD

Reactions

$$\Sigma(M)_B = 0 = 8R_A - 10 \times 5 - 12 \times 2 + 8 \times 3$$

$$\therefore R_A = \frac{50 + 24 - 24}{8} = 6\cdot25 \text{ kN}$$

$$\Sigma(M)_A = 0 = -8R_B + 10 \times 3 + 12 \times 6 + 8 \times 11$$

$$\therefore R_B = \frac{30 + 72 + 88}{8} = 23\cdot75 \text{ kN}$$

Check: $R_A + R_B = 30$ kN = total load
(Note that the moment of the 8 kN load is clockwise (and therefore positive) about both A and B.)

Bending moments

$$M_A = M_B = 0$$
$$M_C = 6\cdot25 \times 3 \qquad\qquad = 18\cdot75 \text{ kNm}$$
$$M_D = 6\cdot25 \times 6 - 10 \times 3 = 37\cdot5 - 30\cdot0 = 7\cdot5 \text{ kNm}$$
$$\text{or } M_D = 23\cdot75 \times 2 - 8 \times 5 = 47\cdot5 - 40\cdot0 = 7\cdot5 \text{ kNm}$$
$$M_B = -8 \times 3 \qquad\qquad = -24\cdot0 \text{ kN}$$

Maximum bending moment

Maximum bending moments occur at the two points of zero shearing force.
Maximum sagging moment at C: $M_C = 18\cdot75$ kNm
Maximum hogging moment at B: $M_B = 24\cdot0$ kNm
The *absolute* maximum of these is the support moment at B.

At section P

$$Q_p = -15\cdot75 \text{ kN}$$
$$M_p = 23\cdot75 \times 1 - 8 \times 4 = -8\cdot25 \text{ kNm}$$

Point of contraflexure

This is the point where the bending moment changes from sagging to hogging, the value of the bending moment at this point is therefore zero. Let the point of contraflexure occur at K, at a distance of k from B.

At K
$$M_k = 0 = 23\cdot75 k - 8(k + 3)$$
$$= 15\cdot75k - 24\cdot0$$

Then
$$k = \frac{24\cdot0}{15\cdot75} = 1\cdot52 \text{ m}$$

EXAMPLE 5.8 Show that, for a simply supported beam, loaded with a uniformly distributed load over the whole length, the maximum bending moment occurs at midspan and has a value equal to $wL^2/8$ (Fig. 5.21).

Reactions

$$R_A = R_B = \frac{wL}{2}$$

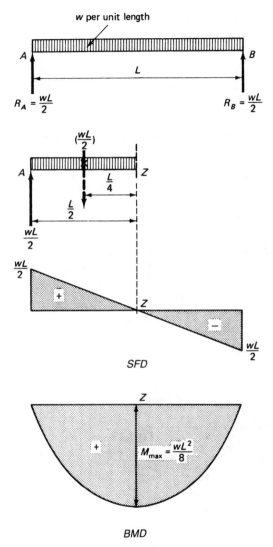

Fig. 5.21

Maximum bending moment

M_{max} occurs at the point of zero shearing force (Section 5.8), which, from the symmetry of the shearing force diagram, is at mid-span. Q.E.D. Referring to the half-span sketch, and taking moments about the point of zero shearing force (z), the maximum bending moment will be:

$$M_{max} = \frac{wL}{2} \times \frac{L}{2} - \frac{wL}{2} \times \frac{L}{4}$$

162

$$= \frac{wL^2}{4} - \frac{wL^2}{8}$$

$$= \frac{wL^2}{8} \qquad\qquad \text{Q.E.D.}$$

EXAMPLE 5.9   (Fig. 5.22).

UDL resultants $(A - C)$ = 40 × 6 = 240 kN

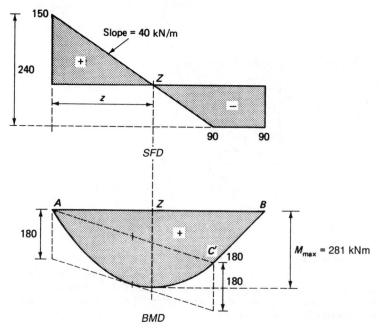

Fig. 5.22

Reactions $R_A = \dfrac{240 \times 5}{8} = 150$ kN

$R_B = \dfrac{240 \times 3}{8} = 90$ kN

Check $R_A + R_B = 240$ kN $=$ total load

Bending moments $M_A = M_B = 0$

$M_C = 90 \times 2 = 180$ kNm

The bending moment diagram is constructed by first plotting the straight-line portion $C - B$ and then drawing a parabola on the resulting base-line $A - C$. The apex height of the parabola above the base $A - C'$ is the FREE BENDING MOMENT value for the *UDL* between A and C. The FREE BENDING MOMENT value of a given length (l) of uniform load (w per unit length) is the maximum moment produced by this load when placed on a simply ('freely') supported beam of the *same length*.

$$FBM = \frac{wl^2}{8}$$

In the case of this example:

$$FBM_{AC} = \frac{40 \times 6^2}{8} = 180 \text{ kNm}$$

The parabola is now sketched, or, if a scale drawing is required, the procedure described in Section 5.7 and shown in Fig. 5.8 may be used.

Maximum bending moment

The maximum bending moment occurs at the point of zero shearing force (Section 5.8), which is at point Z at distance z from A. Then using the similar triangles rule, (i.e. for similar triangles: $\dfrac{\text{base}}{\text{height}} = \text{constant}$)

$$\frac{z}{150} = \frac{6}{240}$$

$$\therefore z = \frac{6 \times 150}{240} = 3 \cdot 75 \text{ m}$$

or, since the slope of the *SFD* is 40 kN/m:

At Z $Q_Z = 0 = 150 - 40z$

$$\therefore z = \frac{150}{40} = 3 \cdot 75 \text{ m}$$

For convenience, the bending effect of the two forces acting on the length of beam AZ may be considered separately (Fig. 5.23).

The reaction causes a sagging moment of $150 \times 3 \cdot 75$ and the *UDL* causes a hogging moment of $40 \times 3 \cdot 75^2 / 2$.

Hence $M_{max} = 150 \times 3 \cdot 75 - 40 \times \dfrac{3 \cdot 75^2}{2}$

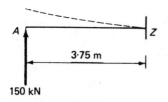

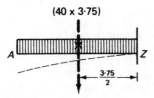

Fig. 5.23

$$= 562 \cdot 5 - 281 \cdot 25$$
$$= 281 \cdot 25 \text{ kNm}$$

EXAMPLE 5.10   (Fig. 5.24).

UDL resultants
$$(A - C) = 20 \times 4 = 80 \text{ kN}$$
$$(C - B) = 10 \times 6 = 60 \text{ kN}$$

Reactions
$$\Sigma(M)_B = 0 = 10R_A - 80 \times 8 - 60 \times 3$$
$$\therefore R_A = \frac{640 + 180}{10} = 82 \text{ kN}$$
$$\Sigma(M)_A = 0 = -10R_B + 80 \times 2 + 60 \times 7$$
$$\therefore R_B = \frac{160 + 420}{10} = 58 \text{ kN}$$

Check: $R_A + R_B = 140 \text{ kN} = \text{total load}$

Bending moments
$$M_A = M_B = 0$$
$$M_C = 82 \times 4 - 80 \times 2$$
$$= 328 - 160 = 168 \quad \text{kNm}$$

The bending moment diagram consists of two parabolas, whose apex heights are the respective FREE BENDING MOMENTS for lengths *AC* and *CB*.

$$FBC_{AC} = \frac{20 \times 4^2}{8} = 40 \text{ kNm}$$

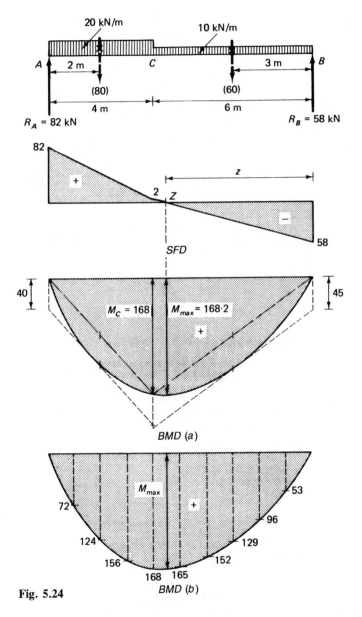

Fig. 5.24

$$FBM_{CB} = \frac{10 \times 6^2}{8} = 45 \text{ kNm}$$

$BMD(a)$ is constructed in this way. An alternative procedure is to calculate the bending moment values at (say) 1 metre intervals and then plot these as ordinates to the BMD curve; $BMD(b)$ is plotted in this way.

BMD Ordinates

From A to C:
$$M_x = 82x - \frac{20x^2}{2}$$

$$= 82x - 10x^2$$

From B to C:
$$M_x = 58x - \frac{10x^2}{2}$$

$$= 58x - 5x^2$$

		A				C	C						B
x	(m)	0	1	2	3	4	6	5	4	3	2	1	0
M_x	(kNm)	0	72	124	156	168	168	165	152	129	96	53	0

Maximum bending moment

The maximum bending moment occurs at the point of zero shearing force Z, at a distance z from B.

$$\text{At } Z: \quad Q_z = 0 = -58 + 10z$$
$$\therefore \quad z = 5 \cdot 8 \text{ m}$$

Then

$$M_{max} = 58 \times 5 \cdot 8 - \frac{10 \times 5 \cdot 8^2}{2}$$

$$= 336 \cdot 4 - 168 \cdot 2$$

$$= 168 \cdot 2 \text{ kNm}$$

EXAMPLE 5.11 (*Fig. 5.25*). *Determine also the point of contraflexure.*

UDL resultants

$$(A - C) = 30 \times 12 = 360 \text{ kN}$$
$$(A - B) = 30 \times 9 = 270 \text{ kN}$$
$$(B - C) = 30 \times 3 = 90 \text{ kN}$$

Reactions

$$R_A = \frac{360 \times 3}{9} = 120 \text{ kN}$$

$$R_B = \frac{360 \times 6}{9} = 240 \text{ kN}$$

Bending moments

$$M_A = M_C = 0$$
$$M_B = -90 \times 1 \cdot 5 = -135 \text{ kNm}$$

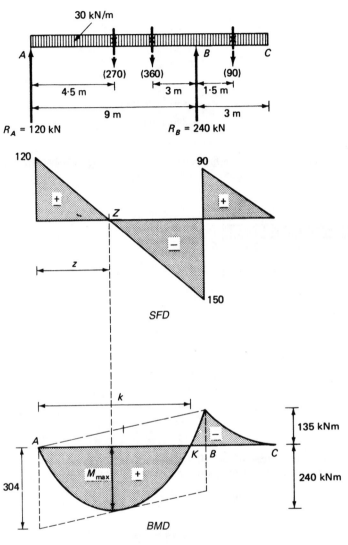

Fig. 5.25

The bending moment diagram consists of two parabolic curves:

     (i) on base-line BC with apex height $= -135$ kNm
    (ii) on base-line AB with apex height $= FBM_{AB}$

$$FBM_{AB} = \frac{30 \times 9^2}{8} = 304 \text{ kNm}$$

Maximum bending moment
The maximum bending moment occurs at the point of zero shearing force Z, at a distance z from A.

At Z:
$$Q_z = 0 = 120 - 30z$$

$$\therefore \quad z = \frac{120}{30} = 4\cdot 0 \text{ m}$$

Then
$$M_{max} = 120 \times 4 - \frac{30 \times 4^2}{2}$$

$$= 480 - 240$$

$$= 240 \text{ kNm}$$

Point of contraflexure
The point of contraflexure occurs at the point where the bending moment is zero. Let this point be K and its distance from A be k metres.

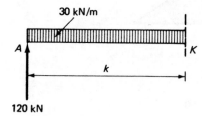

Fig. 5.26

At K: (refer to Fig. 5.26)
$$M_k = 0 = 120k - \frac{30k^2}{2}$$

$$= 120k - 15k^2$$

$$\therefore \quad k = \frac{120}{15} = 8 \text{ m}$$

Alternatively, since the portion of the *BMD* between A and K is equivalent to a FREE BENDING MOMENT DIAGRAM for a beam of length k:

$$\frac{30k^2}{8} = 240 \text{ kNm}$$

$$\therefore \quad k = \sqrt{\frac{240 \times 8}{30}}$$

$$= 8 \text{ m}$$

EXAMPLE 5.12 *Determine also the point of contraflexure.*

UDL resultant

$$(A - C) = 30 \times 4 = 120 \text{ kN}$$

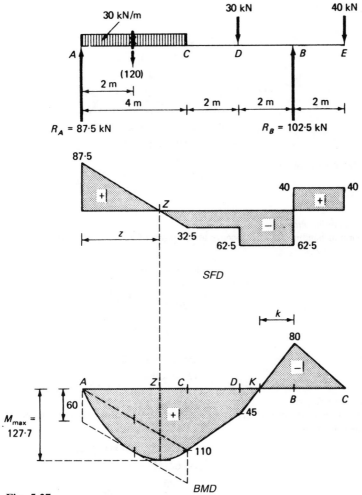

Fig. 5.27

Reactions

$$\Sigma(M)_B = 0 = 8R_A - 120 \times 6 - 30 \times 2 + 40 \times 2$$

$$\therefore R_A = \frac{720 + 60 - 80}{8} = 87 \cdot 5 \text{ kN}$$

$$\Sigma(M)_A = 0 = -8R_B + 120 \times 2 + 30 \times 6 + 40 \times 10$$

$$\therefore R_B = \frac{240 + 180 + 400}{8} = 102 \cdot 5 \text{ kN}$$

Check: $\quad R_A + R_B = 190 \text{ kN} = \text{total load}$

Bending moments

$$M_A = M_E = 0$$
$$M_B = -40 \times 2 = -80 \text{ kNm}$$
$$M_D = 102 \cdot 5 \times 2 - 40 \times 4$$
$$= 205 - 160 = 45 \text{ kNm}$$
$$M_C = 102 \cdot 5 \times 4 - 40 \times 6 - 30 \times 2$$
$$= 410 - 240 - 60 = 110 \text{ kNm}$$

or
$$M_C = 87 \cdot 5 \times 4 - 120 \times 2$$
$$= 350 - 240 = 110 \text{ kNm}$$

For the parabola between *A* and *C*:

$$FBM_{AC} = \frac{30 \times 4^2}{8} = 60 \text{ kNm}$$

Maximum bending moment

The maximum bending moment occurs at the point of zero shearing force Z, at distance *z* from *A*.

At Z:
$$Q_z = 0 = 87 \cdot 5 - 30z$$
$$\therefore \quad z = \frac{87 \cdot 5}{30} = 2 \cdot 92 \text{ m}$$

Then
$$M_{max} = 87 \cdot 5 \times 2 \cdot 92 - \frac{30 \times 2 \cdot 92^2}{2}$$
$$= 255 \cdot 5 - 127 \cdot 8$$
$$= 127 \cdot 8 \text{ kNm}$$

Point of contraflexure

The point of contraflexure occurs at the point of zero bending moment, say point *K* at a distance *k* from *A*.

At K:
$$M_k = 0 = 102 \cdot 5k - 40 (k + 2)$$
$$= 62 \cdot 5k - 80$$
$$\therefore \quad k = \frac{80}{62 \cdot 5} = 1 \cdot 28 \text{ m}$$

5.10 Solutions using computers

The bending behaviour of beams is one of the most difficult concepts that students are required to learn. The repetitious working of many examples is certainly helpful, since the understanding of the underlying behaviour comes from comparative observation. Using a computer to do the numerical 'donkey work', while making repeated small changes in data, will prove to be one of the best approaches to this problem. The author has made good use of purpose-designed spreadsheets to this end. The printout shown in Fig. 5.28(*a*) uses the data given for Example 5.7. A

```
SSB1     SIMPLE BEAM WITH POINT LOADS
------------------------------------------------------------------
                                    | W
                                    |
         <----------- a ----------->|
                                    V
         ----------------------------^----------------------^-------
         <----- al ----->| RL                               |
         <-------------------------------- ar ------------->|  RB
------------------------------------------------------------------
                      RL       RR        LOADS
------------------------------------------------------------------
R or W value:        6.25     23.75      10        12        8
Distance al/ar/a:       0        8        3         6       11
------------------------------------------------------------------
                   W(ar-a)/L              6.25       3       -3        0
                   W(a-al)/L              3.75       9       11        0
------------------------------------------------------------------
    x       M        Component moment due to each load
------------------------------------------------------------------
    0       0        0        0        0        0        0        0
    3    18.75    18.75       0        0        0        0        0
    6     7.5     37.5        0      -30        0        0        0
    8     -24     50          0      -50      -24        0        0
   11      0     68.75     71.25     -80      -60        0        0
```

(*a*)

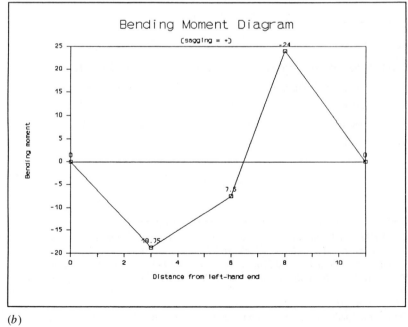

(*b*)

Fig. 5.28

172

variety of different loading and reaction conditions can be entered easily and quickly by the learner and almost instantaneous results obtained in the screen display. Hard-copy printouts can be taken when required and the bending moment diagram drawn using the graph-plotting facility; this too may be printed as hard copy (Fig. 5.28(b)). Other spreadsheets can be easily devised for other types of loading, e.g. uniformly distributed loads, cantilevers, multiple loads, applied moments, etc.

See Section 1.14 for details of sources of spreadsheet and other suitable computer software.

Further worked examples

EXAMPLE 5.13   The cantilever beam shown in Fig. 5.29 carries a distributed load which varies linearly from 60 kNm at A to 20 kNm at B. Determine the shearing force and bending moment at A.

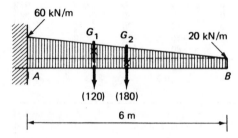

Fig. 5.29

Dividing the total load into two separate portions, the *UDL* resultants are:

for the triangular portion $= (60 - 20) \frac{6}{2} = 120$ kN acting through G_1

for the rectangular portion $= 30 \times 6 = 180$ kN acting through G_2

Then the shearing force at $A, Q_A = 120 + 180 = 300$ kN (Answer)

and the bending moment at $A, M_A = -120 \times 2 - 180 \times 3$

$$= -240 - 360$$

$$= -600 \text{ kNm (Answer)}$$

EXAMPLE 5.14   The beam shown in Fig. 5.30 is simply supported at A and B and carries distributed loads which vary linearly from 12 kNm at A to 40 kNm at B and from zero at C to 50 kNm at B. Sketch the bending moment diagram and determine the magnitude and position of the maximum sagging bending moment.

UDL resultants
Divide the loads into triangles and rectangles, then the *UDL* resultants are:

$$(A - B) \text{ triangle } = (40 - 12) \frac{6}{2} = 84 \text{ kN acting through } G_1$$

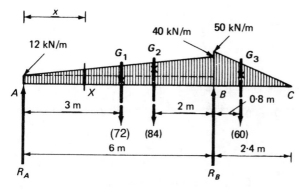

Fig. 5.30

$$(A - B) \text{ rectangle} = 12 \times 6 \qquad\qquad = 72 \text{ kN acting through } G_2$$

$$(B - C) \text{ triangle} = 50 \times \frac{2 \cdot 4}{2} \qquad = 60 \text{ kN acting through } G_3$$

Reactions

$$\Sigma(M)_B = 0 = 6R_A - 72 \times 3 - 84 \times 2 + 60 + 0 \cdot 8$$

$$\therefore \qquad R_A = \frac{216 + 168 - 48}{6} = \frac{336}{6} = 56 \text{ kN}$$

$$\Sigma(M)_A = 0 = -6R_B + 72 \times 3 + 84 \times 4 + 60 \times 6 \cdot 8$$

$$\therefore \qquad R_B = \frac{216 + 336 + 408}{6} = \frac{960}{6} = 160 \text{ kN}$$

Check $\qquad R_A + R_B = 216 \text{ kN} = \text{total load}$

Expression for M_x (Fig. 531)

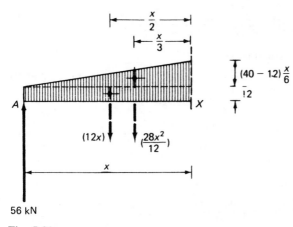

Fig. 5.31

At X:
$$M_x = 56x - \frac{12x^2}{2} - \frac{28x^2}{12}\frac{x}{3}$$

$$= 56x - 6x^2 - \frac{7}{9}x^3$$

At the point of zero shearing force: (equation 5.10)

$$Q_x = 0 = \frac{dM}{dx}$$

$$= 56 - 12x - \frac{7}{3}x^2$$

Dividing by $\frac{7}{3}$:

$$0 = x^2 + 5 \cdot 14x - 24 \cdot 0$$

$$= (x + 8 \cdot 10)(x - 2 \cdot 96)$$

Discarding the negative root: $x = 2 \cdot 96$ m (Answer)

Then

$$M_{max} = 56 \times 2 \cdot 96 - 6 \times 2 \cdot 96^2 - \frac{7 \times 2 \cdot 96^3}{9}$$

$$= 165 \cdot 8 - 52 \cdot 6 - 20 \cdot 2$$
$$= 93 \cdot 0 \text{ kNm (Answer)}$$

EXAMPLE 5.15   The beam shown in Fig. 5.32 is simply supported at A, B and C and carries a uniformly distributed load of 6 kNm over its entire length. At D there is a hinge which transmits shearing force but not bending moment. Draw the shearing force and bending moment diagrams.

UDL resultants
$$(A - C) = 6 \times 18 = 108 \text{ kN}$$
$$(A - B) = 6 \times 10 = 60 \text{ kN}$$
$$(B - D) = 6 \times 4 = 24 \text{ kN}$$
$$(D - C) = 6 \times 4 = 24 \text{ kN}$$

Reactions
Since the bending moment at D is zero:

$$\therefore \quad \Sigma(M)_D = 0 = 24 \times 2 - 4R_c \qquad \text{(moments to the right of } D\text{)}$$

$$\therefore \qquad R_c = \frac{24 \times 2}{4} = 12 \text{ kN}$$

$$\Sigma(M)_A = 0 = -10RB - 18RC + 108 \times 9$$

$$\therefore \qquad R_B = \frac{972 \times 216}{10} = 75 \cdot 6 \text{ kN}$$

$$\Sigma(M)_B = 10R_A - 108 \times 1 - 8R_c$$

$$\therefore \qquad R_A = \frac{108 + 96}{10} = 20 \cdot 4 \text{ kN}$$

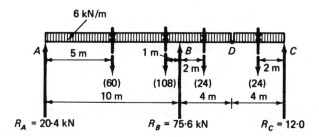

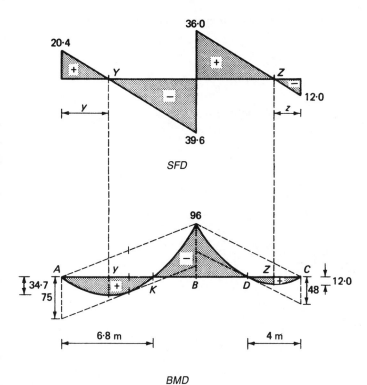

SFD

BMD

Fig. 5.32

Check:
$$R_A + R_B + R_C = 108 \text{ kN} = \text{total load}$$

Bending moments

$$
\begin{aligned}
M_A &= M_C = M_D = 0 \\
M_B &= 12 \times 8 - 24 \times 2 - 24 \times 6 \\
&= 96 - 48 - 144 \\
&= -96 \text{ kNm}
\end{aligned}
$$

or
$$M_B = 20 \cdot 4 \times 10 - 60 \times 5$$

$$= 204 - 300$$
$$= -96 \text{ kNm}$$

Free bending moments

$$FBM_{AB} = \frac{6 \times 10^2}{8} = 75 \text{ kNm}$$

$$FBM_{BC} = \frac{6 \times 8^2}{8} = 48 \text{ kNm}$$

Maximum bending moments
Three points of zero shearing force occur at Y, B and Z. The maximum hogging moment is at B: $M_B = -96$ kNm.

At Y:

$$Q_y = 0 = 20 \cdot 4 - 6y$$
$$y = \frac{20 \cdot 4}{6} = 3 \cdot 4 \text{ m}$$

Then

$$M_y = 20 \cdot 4 \times 3 \cdot 4 - 6 \times \frac{3 \cdot 4^2}{2}$$
$$= 69 \cdot 4 - 34 \cdot 7$$
$$= 34 \cdot 7 \text{ kNm}$$

Alternatively,

$$M_y = \frac{6 \times 6 \cdot 8^2}{8} = 34 \cdot 7 \text{ kNm}$$

At Z:

$$Q_z = 0 = 12 - 6z$$
$$z = \frac{12}{6} = 2 \cdot 0 \text{ m}$$

Then

$$M_z = 12 \times 2 - \frac{6 \times 2^2}{2}$$
$$= 24 \cdot 0 - 12 \cdot 0$$
$$= 12 \cdot 0 \text{ kNm}$$

Alternatively,

$$M_z = \frac{6 \times 4^2}{8} = 12 \cdot 0 \text{ kNm}$$

EXERCISES   CHAPTER 5

1–10. Calculate the reactions and draw the shearing force and bending moment diagrams for the beams shown in Figs 5.33–5.42. Also calculate the magnitude of the shearing force and bending moment at point K in each case.

11–22. Draw the shearing force and bending moment diagrams for the beams shown in Figs 5.43–5.54. Calculate (a) the position and magnitude of the maximum sagging bending moment and (b) the magnitude of the shearing force and the bending moment at point K.

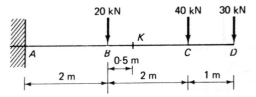

Fig. 5.33

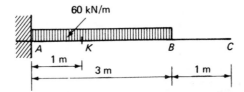

Fig. 5.34

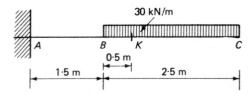

Fig. 5.35

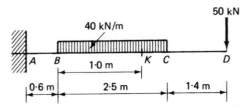

Fig. 5.36

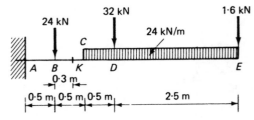

Fig. 5.37

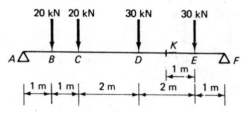

Fig. 5.38

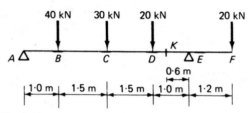

Fig. 5.39

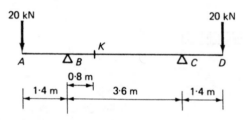

Fig. 5.40

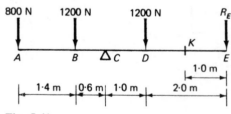

Fig. 5.41

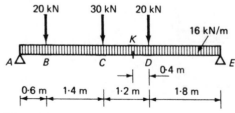

Fig. 5.42

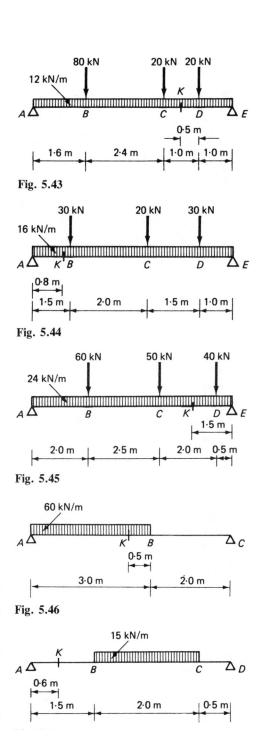

Fig. 5.43

Fig. 5.44

Fig. 5.45

Fig. 5.46

Fig. 5.47

180

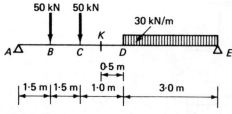

Fig. 5.48

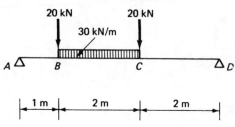

Fig. 5.49

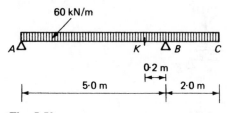

Fig. 5.50

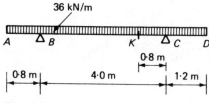

Fig. 5.51

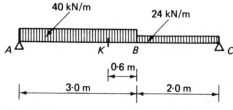

Fig. 5.52

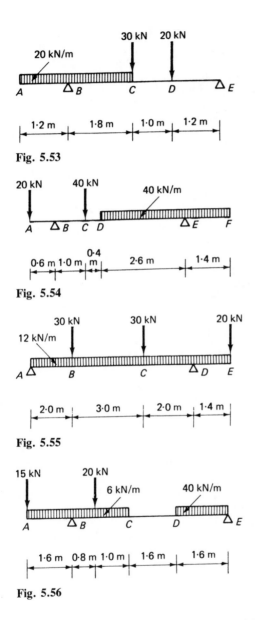

Fig. 5.53

Fig. 5.54

Fig. 5.55

Fig. 5.56

23–28. Calculate the position and magnitude of the maximum bending moment, both sagging and hogging where appropriate, occurring in the beams shown in Figs 5.55–5.60.

29 and 30. For the beams shown in Figs 5.61 and 5.62, calculate the distance a in the terms of the span L that will give (a) the lowest value of bending moment and (b) zero bending moment at a point mid-way between B and C.

182

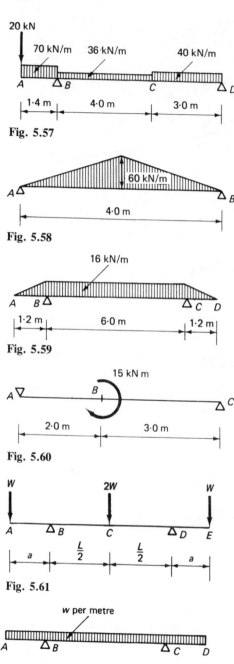

Fig. 5.57

Fig. 5.58

Fig. 5.59

Fig. 5.60

Fig. 5.61

Fig. 5.62

31 and 32. For the cantilever beams shown in Figs 5.63 and 5.64, calculate the magnitude of the shearing force and the bending moment occurring at point A.

33−40. Calculate the position and magnitude of the maximum bending moment, both sagging and hogging where appropriate, occurring in the beams shown in Figs 5.65−5.72.

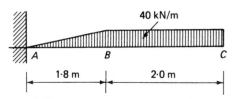

Fig. 5.63

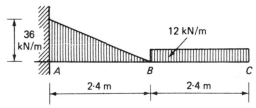

Fig. 5.64

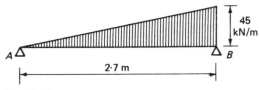

Fig. 5.65

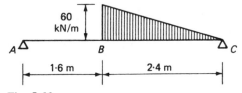

Fig. 5.66

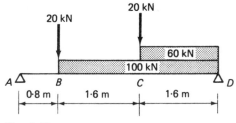

Fig. 5.67

184

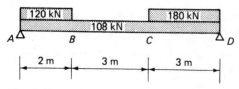

Fig. 5.68

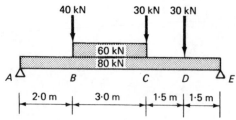

Fig. 5.69

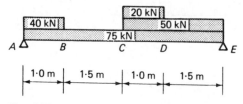

Fig. 5.70

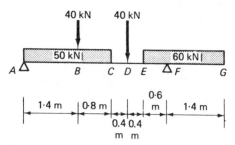

Fig. 5.71

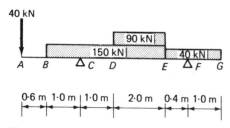

Fig. 5.72

41. In the beam shown in Fig. 5.73, the bending moment occurring at *B* is 700 kNm (hogging). Calculate the value of *W* and also the reactions at *A* and *B*.

42. For the beam shown in Fig. 5.74, calculate the position and magnitude of the maximum bending moment and sketch the shearing force and bending moment diagrams.

43. For the beam shown in Fig. 5.75, calculate the position and magnitude of the maximum bending moment and also the position of the point of contraflexure. What is the magnitude of the bending moment at mid-span?

44. For the beam shown in Fig. 5.76, (a) draw the shearing force and bending moment diagrams, (b) calculate the magnitude of the bending moment at mid-span and (c) calculate the positions of the points of contraflexure.

45. For the beam shown in Fig. 5.77, calculate the position of the points of contraflexure.

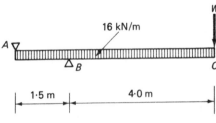

Fig. 5.73

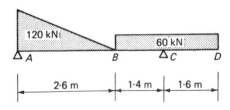

Fig. 5.74

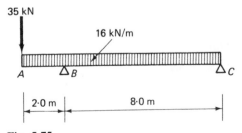

Fig. 5.75

186

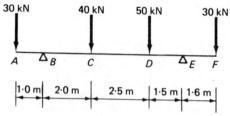

Fig. 5.76

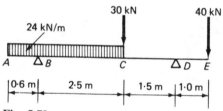

Fig. 5.77

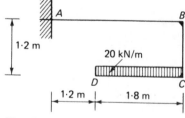

Fig. 5.78

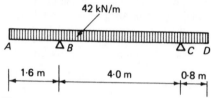

Fig. 5.79

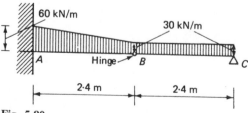

Fig. 5.80

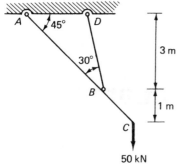

Fig. 5.81

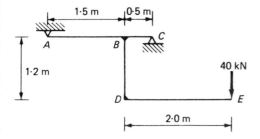

Fig. 5.82

46. For the beam shown in Fig. 5.78, calculate the position and magnitude of the maximum sagging bending moment and also the position of the points of contraflexure.

47. Fig. 5.79 shows a cantilever bent which is encastré at point A and contains fully rigid joints at B and C which are capable of transmitting bending moments. Draw the complete bending moment diagram for the bent.

48. Fig. 5.80 shows a loaded beam which is encastré at A, is simply supported at C and contains a hinge at B. Draw the bending moment diagram for the beam.

49. Draw the bending moment diagram for the inclined beam shown in Fig. 5.81.

50. Fig. 5.82 shows a rigid-jointed framework the members of which have constant cross-section. Draw the bending moment diagram for the complete framework.

6 Stresses in beams

Symbols

A	$=$	area
B, b	$=$	breadth of section
D, d	$=$	depth of section
c	$=$	distance to centroid of an element of area
E	$=$	Young's modulus of elasticity
f, f_b	$=$	longitudinal stress due to bending
f_y	$=$	yield strength
I	$=$	second moment of area (about a given axis)
I_{xy}	$=$	product moment of area
L	$=$	length, span
M	$=$	bending moment, moment of resistance
M_{ep}	$=$	elasto-plastic moment of resistance
M_p	$=$	fully plastic moment of resistance
m	$=$	modular ratio
p_y	$=$	yield strength
Q	$=$	shearing force
R	$=$	radius of curvature
R	$=$	reaction force at a given point
S	$=$	plastic section modulus (about a given axis)
W	$=$	concentrated load, total uniform load
w	$=$	uniform load per unit length
y	$=$	distance from neutral axis to any fibre
y_c	$=$	distance from neutral axis to extreme compression fibre

y_t	= distance from neutral axis to extreme tension fibre
$\bar{y}$	= distance from neutral axis to the centroid of a component area
Z	= elastic section modulus (about a given axis)
Z_c	= section modulus with respect to the extreme compression fibre
Z_t	= section modulus with respect to the extreme tension fibre
z	= distance to the point of zero shearing force
α	= shearing stress distribution coefficient
γ	= partial factor of safety density
θ	= angle
ϵ	= strain
τ	= shearing stress

6.1 Stresses induced by bending

When a beam is subjected to lateral loads it bends until either a state of equilibrium is reached, or the beam fails. One of the main aims in designing beams is to ensure that, under the application of the full working loads, the beam reaches a state of equilibrium at which the stresses induced in the fibres are just below the maximum permitted values. The work of this chapter will be concerned with the analysis of the stress states induced in beams due to lateral loading. Stresses resulting from bending moments will be considered first, followed later in the chapter by stresses resulting from shearing forces.

A definition of terms is useful at this point:

Bending stresses are longitudinal stresses, i.e. acting parallel to the longitudinal axis of the beam, which may be tensile or compressive, and are induced by the application of a bending moment.

Shearing stresses are lateral or longitudinal stresses, acting tangentially to the plane of reference, and are induced by the application of shearing forces.

6.2 Pure bending

A beam is said to be in a state of pure bending if, along its length, the induced bending moment is constant and the shearing force is zero. For example, the length of beam between supports A and B in Fig. 6.1 will be in *pure bending*; in Fig. 6.2, equal couples applied at the ends of the beam produce the same effect. Although *pure* bending only occurs occasionally in practice, it is useful to consider the analysis of bending stresses for this condition since the effects of shearing stresses and strains can be eliminated.

6.3 The theory of simple bending

An analysis will now be made of the distribution of stress and the various relationships that exist at a given section of a beam subject to pure

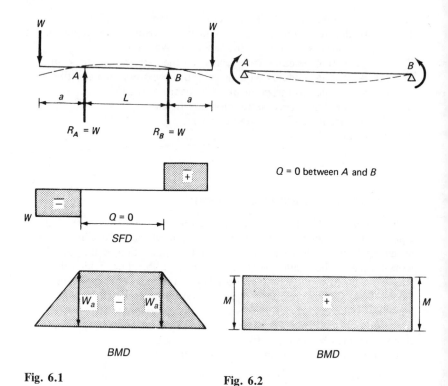

Fig. 6.1   Fig. 6.2

bending. In order that correct mathematical methods may be employed certain assumptions must be made, as follows:

1. The beam is in a state of pure bending.
2. The beam is of homogeneous material, which is elastically isotropic and has the same value for Young's modulus in both tension and compression.
3. The beam is initially straight and all longitudinal fibres bend in the form of circular arcs with a common centre of curvature and a radius which is large compared with the dimensions of the cross-section.
4. Lateral cross-sections remain plane and perpendicular to the neutral axis after bending.

Allowing these assumptions, consider the portion of beam shown in Fig. 6.3. In the elemental length AB, the sections at A and B remain perpendicular to the axis after bending, so that a fibre CD, having a cross-sectional area δA and at distance y from the neutral axis, undergoes a change in length from CD to $C'D'$.

Therefore the strain in $CD = \dfrac{C'D' - CD}{CD} = \dfrac{C'D' - AB}{AB}$

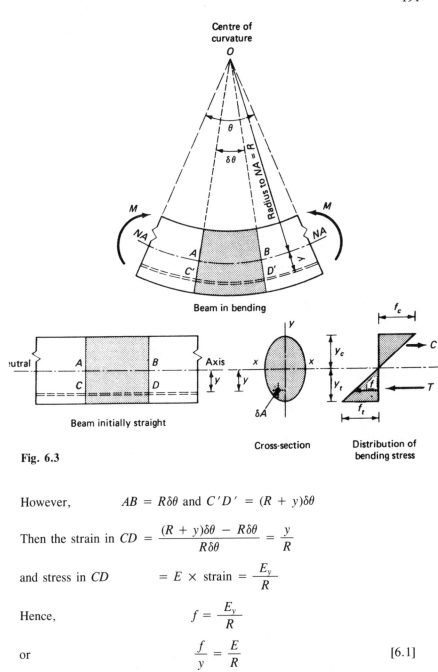

Fig. 6.3

However, $AB = R\delta\theta$ and $C'D' = (R + y)\delta\theta$

Then the strain in $CD = \dfrac{(R + y)\delta\theta - R\delta\theta}{R\delta\theta} = \dfrac{y}{R}$

and stress in $CD = E \times \text{strain} = \dfrac{E_y}{R}$

Hence, $f = \dfrac{E_y}{R}$

or $\dfrac{f}{y} = \dfrac{E}{R}$ [6.1]

(*Note:* it is common practice to use the symbol f for bending stress rather than the more general symbol σ.)

Now consider the longitudinal force transmitted by the element *CD*.

Since the stress in the element $= f$

the force in the element $= f\delta A = \dfrac{Ey}{R}\,\delta A$

However, for equilibrium in pure bending, the total horizontal force is equal to zero, i.e. the total tension (T) = the total compression (C).

Therefore,
$$\Sigma \left(\dfrac{E}{R}\, y\,\delta A \right) = 0$$

or between the limits of y_c and y_t:

$$\dfrac{E}{R} \int_{y_t}^{y_c} y\,\mathrm{d}A = 0$$

Hence *the neutral axis passes through the centroid of the cross-section.*

Consider again the longitudinal force in the element, i.e. $\dfrac{E}{R}\, y\,\delta A$.

The moment of this force about the neutral axis $= \dfrac{E}{R}\, y^2\,\delta A$.

The sum of the moments of all such elements about the neutral axis is called the MOMENT OF RESISTANCE of the beam section. For equilibrium, the moment of resistance must be equal to the externally applied bending moment.

Bending moment = moment of resistance $= M = \Sigma \left(\dfrac{E}{R}\, y^2\,\delta A \right)$

Since E/R is constant for a given bending moment, then between the limits y_c and y_t:

$$M = \dfrac{E}{R} \int_{y_t}^{y_c} y^2\,\mathrm{d}A$$

But from equation [4.6] p. 121.
$$\int_{y_t}^{y_c} y^2\,\mathrm{d}A = I$$

Hence
$$M = \dfrac{E}{R} \times I$$

or
$$\dfrac{M}{I} = \dfrac{E}{R} \qquad\qquad [6.2]$$

If equations [6.1] and [6.2] are taken together the resulting expression gives the general relationships for simple bending and is termed the GENERAL EXPRESSION FOR SIMPLE BENDING.

$$\dfrac{M}{I} = \dfrac{f}{y} = \dfrac{E}{R} \qquad\qquad [6.3]$$

It is from this genereal expression that the formulae used for design are derived. Equation [6.3] is derived, of course, for the condition of *pure* or *simple* bending; a more rigorous analysis is required to produce a similar expression for *ordinary* bending, i.e. where the effects of shearing strains and varying bending moments occur. However, the expression for simple bending is generally accepted as being of sufficient accuracy for the majority of cases of ordinary bending. Further justification lies in the fact that, in many cases, the maximum bending moment occurs where the shearing force is zero. Also the factors of safety adopted are more than adequate to take care of the small errors involved, even in the more extreme cases.

The following important conclusions can now be drawn from the foregoing analysis:

1. It will be seen from equations [6.1] and [6.3] that the bending stress in any given longitudinal fibre is proportional to the distance of that fibre from the neutral axis.

$$f = \frac{E}{R} \times y \qquad \left(\frac{E}{R} = \text{constant} \right)$$

The distribution of bending stress (Fig. 6.3) is therefore triangular, the maximum values occurring at the extreme fibres.

2. It will be seen from equation [6.3] that the bending stress in any given longitudinal fibre is proportional to the applied bending moment.

$$f = M \times \frac{I}{y}$$

The value of I/y will be constant for a given fibre.

6.4 Section modulus

In conclusion 2 in the preceding section, the bending stress is shown to be proportional to the bending moment; and in conclusion 1, the bending stress is shown to have a maximum value at the extreme fibre. The ratio of the bending moment to the maximum stress at a given section is called the ELASTIC SECTION MODULUS of the section (Z).

$$\frac{M}{f} = Z \qquad\qquad [6.4]$$

In Section 4.11 it was established that the section modulus is also a purely geometrical function.

$$Z_c = \frac{I}{y_c}$$

and

$$Z_t = \frac{I}{y_t}$$

Also, from equation [6.3] $\quad \dfrac{M}{f} = \dfrac{I}{y}$

Thus for a given geometrical section, a value for the section modulus can be determined and the stress state due to bending evaluated.

$$\text{Maximum compressive stress,} \quad f_c = \frac{M}{Z_c} \qquad [6.5]$$

$$\text{Maximum tensile stress,} \quad f_t = \frac{M}{Z_t} \qquad [6.6]$$

Equations [6.5] and [6.6] are basic expressions used in the design of beams.

EXAMPLE 6.1   A strip of steel 1·4 mm thick has to pass over a pulley. Determine the diameter of the pulley required if the bending stress in the steel must not exceed 120 kN/mm². E = 120 kN/m²

Radius of pulley = radius of bending curvature

Then from equation [6.1], $\quad R = \dfrac{E_y}{f} = \dfrac{210 \times 10^3 \times 1\cdot4}{120 \times 2}$

$$= 1\cdot23 \times 10^3 \text{ mm}$$

$$= 1\cdot23 \text{ mm}$$

Hence diameter of pulley $\quad = 2\cdot46$ m (Answer)

EXAMPLE 6.2   Determine the maximum bending stress that will occur in a timber beam of rectangular section 150 mm × 50 mm (Fig. 6.4), when a bending moment of 600 Nm is applied about the x−x axis.

For a rectangle, $\quad I_{xx} = \dfrac{bd^3}{12}$ (Table 4.1)

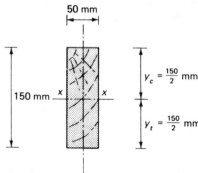

Fig. 6.4

Since
$$y_c = y_t$$
$$Z_{xx(c)} = Z_{xx(t)} = Z_{xx}$$
$$Z_{xx} = \frac{50 \times 150^3}{12} \times \frac{2}{150}$$
$$= \frac{50 \times 22\,500}{6}$$
$$= 187\,500 \text{ mm}^3$$

Hence,
$$f_{max} = \frac{M_{max}}{Z_{xx}}$$
$$= \frac{600 \times 10^3}{187\,500} = 3 \cdot 2 \text{ N/mm}^2 \text{ (Answer)}$$

EXAMPLE 6.3   A timber beam of rectangular section 240 mm × 80 mm is simply supported over a span of 4 m. If the permissible bending stress is 5 N/mm², determine the maximum allowable uniformly distributed load that the beam may carry, (a) applied perpendicular to the x−x axis, and (b) applied perpendicular to the y−y axis.

For the maximum stress condition (at the extreme fibres) the section moduli

are:
$$Z_{xx} = \frac{I_{xx}}{y_{xx}} = \frac{80 \times 240^3 \times 2}{12 \times 240} = 768 \times 10^3 \text{ mm}^3$$
$$Z_{yy} = \frac{I_{yy}}{y_{yy}} = \frac{240 \times 80^3 \times 2}{12 \times 80} = 256 \times 10^3 \text{ mm}^3$$

(see Table 4.1)

Therefore the allowable bending moments about the x−x and y−y axes respectively will be:
$$M_{xx} = 768 \times 10^3 \times 5 = 3 \cdot 84 \times 10^6 \text{ Nmm}$$
$$= 3 \cdot 84 \qquad \text{kNm}$$
$$M_{yy} = 256 \times 10^3 \times 5 = 1 \cdot 28 \times 10^6 \text{ Nmm}$$
$$= 1 \cdot 28 \qquad \text{kNm}$$

Now, for a uniform load, the bending moment $M = wL^2/8$ so that, rearranging:
$$w = \frac{8M}{L^2}$$

The allowable loads are therefore:
$$w_{xx} = \frac{8 \times 3 \cdot 84}{4^2} = 1 \cdot 92 \text{ kNm (Answer)}$$
$$w_{yy} = \frac{8 \times 1 \cdot 28}{4^2} = 0 \cdot 64 \text{ kNm (Answer)}$$

*EXAMPLE 6.4   Calculate the percentage reduction in the second moment of area
(I_{xx}) that will oₒcuₙ at a section in a universal beam where a hole of diameter
200 mm has Leᵉ₁ cut through the web as shown in Fig. 6.5.*

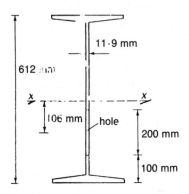

Fig. 6.5

The sectional properties of the universal beam area as follows:

Overall depth = 612 mm   Web thickness = 11·9 mm

Cross-sectional area = 15 960 mm²

Second moment of area (I_{xx}) = 985·8 × 10⁶ mm⁴

The cross-sectional area and second moment of area of the hole are calculated as
they would be for solid material; they are simply geometrical properties.

The position of the neutral axis of the section will be shifted because of the
hole. The amount of this shift can be calculated by taking (first) moments of area
about the original neutral axis.

The distance from the neutral axis to the centre of the hole

$$= \tfrac{1}{2} \times 612 - 100 - \tfrac{1}{2} \times 200 = 106 \text{ mm}$$

Shift in netural axis, $\Delta y = \dfrac{\text{moment of area of hole}}{\text{original area} - \text{area of hole}}$

$$= \dfrac{-200 \times 11 \cdot 9 \times 106}{15\ 960 - 200 \times 11 \cdot 9} = -18 \cdot 58 \text{ mm}$$

The second moment of area of the hole about the neutral axis will be:

$$\Delta I_{xx} = \dfrac{11 \cdot 9 \times 200^3}{12} + 11 \cdot 9 \times 200 (10 \cdot 6 - 18 \cdot 58)^2$$

$$= (7 \cdot 93 + 8 \cdot 19)10^6 = 26 \cdot 12 \times 10^6 \text{ mm}^4$$

Percentage reduction in $I_{xx} = \dfrac{26 \cdot 12 \times 100}{985 \cdot 8} = 2 \cdot 6\% \text{ (Answer)}$

EXAMPLE 6.5 *The section shown in Fig. 6.6 is that of a cantilever 4 m long carrying a uniform load of 3 kN/m which is applied perpendicular to the x—x axis. Calculate (a) the maximum bending stress in the beam under this loading, (b) the maximum concentrated load that may be carried at the free end of the cantilever in addition to the uniform load, if the permissible bending stress is 120 N/mm².*

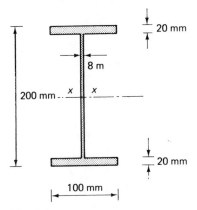

Fig. 6.6

The second moment area will be the difference between the values for two rectangles.

$$I_{xx} = \frac{100 \times 200^3}{12} - \frac{92 \times 160^3}{12}$$

$$= \frac{800 - 377}{12} \times 10^6$$

$$= 35 \cdot 25 \times 10^6 \text{ mm}^4$$

(a)
$$M_{max} = 3 \times 4 \times \frac{4}{2} = 24 \text{ kNm}$$

$$= 24 \times 10^6 \text{ Nmm}$$

$$f = \frac{M}{I} y = \frac{24 \times 10^6 \times 100}{35 \cdot 25 \times 10^6}$$

$$= 68 \cdot 1 \text{ N/mm}^2 \text{ (Answer)}$$

(b) Since the total stress must not exceed 120 N/mm², the stress due to the concentrated load only must be limited to

$$f_{cl} = 120 \cdot 0 - 68 \cdot 1 = 51 \cdot 9 \text{ N/mm}^2$$

Then the allowable additional bending moment due to the concentrated load will be

$$M_{cl} = \frac{fI}{y} = \frac{51 \cdot 9 \times 35 \cdot 25 \times 10^6}{100}$$

$$= 18\cdot29 \times 10^6 \text{ Nmm}$$
$$= 18\cdot29 \text{ kNm}$$

Now for a concentrated load on the free end of a cantilever $M = W \times L$. Then the allowable concentrated load, $W_{cl} = 18\cdot29/4 = 4\cdot57$ kN (Answer)

EXAMPLE 6.6   A rectangular steel box section is shown in Fig. 6.7. Determine the average compressive stress in the walls of the section parallel to the x−x axis, when the applied bending moment about this axis is 15 kNm.

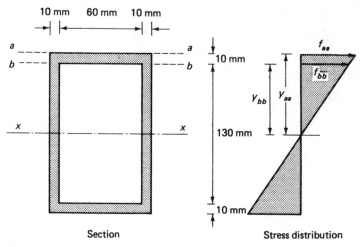

Fig. 6.7

Section   **Stress distribution**

The second moment of area will be the difference between the values for two rectangles.

$$I_{xx} = \frac{80 \times 150^3}{12} - \frac{60 \times 130^3}{12}$$

$$= \frac{270 - 132}{12} \times 10^6$$

$$= 11\cdot5 \times 10^6 \text{ mm}^4$$

Extreme fibre stress, $$f_{aa} = \frac{My_a}{I_{xx}}$$

$$= \frac{15 \times 10^6 \times 75}{11\cdot5 \times 10^6}$$

$$= 97\cdot8 \text{ N/mm}^2$$

Stress at inside face, $$f_{bb} = \frac{My_b}{I_{xx}}$$

$$= \frac{15 \times 10^6 \times 65}{11 \cdot 5 \times 10^6}$$

$$= 84 \cdot 8 \text{ N/mm}^2$$

Average stress in wall of section $\quad = \dfrac{97 \cdot 8 + 84 \cdot 8}{2}$

$$= 91 \cdot 3 \text{ N/mm}^2 \text{ (Answer)}$$

---

*EXAMPLE 6.7 A timber beam having a rectangular cross-section 240 mm ×
85 mm is loaded as shown in Fig. 6.8. Determine (a) the maximum bending stress
in the beam, (b) the bending stress in the beam at a point situated 0·2 m to the
left of B and 30 mm below the upper edge of the section.*

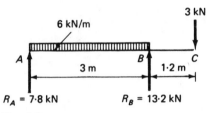

Fig. 6.8

(a) Reactions

$$\Sigma(M)_B \quad = 0 = 3R_A - 6 \times 3 \times 1 \cdot 5 + 3 \times 1 \cdot 2$$

$$\therefore \ R_A \quad = \frac{27 \cdot 0 - 3 \cdot 6}{3} = \frac{23 \cdot 4}{3} = 7 \cdot 8 \text{ kN}$$

$$\Sigma(M)_A \quad = 0 = -3R_B + 6 \times 3 \times 1 \cdot 5 + 3 \times 4 \cdot 2$$

$$\therefore \ R_B \quad = \frac{27 \cdot 0 + 12 \cdot 6}{3} = \frac{39 \cdot 6}{3} = 13 \cdot 2 \text{ kN}$$

Check: $\quad R_A + R_B \ = 21 \cdot 0 \text{ kN} = \text{total load}$

Maximum bending moment
The maximum sagging moment occurs at the point of zero shear, say z m from
A.

At z $\qquad\qquad \therefore \ Q_z = 0 = 7 \cdot 8 - 6z$

$$\therefore \ z \ = \frac{7 \cdot 8}{6} = 1 \cdot 3 \text{ m}$$

Then $\qquad M_{\max} \text{ (sagging)} \ = 7 \cdot 8 \times 1 \cdot 3 - 6 \times \dfrac{1 \cdot 3^2}{2}$

$$= 10 \cdot 14 - 5 \cdot 07$$

$$= 5 \cdot 07 \text{ kNm}$$

At B $\qquad M_{max}$ (hogging) $\quad = -3 \times 1 \cdot 2$

$$= -3 \cdot 6 \text{ kNm}$$

Hence $\qquad M_{max} \qquad\qquad = 5 \cdot 07 \times 10^6 \text{ Nmm}$

$$I_{xx} = \frac{85 \times 240^3}{12} = 97 \cdot 92 \times 10^6 \text{ mm}^4$$

Then the maximum bending stress (occurring at z) is

$$f_{max} = \frac{M_{max} \times y}{I_{xx}}$$

$$= \frac{5 \cdot 07 \times 10^6 \times 120}{97 \cdot 92 \times 10^6}$$

$$= 6 \cdot 21 \text{ N/mm}^2 \text{ (Answer)}$$

(b) At a point $0 \cdot 2$ m to the left of B,

$$M = 13 \cdot 2 \times 0 \cdot 2 - 6 \times \frac{0 \cdot 2^2}{2} - 3 \times 1 \cdot 4$$

$$= 2 \cdot 64 - 0 \cdot 12 - 4 \cdot 20$$

$$= -1 \cdot 66 \text{ kNm}$$

$$= -1 \cdot 66 \times 10^6 \text{ Nmm}$$

The extreme fibre stress at this point will be,

$$f = \frac{My}{I} = \frac{1 \cdot 66 \times 10^6 \times 120}{97 \cdot 92 \times 10^6}$$

$$= 2 \cdot 04 \text{ N/mm}^2$$

This will be tension at the top edge and compression at the bottom edge. The stress required is f' shown in Fig. 6.9. By similar triangles,

$$f' = 2 \cdot 04 \times \frac{90}{120}$$

$$= 1 \cdot 53 \text{ N/mm}^2, \text{ tension (Answer)}$$

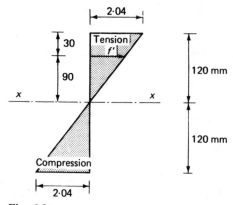

Fig. 6.9

6.5 Stresses due to simultaneous or asymmetric moments

Structural members may often be subject to bending moments about two orthogonal axes simultaneously, or to a single bending moment applied about an axis that is not an axis of symmetry. In such cases, an asymmetrical form of bending occurs in which the resulting radius of curvature in a given plane is affected by all the component bending moments.

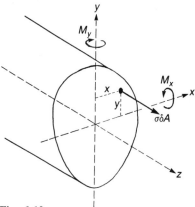

Fig. 6.10

In Fig. 6.10 a beam is shown to be subject to simultaneously applied bending moments M_x and M_y. If, in a fibre at co-ordinates (x,y), the resulting longitudinal stress is f, then for equilibrium:

in the yz plane: $M_x = f y \, dA$

$M_y = f x \, dA$

in the xz plane $M_x = f x \, dA$

$M_y = f y \, dA$

Now, from eqn [6.3], for compatibility:

in the yz plane: $f = yE/R_y$

in the xz plane: $f = xE/R_x$

Substituting:

in the yx plane: $M_x = \dfrac{E}{R_y} \, y^2 \, dA = \dfrac{E \, I_{xx}}{R_y}$

$M_y = \dfrac{E}{R_y} \, xy \, dA = \dfrac{E \, I_{xy}}{R_v}$

in the xz plane:
$$M_x = \frac{E}{R_x} \; xy \, dA \; = \frac{E \, I_{xy}}{R_x}$$

$$M_y = \frac{E}{R_x} \; x^2 \, dA \; = \frac{E \, I_{yy}}{R_x}$$

For simultaneous bending in the yz and xz planes the equations above are added together:

$$M_x = \frac{E \, I_{xx}}{R_y} + \frac{E \, I_{xy}}{R_x} \qquad\qquad [6.7(a)]$$

$$M_y = \frac{E \, I_{yy}}{R_x} + \frac{E \, I_{xy}}{R_y} \qquad\qquad [6.7(b)]$$

and from these equations it can be shown that

$$\frac{E}{R_y} = \frac{M_x \, I_{yy} - M_y \, I_{xy}}{I_{xx} \, I_{yy} - I_{xy}^{\,2}}$$

$$\frac{E}{R_x} = \frac{M_y \, I_{xx} - M_x \, I_{xy}}{I_{xx} \, I_{yy} - I_{xx}^{\,2}}$$

So that the bending stress resulting from the simultaneous application of M_x and M_y is given by:

$$f = \frac{yE}{R_y} + \frac{xE}{R_x}$$

$$= \frac{y(M_x I_{yy} - M_y I_{xy}) + x(M_y I_{xx} - M_x I_{xy})}{I_{xx} I_{yy} - I_{xy}^2} \qquad [6.8]$$

In the case of the x−x and y−y axes being principal axes, $I_{xy} = 0$, so that:

$$f = \frac{y \, M_x}{I_{xx}} + \frac{x \, M_y}{I_{yy}} \qquad\qquad [6.9]$$

In the case of a single bending moment (M) being applied at an oblique angle (θ) to the major principal axis, the component moments with respect to the two principal axes will be (Fig. 6.11):

$$M_x = M \cos\theta \qquad\qquad [6.10(a)]$$

$$M_y = M \sin\theta \qquad\qquad [6.10(b)]$$

and, substituting in eqn [6.9]:

$$f = \frac{y \, M \cos\theta}{I_{xx}} + \frac{x \, M \sin\theta}{I_{yy}} \qquad\qquad [6.11]$$

The neutral axis (neutral plane) will not necessarily coincide with either

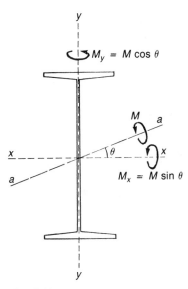

Fig. 6.11

the orthogonal axes or the plane of the bending. Since, at the neutral axis, $f = 0$, then substituting in eqn [6.8]:

$$f = 0 = y(M_x I_{yy} - M_y I_{xy}) + x(M_y I_{xx} - M_x I_{xy}) \qquad [6.12]$$

Thus,
$$\frac{y}{x} = -\frac{M_y I_{xx} - M_x I_{xy}}{M_x I_{yy} - M_y I_{xy}} \qquad [6.13]$$

or where $x-x$ and $y-y$ are principal axes:

$$\frac{y}{x} = -\frac{M_y I_{xx}}{M_x I_{xy}}$$

or substituting from eqn [6.11]:

$$\frac{x}{y} = -\frac{\sin \theta \, I_{xx}}{\cos \theta \, I_{yy}} = -\tan \theta \, \frac{I_{xx}}{I_{yy}} \qquad [6.14]$$

EXAMPLE 6.8   Figure 6.12 shows the cross-section of a gantry girder which is subject to bending about both principal axes simultaneously. Determine the stresses induced at a, b, c and d if the moments are $M_{xx} = 150$ kNm and $M_{yy} = 10$ kNm.

The component sections have the following properties:

Channel
$$I_{xx} = 33 \cdot 67 \times 10^6 \text{ mm}^4$$

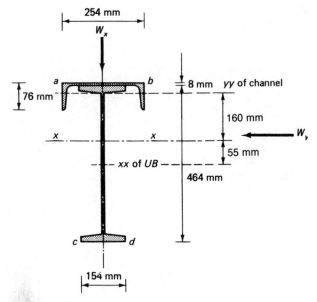

Fig. 6.12

$$I_{yy} = 1·63 \times 10^6 \text{ mm}^4$$
$$A = 3\ 600 \text{ mm}^2$$

Universal beam

$$I_{xx} = 361·6 \times 10^6 \text{ mm}^4$$
$$I_{yy} = 10·93 \times 10^6 \text{ mm}^4$$
$$A = 10\ 440 \text{ mm}^2$$

Second moments of area

$$I_{xx} = (361·6 + 1·6)\ 10^6 + 10\ 440 \times 55^2 + 3\ 600 \times 160^2$$
$$= (363·2 + 29·3 + 94·5)\ 10^6$$
$$= 487 \times 10^6 \text{ mm}^4$$
$$I_{yy} = (10·93 + 33·67)\ 10^6$$
$$= 44·6 \times 10^6 \text{ mm}^4$$

Extreme fibre distances

$$y_a = y_b = 232 - 55 + 8 = 185 \text{ mm}; \quad x_a = x_b = 254/_2 = 127 \text{ mm}$$
$$y_c = y_d = 232 + 55 \qquad = 287 \text{ mm}; \quad x_c = x_d = 154/_2 = 77 \text{ mm}$$

Stresses (compressive = + ve):

$$f_a = \frac{M_x y_a}{I_{xx}} - \frac{M_y x_a}{I_{yy}}$$

$$= \frac{150 \times 10^6 \times 185}{487 \times 10^6} - \frac{10 \times 10^6 \times 127}{44 \cdot 6 \times 10^6} = \quad \begin{array}{l} 57 \cdot 0 - 28 \cdot 5 = 28 \cdot 5 \\ \text{N/mm}^2 \text{ (compression)} \end{array}$$

$$f_b = \frac{M_x y_b}{I_{xx}} + \frac{M_y x_b}{I_{yy}} \qquad\qquad = \quad \begin{array}{l} 57 \cdot 0 + 28 \cdot 5 = 85 \cdot 5 \\ \text{N/mm}^2 \text{ (compression)} \end{array}$$

$$f_c = - \frac{M_x y_c}{I_{xx}} - \frac{M_y x_c}{I_{yy}}$$

$$= - \frac{150 \times 10^6 \times 287}{487 \times 10^6} - \frac{10 \times 10^6 \times 77}{44 \cdot 6 \times 10^6} = \quad \begin{array}{l} - 88 \cdot 4 - 17 \cdot 3 = 105 \cdot 7 \\ \text{N/mm}^2 \text{ (compression)} \end{array}$$

$$f_d = - \frac{M_x y_d + M_y x_d}{I_{xx} \, I_{yy}} \qquad\qquad = \quad \begin{array}{l} - 88 \cdot 4 + 17 \cdot 3 = 71 \cdot 1 \\ \text{N/mm}^2 \text{ (tension)} \end{array}$$

EXAMPLE 6.9 The angle section shown in Fig. 4.21 is subject to a bending moment of 10 kNm applied about an axis p−p which is inclined at an angle of 40° to the x−x axis (Fig. 6.13). Calculate the longitudinal stress at point A in the section and determine the orientation of the neutral axis.

The section properties of this angle section were determined in Example 4.10:

$$I_{xx} = 4 \cdot 600 \times 10^6 \text{ mm}^4$$
$$I_{yy} = 3 \cdot 269 \times 10^6 \text{ mm}^4$$
$$I_{xy} = 2 \cdot 240 \times 10^6 \text{ mm}^4$$

The component moments about the reference axes are:

$$M_x = 10 \cos 40° = 7 \cdot 66 \text{ kNm} = 7 \cdot 66 \times 10^6 \text{ Nmm}$$
$$M_y = 10 \sin 40° = 6 \cdot 43 \text{ kNm} = 6 \cdot 43 \times 10^6 \text{ Nmm}$$

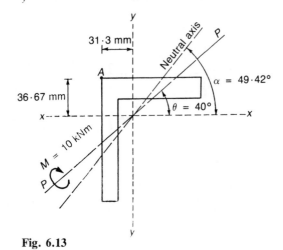

Fig. 6.13

The bending stress may be obtained using eqn [6.8]:

$$f = \frac{y(M_x I_{yy} - M_y I_{xy}) + x(M_y I_{xx} - M_x I_{xy})}{I_{xx}I_{yy} - I_{xy}^2}$$

$$= \frac{y(7\cdot66 \times 3\cdot269 - 6\cdot43 \times 2\cdot240) + x(6\cdot43 \times 4\cdot600 - 7\cdot66 \times 2\cdot240)}{4\cdot600 \times 3\cdot269 - 2\cdot240^2}$$

$$= 1\cdot062y + 1\cdot239x$$

Then, at point A, where $x = 31\cdot33$ mm and $y = 36\cdot67$ mm:

$$f = 1\cdot062 \times 36\cdot67 + 1\cdot239 \times 31\cdot33 = 77\cdot8 \text{ N/mm}^2 \text{ (Answer)}$$

The orientation of the neutral axis is obtained from eqn [6.13]:

$$\frac{y}{x} = \tan \alpha = -\frac{M_y I_{xx} - M_x I_{xy}}{M_x I_{yy} - M_y I_{xy}}$$

$$= -\frac{6\cdot43 \times 4\cdot600 - 7\cdot66 \times 2\cdot240}{7\cdot66 \times 3\cdot269 - 6\cdot43 \times 2\cdot240} = -1\cdot1675$$

Giving, $\qquad\qquad \alpha = 49\cdot42°$ (Answer)

6.6 Bending due to oblique loading and moments

Oblique loads and moments are those applied in a plane inclined to the principal axes of the section (Fig. 6.14). A convenient method of dealing with this situation is to replace the oblique system with a series of loads and moments aligned with the principal axes, but producing equivalent effects. The combined-stress distribution is then analysed.

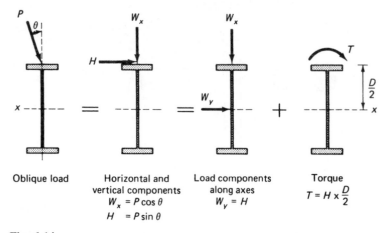

| Oblique load | Horizontal and vertical components $W_x = P \cos \theta$ $H = P \sin \theta$ | Load components along axes $W_y = H$ | Torque $T = H \times \dfrac{D}{2}$ |

Fig. 6.14

It should be appreciated that methods such as the one shown in Fig. 6.14 are only approximate and that more rigorous analytical procedures are required if an accurate stress distribution is necessary.

6.7 Beams of two materials — composite beams

If a beam is made up of two materials with different elastic moduli and the components are joined together in such a way that the bending strains at the material interfaces are the same in both, composite action takes place when the beam is subject to bending.

Consider the timber and steel beam shown in Fig. 6.15(a). The timber

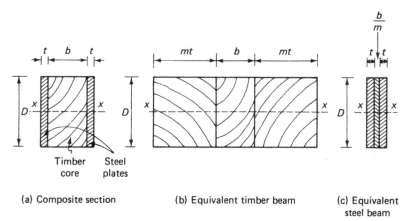

(a) Composite section   (b) Equivalent timber beam   (c) Equivalent steel beam

Fig. 6.15

joist section is reinforced with steel plates attached securely along the whole length. A given applied bending moment must therefore be taken partly by the timber and partly by the steel.

$$M = M_t + M_s \qquad [6.15]$$

However, because of composite action, the radius of curvature must be the same for both timber and steel.

Then from equation [6.3],

$$M = \frac{E_t I_t}{R} + \frac{E_s I_s}{R}$$

But

$$\frac{E_s}{E_t} = m \text{ (the } modular\ ratio)$$

Therefore,

$$M = (I_t + mI_s)\frac{E_t}{R}$$

208

Also $$\frac{E_t}{R} = \frac{f_t}{y}$$

Hence, $$M = (I_t + mI_s)\frac{f_t}{y} \qquad [6.16]$$

The value in the bracket is called the *equivalent second moment of area* of the composite section. In the case of equation [6.16], this represents the second moment of area of an *equivalent timber beam* — see Fig. 6.15(*b*). Thus the calculations are dealt with as if for an all-timber beam. Similarly, the calculations can be done in terms of an *equivalent steel beam* — see Fig. 6.15(*c*) and use equation [6.17].

$$M = \left(\frac{I_t}{m} + I_s\right)\frac{f_s}{Y} \qquad [6.17]$$

Equations [6.16] and [6.17] may be used for bending moments about either axis — see Fig. 6.16.

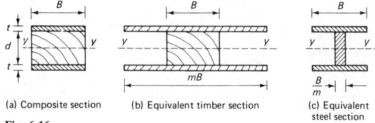

(a) Composite section   (b) Equivalent timber section   (c) Equivalent steel section

Fig. 6.16

Examples of composite beams may be found in reinforced timber joists, flitched beams (steel plate sandwiched between two timbers) and filled steel tubes.

EXAMPLE 6.10 A composite beam consists of a rectangular timber core 150 mm × 100 mm which is secured along its entire length between steel plates 150 mm × 9 mm (Fig. 6.17). Determine the maximum bending stresses induced in the timber and steel by a bending moment applied about the x–x axis of 6 kNm. $E_{timber} = 10 \cdot 5$ kN/mm²; $E_{steel} = 210$ kN/mm².

9 mm   100 mm   9 mm

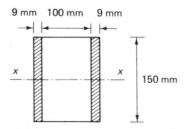

150 mm

Fig. 6.17

Modular ratio, $\quad m = \dfrac{E_s}{E_t} = \dfrac{210}{10 \cdot 5} = 20$

Working in timber units,

$$
\begin{aligned}
I_{xx(t)} &= I_t + mI_s \\
&= \frac{100 \times 150^3}{12} + \frac{20 \times 9 \times 150^3 \times 2}{12} \\
&= \frac{460 \times 150^3}{12} \\
&= 135 \times 10^6 \text{ mm}^4
\end{aligned}
$$

From equation [6.16],

$$
M = (I_t + mI_s) \frac{f_t}{y} = \frac{I_{xx(t)} f_t}{y}
$$

Therefore the stress in the timber is,

$$
f_t = \frac{M_y}{I_{xx(t)}} = \frac{6 \times 10^6 \times 75}{135 \times 10^6}
$$

$$
= 3 \cdot 33 \text{ N/mm}^2 \text{ (Answer)}
$$

and the stress in the steel (equation [3.9]) is,

$$
\begin{aligned}
f_s = m f_t &= 20 \times 3 \cdot 33 \\
&= 66 \cdot 7 \text{ N/mm}^2 \text{ (Answer)}
\end{aligned}
$$

EXAMPLE 6.11   A composite beam section consists of a timber joist and a steel plate of the same length as shown in Fig. 6.18(a). The beam when loaded is subject to a sagging bending moment in the plane of the y—y axis which has a maximum value of $1 \cdot 8$ kNm. Determine the maximum compressive and tensile bending stresses induced in the section by this moment when: (a) the two sections remain in contact with each other but are not bolted together, (b) the two sections are firmly bolted together along their length.
   Young's moduli: timber = $10 \cdot 5$ kN/mm²; steel = 210 kN/mm².

(a)   If the timber and steel are not bolted together they will bend separately without composite action, but if they also remain in contact their radii of curvature will be equal.

Then $\qquad \dfrac{f}{E_y} = \dfrac{I}{R} = \text{constant}$

Giving $\qquad \dfrac{f_s}{f_t} = \dfrac{E_s y_s}{E_t y_t} = \dfrac{210 \times 10}{10 \cdot 5 \times 75} = 2 \cdot 67$

Now $\quad \dfrac{\text{Total}}{BM} = \dfrac{\text{Total moment}}{\text{of resistance}} = MR \text{ in steel } + \text{ } MR \text{ in timber}$

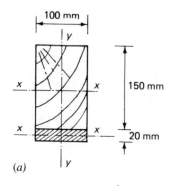

(a)

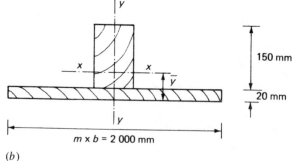

(b)

Fig. 6.18

$$= \left(\frac{fI}{y}\right)_s + \left(\frac{fI}{y}\right)_t$$

$$= \frac{2 \cdot 67 f_t \times 100 \times 20^3 \times 2}{12 \times 20} + \frac{f_t \times 100 \times 150^3 \times 2}{12 \times 150}$$

Therefore

$$1 \cdot 8 \times 10^6 = \frac{100 f_t}{6} (2 \cdot 67 \times 20^2 + 150^2)$$

$$= \frac{100 f_t \times 23\,567}{6}$$

Hence

$$f_t = \frac{1 \cdot 8 \times 10^6 \times 6}{100 \times 23\,567} = 4 \cdot 58 \text{ N/mm}^2 \text{ (Answer)}$$

and

$$f_s = 2 \cdot 67 \times 4 \cdot 58 = 12 \cdot 24 \text{ N/mm}^2 \text{ (Answer)}$$

(b) If the steel and timber components are firmly fixed together composite action will be developed and the equivalent timber section will be as shown in Fig. 6.18(b)

Then

$$\frac{f_s}{f_t} = \frac{E_s}{E_t} = \frac{210}{10 \cdot 5} = 20 = m \text{ (modular ratio)}$$

Position of neutral axis in the equivalent timber section

$$\bar{y} = \frac{2\,000 \times 20 \times 10 + 100 \times 150 \times 95}{2\,000 \times 20 + 100 \times 150}$$

$$= \frac{40 \cdot 0 + 142 \cdot 5}{4 \cdot 0 + 1 \cdot 5} = 33 \cdot 18 \text{ mm}$$

Second moment of area of the equivalent timber section

$$I_{xx} = \frac{2000 \times 20^3}{12} + \frac{100 \times 150^3}{12} + 2000 \times 20\,(33 \cdot 18 - 10 \cdot 00)^2 \\ + 100 \times 150\,(95 \cdot 00 - 33 \cdot 18)^2$$

$$= (1 \cdot 33 + 28 \cdot 13 + 21 \cdot 49 + 57 \cdot 33)\,10^6$$

$$= 108 \cdot 28 \times 10^6 \text{ mm}^4 \text{ (timber units)}$$

Hence

$$\frac{My_s}{I_t} = \frac{1 \cdot 8 \times 10^6 (170 \cdot 00 \times 33 \cdot 18)}{77 \cdot 18 \times 10^6} = 2 \cdot 27 \text{ N/mm}^2 \text{ (Answer)}$$

$$\frac{My_s}{I_s} = \frac{1 \cdot 8 \times 10^6 \times 33 \cdot 18 \times 20}{108 \cdot 28 \times 10^6} = 11 \cdot 03 \text{ N/mm}^2 \text{ (Answer)}$$

$$\left[I_s = \frac{I_t}{m} \right]$$

6.8 Reinforced concrete beams — elastic theory

Reinforced concrete beams (Fig. 6.19) are a special case of beams in which composite action is assumed to take place between different materials, in this case concrete and steel. As with the theory of simple bending, certain assumptions have to be made regarding elastic behaviour. These may be summarized as follows:

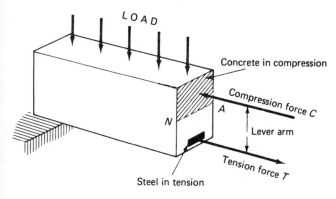

Fig. 6.19

1. Lateral cross-sections remain plane and perpendicular to the neutral axis after bending.
2. Concrete behaves elastically so that the stress is proportional to the strain.
3. No tensile stress will develop in the concrete.
4. The tensile stress in the steel reinforcement is uniform.
5. There is perfect bond (adhesion) between the concrete and the steel reinforcement.

Therefore, the compressive stress in the concrete is assumed to have a triangular distribution varying from zero at the neutral axis to a maximum at the extreme compression fibre (see Fig. 6.20). The combined effect of assumptions (3), (4) and (5) results in assuming that the tension force is taken entirely by the steel reinforcing bars and that, although the stress in the adjacent concrete is zero (i.e the concrete is assumed to crack), the strain in the adjacent concrete is equal to the strain in the steel.

The moment of resistance, which is equal to the bending moment, will be given by the couple set up by the compression force in the concrete and the tension force in the steel (Fig. 6.19). Consider the equivalence of strains shown in the strain diagram in Fig. 6.20.

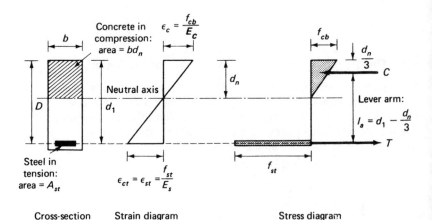

Cross-section   Strain diagram   Stress diagram

Fig. 6.20

$$\text{Compressive strain in the concrete} = \epsilon_c = \frac{f_{cb}}{E_c}$$

$$\text{Tensile strain in the concrete} = \text{tensile strain in the steel}$$

$$\epsilon_{ct} = \epsilon_{st}$$

$$= \frac{f_{st}}{E_s}$$

Then since the compression and tension portions of the diagram are similar triangles:

$$\frac{d_n}{f_{cb}/E_c} = \frac{d_1 - d_n}{f_{st}/E_s}$$

rearranging

$$\frac{d_n}{d_1 - d_n} = \frac{f_{cb}}{f_{st}} \frac{E_s}{E_c}$$

but

$$\frac{E_s}{E_c} = m$$

i.e. the MODULAR RATIO (see equation [3.9]), the value of which is usually taken as 15 for ordinary reinforced concrete.

Then

$$\frac{d_n}{d_1 - d_n} = \frac{15 f_{cb}}{f_{st}} \qquad [6.18]$$

From this expression the value of d_n in terms of d_1 can be obtained. The moment of resistance of the section is provided by the compression–tension force couple, the lever arm of which is given by:

$$l_a = d_1 - \frac{d_n}{3} \qquad [6.19]$$

Now the total compression force, $C = \dfrac{\text{area in}}{\text{compression}} \times \dfrac{\text{average compressive}}{\text{stress}}$

$$= b \times d_n \times \tfrac{1}{2} \times f_{cb}$$

$$= \tfrac{1}{2} f_{cb} b d_n \qquad \begin{array}{l}\text{acting at a depth} \\ \text{of } \tfrac{1}{3} d_n \text{ below the} \\ \text{top edge}\end{array}$$

and the total tension force, $T = \dfrac{\text{area of steel}}{\text{in tension}} \times \dfrac{\text{tensile stress}}{\text{in steel}}$

$$= A_{st} f_{st} \qquad \begin{array}{l}\text{acting at a depth} \\ \text{of } d_1 \text{ below the} \\ \text{top edge}\end{array}$$

The moment of resistance of the section with respect to the concrete in compression is now given by:

$$M_c = \tfrac{1}{2} f_{cb} b d_n \times l_a$$

or

$$M_c = K b d_1^2 \qquad [6.20]$$

where

$$K = \tfrac{1}{2} f_{cb} \cdot \frac{d_n}{d_1} \cdot \frac{l_a}{d_1}$$

$$\frac{d_n}{d_1} = \text{the neutral axis factor}$$

$$\frac{l_a}{d_1} = \text{the lever arm factor}$$

The moment of resistance of the section with respect to the steel in tension is given by:

$$M_s = A_{st} f_{st} l_a \qquad [6.21]$$

The four expressions established here may be regarded as the primary design equations for the *elastic* design of reinforced concrete beams and slabs. They may be used directly, as illustrated in the following examples, or indirectly by using tables which have been derived from them.

EXAMPLE 6.12 *Calculate the elastic design factors d_n, l_a, and K for a reinforced concrete beam when the permissible stresses are*

$$p_{cb} = \quad 7 \text{ N/mm}^2$$
$$p_{st} = 140 \text{ N/mm}^2$$

and

$$m = \quad 15$$

Using equation [6.18] and putting $f_{cb} = p_{cb}$ and $f_{st} = p_{st}$,

$$\frac{d_n}{d_1 - d_n} = \frac{15 \times 7}{140} = 0 \cdot 750$$

$$d_n = 0 \cdot 750 \, (d_1 - d_n)$$

$$= \frac{0 \cdot 750 \, d_1}{1 + 0 \cdot 750}$$

$$= 0 \cdot 429 \, d_1 \text{ (Answer)}$$

Using equation [6.19]

$$l_a = d_1 - \frac{0 \cdot 429 \, d_1}{3}$$

$$= (1 \cdot 00 - 0 \cdot 143) \, d_1$$

$$= 0 \cdot 857 \, d_1 \text{ (Answer)}$$

Using equation [6.20],

$$M_c = K \, b \, d_1^2$$

and

$$K = \tfrac{1}{2} \times 7 \times$$
$$0 \cdot 429 \times$$
$$0 \cdot 857$$

$$= 1 \cdot 29$$

Hence

$$M_c = 1 \cdot 29 \, b \, d_1^2 \text{ (Answer)}$$

EXAMPLE 6.13 *A reinforced concrete beam of rectangular section is required to transmit a bending moment of 40 kNm. The breadth of the beam is to be 250 mm and the permissible stresses 7 N/mm² for the concrete and 140 N/mm² for the steel. Determine the effective depth (d_1) and the area of steel reinforcement (A_{st}) required.*

Using the elastic design factors obtained in Example 6.12 i.e.

$$d_n = 0 \cdot 429 \, d_1$$
$$l_a = 0 \cdot 875 \, d_1$$
$$K = 1 \cdot 29$$

and rearranging equation [6.20], the effective depth required is

$$d_1 = \sqrt{\frac{M}{Kb}}$$
$$= \sqrt{\frac{40 \times 10^6}{1 \cdot 29 \times 25}}$$
$$= \sqrt{124\,000} = 352 \text{ mm. Say 360 mm (Answer)}$$

Using equation [6.21] the area of tension reinforcement required is

$$A_{st} = \frac{M}{p_{st} l_a}$$
$$= \frac{40 \times 10^6}{140 \times 0 \cdot 857 \times 360}$$
$$= 926 \text{ mm}^2 \text{ (Answer)}$$

EXAMPLE 6.14 *A reinforced concrete beam of rectangular section has a breadth of 200 mm and an effective depth of 420 mm. Determine the moment of resistance of the beam and the area of steel reinforcement required. What is the maximum uniformly distributed load that the beam may safely carry when simply supported over a span of 5 m?*

Permissible stresses:

$$p_{cb} = 8 \cdot 5 \text{ N/mm}^2 \qquad m = 15$$
$$p_{st} = 140 \text{ N/mm}^2$$

First of all establish the elastic design factors.

$$\frac{d_n}{d_1 - d_n} = \frac{15 \times 8 \cdot 5}{140} = 0 \cdot 911$$

$$\therefore \qquad d_n = 0 \cdot 911 \, (d_1 - d_n)$$
$$= \frac{0 \cdot 911 \, d_1}{1 \cdot 911}$$
$$= 0 \cdot 477 \, d_1$$

Using equation [6.19]:

$$l_a = d_1 - \frac{0 \cdot 477\, d_1}{3}$$

$$= (1 \cdot 000 - 0 \cdot 159)\, d_1$$

$$= 0 \cdot 841\, d_1$$

Using equation [6.20]:

$$K = \tfrac{1}{2} \times 8 \cdot 5 \times 0 \cdot 477 \times 0 \cdot 841$$

$$= 1 \cdot 70$$

Then the moment of resistance of the beam is:

$$M = K\, b\, d_1^2$$

$$= 1 \cdot 70 \times 200 \times 420^2$$

$$= 60 \cdot 0 \times 10^6 \text{ N mm}$$

$$= 60 \cdot 0 \text{ kNm (Answer)}$$

The area of steel required is:

$$A_{st} = \frac{M}{p_{st} l_a}$$

$$= \frac{60 \times 10^6}{140 \times 0 \cdot 841 \times 420}$$

$$= 1\,213 \text{ mm}^2 \text{ (Answer)}$$

In calculating the allowable load that the beam may carry, the self-mass of the beam must be taken into account.

The bending moment due to a uniform load is given by $M = WL/8$

Therefore the total allowable load will be, $W = \dfrac{8 \times 60}{5}$

$$= 96 \cdot 0 \text{ kN}$$

If the overall depth is assumed to be 450 mm the self-mass of the beam will be $2\,400 \times 0 \cdot 450 \times 0 \cdot 200 = 216$ kg/m

Therefore the uniform load due to the self-mass $= 216 \times 5 \times \dfrac{9 \cdot 81}{1\,000}$

$$= 10 \cdot 6 \text{ kN}$$

Hence the allowable load that the beam may safely carry will be

$$96 \cdot 0 - 10 \cdot 6 = 85 \cdot 4 \text{ kN (Answer)}$$

6.9 Plastic bending theory

In the *elastic method* of analysis and design, the stresses are assumed to be proportional to the strains, i.e. Hooke's law is obeyed; an assumption that only remains valid so long as the stresses are below the yield strength of the material. In materials such as concrete, brick, rock, etc., *brittle failure* will result at a tensile stress just beyond the yield strength. In *ductile* materials, such as mild steel, distortion will continue with increasing stress well beyond the yield strength before brittle failure occurs (see Section 3.9). However, at stresses beyond the yield strength *plastic* (irrecoverable) strain takes place and Hooke's law no longer applies. Furthermore, the bending stress distribution over the depth of the beam will no longer be linear. In the case of mild steel, the yield strength and modulus of elasticity are assumed to be the same in compression as in tension. Figure 6.21 shows an idealized stress/strain curve illustrating the *elasto-plastic* behaviour of mild steel up to and beyond the yield strength (p_y).

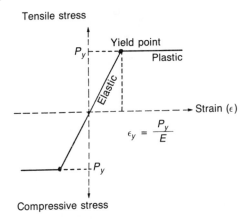

Fig. 6.21

When the stress in the extreme fibres in a beam subject to bending stress reaches the yield strength the material will yield rather than sustain a further increase in stress. If the bending moment at this section is further increased, the stress in those fibres that are yielding will remain constant (at the yield strength), while the stress in those below the yield strength will increase until a *plastic hinge* will form and the beam will cease to bend. Failure may then arise due to rotation at the plastic hinge (Fig. 6.22).

The elastic and plastic stress distributions and the corresponding moments of resistance will now be examined for rectangular and I-section beams.

Beam of rectangular section

Figure 6.23 shows the stress distribution during the stages of elastic, elasto-plastic and plastic bending. The value of the bending moment (M_y)

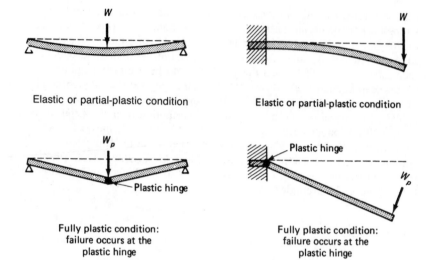

Elastic or partial-plastic condition

Elastic or partial-plastic condition

Fully plastic condition:
failure occurs at the
plastic hinge

Fully plastic condition:
failure occurs at the
plastic hinge

(a) Simply supported beam

(b) Cantilever

Fig. 6.22

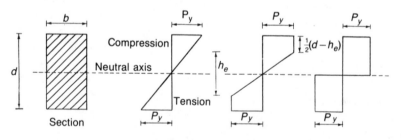

(a) Fully elastic   (b) Elasto-plastic   (c) Fully plastic

Fig. 6.23

required to produce the initial yielding condition (Fig. 6.23(a)) is given by:

$$M_y = \frac{p_y I_{xx}}{d/2}$$

and since

$$I_{xx} = bd^3/12,$$

$$M_y = p_y \frac{bd^2}{6} \qquad [6.22]$$

i.e. the result identical to that obtained for fully elastic bending, except that the extreme fibre stress is p_y.

In the fully plastic state, assuming that the longitudinal stress at all depths in the section is equal to p_y (Fig. 6.23(c)) and that $p_{y(\text{compression})} = p_{y(\text{tension})}$, the *fully plastic moment of resistance* or *moment capacity* (M_p) is given by:

$$M_p = p_y \times \frac{bd}{2} \times \frac{d}{2} = p_y \frac{bd^2}{4} \qquad [6.23]$$

Thus the fully plastic moment, $M_p = 1 \cdot 5\ M_y$ $\qquad [6.24]$

In an intermediate *elasto-plastic* state, wherein the elastic core has a thickness of h_e (Fig. 6.23(b)), the moment of resistance (M_{ep}) is given by:

$$M_{ep} = p_y \left[\frac{1}{2} b \frac{h_e}{2} \frac{2h_e}{3} + b \left(\frac{d}{2} - \frac{h_e}{2} \right) \left(\frac{d}{2} + \frac{h_e}{2} \right) \right]$$

$$= p_y \left[\frac{bh_e^2}{6} + b \left(\frac{d^2}{4} - \frac{h_e^2}{4} \right) \right]$$

$$= p_y \left(\frac{bd^2}{4} - \frac{bh_e^2}{12} \right)$$

$$= p_y \frac{bd^2}{4} \left(1 - \frac{h_e^2}{3d^2} \right) \qquad [6.25]$$

$$= M_p \left(1 - \frac{h_e^2}{3d^2} \right) \qquad [6.26]$$

From the elastic theory of bending (eqn [6.1]), the radius of curvature of the beam is given by:

$$\frac{1}{R} = \frac{f}{E_y}$$

and at the boundary between the elastic and plastic stress zones (Fig. 6.23(b)), $f = p_y$ and $y = h_e$.

Hence, $\qquad\qquad \dfrac{1}{R} = \dfrac{p_y}{Eh_e}$ $\qquad [6.27]$

Fig. 6.24 shows a plot of values of M against $1/R$ using eqns [6.25] and [6.27].

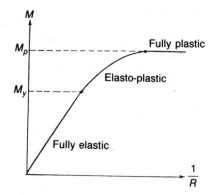

Fig. 6.24

Steel I-section beam

In the case of a steel I-section beam, the plastic moment of resistance is given by the flange—force and web—force couples. The elastic, elasto-plastic and fully plastic stress distributions are shown in Fig. 6.25. For the *fully plastic state*:

$$\text{Force in the flanges,} \quad F_f = p_y BT$$
$$\text{Flange—force lever arm,} \quad l_f = \tfrac{1}{2}(d - T)$$
$$\text{Force in half-web,} \quad F_w = p_y t(\tfrac{1}{2}d - T)\cdot$$
$$\text{Web force lever arm,} \quad l_w = \tfrac{1}{2}(\tfrac{1}{2}D - T)$$

Thus, the *fully plastic moment of resistance* or *plastic moment capacity* (M_p) is given by:

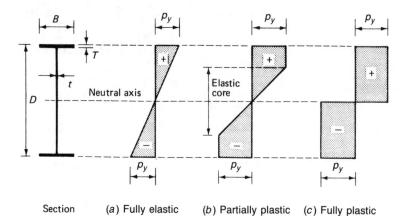

| Section | (a) Fully elastic | (b) Partially plastic | (c) Fully plastic |

Fig. 6.25

$$M_p = 2[F_f l_f + F_w l_w]$$
$$= 2[p_y BT \times \tfrac{1}{2}(D - T) + p_y \tfrac{1}{2} t(\tfrac{1}{2} D - T)^2]$$
$$= p_y[BT(D - T) + t(\tfrac{1}{2} D - T)^2] \qquad [6.28]$$

or, defining the *plastic section modulus* as

$$S = BT(D - T) + t(\tfrac{1}{2} D - T)^2 \qquad [6.29]$$

Then, $$M_p = p_y S \qquad [6.30]$$

For Universal Beam and other standard beam sections, values for the plastic section modulus may be obtained from BCSA tables in the same way as the elastic section modulus.

It is useful to compare the fully plastic moment of resistance (M_p) with the moment of resistance at initial yielding (M_y).

When the extreme fibre stress has just reached p_y:

$$M_y = p_y Z \qquad [6.31]$$

Then, $$\frac{M_p}{M_y} = \frac{S}{Z} \qquad [6.32]$$

The ratio S/Z is a dimensionless quantity, sometimes called the *shape factor*, which indicates the increase in bending moment required in excess of that required to cause the 'just yielding' state to produce a fully plastic hinge. In steelwork design, it is recommended that the *moment capacity* of a section may be taken as $M_p = p_y S$, but not greater than $1 \cdot 2 p_y Z$.

EXAMPLE 6.15 Determine the elasto-plastic and fully plastic moments of resistance of a Universal Beam having the following properties and comment on the plastic moment capacity of the section. $Z_{xx} = 1 \cdot 40 \times 10^6$ mm^4, $S_{xx} = 1 \cdot 62 \times 10^6$ mm^4. Yield strength, $p_y = 275$ N/mm^2.

The elasto-plastic moment of resistance occurs at 'first yield' when $f = p_y$.

$$M_y = 275 \times 1 \cdot 40 \times 10^6 = 385 \times 10^6 \text{ Nmm}$$
$$= 385 \text{ kNm} \quad \text{(Answer)}$$

The fully plastic moment of resistance is:

$$M_p = 275 \times 1 \cdot 62 \times 10^6 = 445 \cdot 5 \times 10^6 \text{ Nmm}$$
$$= 445 \cdot 5 \text{ kNm} \quad \text{(Answer)}$$

The ratio $$\frac{M_p}{M_x} = \frac{445 \cdot 5}{385} = 1 \cdot 19$$

It is recommended that the moment capacity (M_c) should not exceed $1 \cdot 2 \, M_y$, so that the value of $445 \cdot 5$ kNm may be taken for M_c.

6.10 Reinforced concrete beams — plastic theory

The plastic bending theory as described in Section 6.9 can be applied to the yield condition occurring at the failure of a concrete beam in only an approximate form. This is because concrete, being an essentially brittle material, does not display exactly the same plastic yield properties as steel which is a ductile material. Experimental evidence has shown, however, that as the failure condition is approached the distribution of compressive stress in the concrete is much nearer to being rectangular than triangular. The results of many tests have supported the view that a simplified rectangular distribution of stress is sufficiently accurate for design purposes, providing that suitably factored values are used.

The compressive stresses used in the design analysis are based on the 28-day cube strength of concrete (f_{cu}) and the guaranteed yield strength of the steel reinforcement (f_y). The appropriate values of f_{cu} and f_y are divided by a *partial factor of safety* (γ_m), values of which are recommended in BS 8110 *The structural use of concrete*: $\gamma_m = 1 \cdot 5$ for concrete, $\gamma_m = 1 \cdot 15$ for steel reinforcement. Hence the stresses which may be used in design are:

Compressive stress for concrete $= 0 \cdot 6 \, f_{cu}/1 \cdot 5 = 0 \cdot 4 \, f_{cu}$

Tensile stress for steel reinforcement $= f_y/1 \cdot 15 \qquad = 0 \cdot 87 \, f_y$

In the simplest case, usually referred to as a *balanced section*, the depth to the neutral axis is taken to be half-way between the centre of the steel area and extreme compression fibre, i.e. $d_n = \frac{1}{2} d$ (see Fig. 6.26). The moments of resistance will then be as follows:

$$M_{uc} = 0 \cdot 4 \, f_{cu} \times b \times \tfrac{1}{2} d \times \tfrac{3}{4} d$$
$$= 0 \cdot 15 \, f_{cu} \, bd^2 \qquad\qquad [6.33]$$

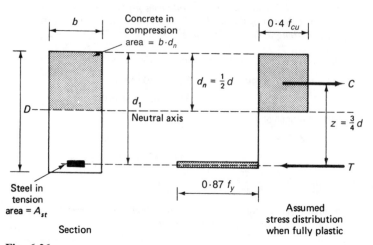

Fig. 6.26

EXAMPLE 6.16 Do again Example 6.13 using the plastic method of design.
28-day cube strength of concrete = 20 N/mm²
Yield strength of reinforcing steel = 250 N/mm²

Using eqn [6.33]: $d = \sqrt{\left(\dfrac{M}{0\cdot15 \times f_{cu} \times b}\right)}$

$= \sqrt{\left(\dfrac{40 \times 10^6}{0\cdot15 \times 20 \times 250}\right)} = 231$ mm (Answer)

Using eqn [6.34]: $A_{st} = \dfrac{M}{0\cdot87 \times f_y \times z}$

$= \dfrac{40 \times 10^6}{0\cdot87 \times 250 \times 0\cdot75 \times 231} = 1\ 062$ mm²

(Answer)

This method has produced a beam of lesser depth, but with an increased area of
tension steel. The plastic method is of advantage when the size of beam is
critical, but this is gained at the expense of increasing the amount of reinforcing
steel required.

6.11 Shearing stresses in beams

The shearing force (Q) at a given section in a loaded beam induces a
shearing stress (τ) which acts tangentially to the cross-sectional plane
(Fig. 6.27). The average value of this shearing stress is generally taken
as

$$\tau = \frac{Q}{A} \qquad [6.34]$$

where A = area of the cross-section.

In addition to this lateral stress, a *complementary longitudinal
tangential stress* is induced along planes parallel to the longitudinal axis

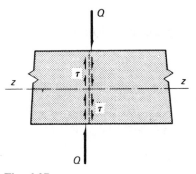

Fig. 6.27

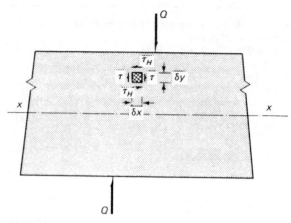

Fig. 6.28

$(z-z)$ of the beam. It can be shown that, at a given point in a beam, the lateral and longitudinal shearing stresses are numerically equal. Consider the element shown in Fig. 6.28, having unit width, of length δz and depth δy.

Let the vertical shearing stress on the element $= \tau$
Therefore the vertical tangential force on each side of the element $= \tau \delta y$
This produces a clockwise couple $= \tau \delta y \delta z$

For equilibrium, an equal and opposite couple is required, and this is provided by the horizontal shearing stress (τ_H) acting on the element

$$\text{Horizontal force on element} \quad = \tau_H \delta z$$
$$\text{Horizontal couple} \quad = \tau_H \delta z \delta y$$

Then $\qquad \tau_H \delta y \delta z = \tau \delta y \delta z$

Hence $\qquad \tau_H = \tau$ $\hfill$ [6.35]

6.12 A general expression for the distribution of shearing stress

Consider two parallel sections of a loaded beam, AB and CD, which are a very small distance apart, δx (Fig. 6.29). Let the bending moments at these sections be M and $M + \delta M$ respectively. Then also consider an elemental fibre between these two sections of area δA and situated at distance y from the neutral axis. The stresses at each end of this fibre will be f and $f + \delta f$ respectively.

The resultant direct force on the fibre is therefore

$$(f + \delta f)\delta A - f\delta A = \delta f \delta A$$

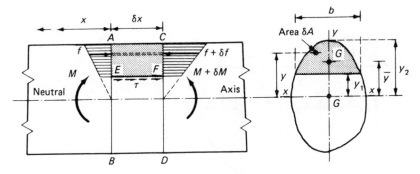

Fig. 6.29

Let τ be the shearing stress on a horizontal plane (EF) situated at distance y_1 from the neutral axis. The total out-of-balance direct force on the prism $ACFE$ must be balanced by the shearing force acting tangential to plane EF.

$$\tau b \,\delta x = \sum_{y_1}^{y_2} (\delta f \delta A)$$

Now,
$$f = \frac{M_y}{I} \text{ and } f + \delta f = \frac{(M + \delta M)y}{I}$$

Then
$$\delta f = \frac{\delta M \, y}{I}$$

Therefore
$$\tau = \frac{\delta M}{\delta x} \frac{\Sigma(\delta A \, y)}{bI}$$

But,
$$\frac{\delta M}{\delta x} = Q \text{ and } \Sigma(\delta A \, y) = A\bar{y}$$

Hence
$$\tau = \frac{QA\bar{y}}{bI} \qquad [6.36]$$

Since, at a given value of y_1, the vertical and horizontal shearing stresses are numerically equal, this expression can be used to give the distribution of vertical shearing stress across the section.

6.13 Distribution of shearing stress in a rectangular section

The cross-section of a rectangular beam is shown in Fig. 6.30. Consider the shearing stress τ at a distance y_1 from the neutral axis.

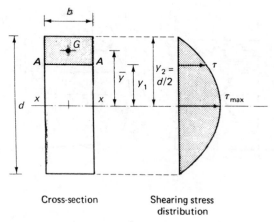

Cross-section

Shearing stress
distribution

Fig. 6.30

For the shaded area above AA: $A = b\left(\dfrac{d}{2} - y_1\right)$

$$\bar{y} = \tfrac{1}{2}\left(\dfrac{d}{2} + y_1\right)$$

Then, from equation [6.21]:

$$\tau = \frac{Qb\left(\dfrac{d}{2} - y_1\right) \cdot \tfrac{1}{2}\left(\dfrac{d}{2} + y_1\right) 12}{b.bd^3}$$

$$= \frac{6Q\left(\dfrac{d^2}{4} - y_1^{\,2}\right)}{bd^3} \qquad [6.37]$$

Since this is a second-order expression, the shearing stress distribution diagram will be in the form of a parabola having the following limits:

$$\tau = 0 \qquad \text{when } y_1 = \frac{d}{2} \text{ or } -\frac{d}{2}$$

$$\tau = \tau_{\max} \quad \text{when } y_1 = 0$$

Then
$$\tau_{\max} = \frac{6Q\,d^2}{bd^3.4}$$

$$= \frac{1 \cdot 5\,Q}{bd} \qquad [6.38]$$

and $$\tau_{average} = \frac{1 \cdot 5 \, Q}{bd} \times \frac{2}{3}$$

$$= \frac{Q}{bd} \qquad \qquad [6.39]$$

which agrees with equation [6.35].

EXAMPLE 6.17 *A timber beam of rectangular section 200 mm × 84 mm is simply supported over a span of 3·2 m and carries a uniformly distributed load. Determine the maximum shearing stress in the beam when the maximum bending stress is 5 N/mm².*

Maximum bending stress occurs at midspan and is induced by a bending moment M where

$$M = \frac{fI}{y}$$

$$= \frac{5 \times 56 \times 10^6}{100}$$

$$\left[I = \frac{84 \times 200^3}{12} \\ = 56 \times 10^6 \text{ mm}^4 \right]$$

$$= 2 \cdot 8 \times 10^6 \text{ Nmm}$$

$$= 2 \cdot 8 \qquad \text{kNm}$$

The load required to produce this moment is

$$W = \frac{2 \cdot 8 \times 8}{3 \cdot 2} = 7 \text{ kN}$$

The maximum shearing force occurs at the supports: $Q = 3 \cdot 5$ kN
Then the maximum shearing stress is:

$$\tau_{max} = \frac{1 \cdot 5 \, Q}{bd}$$

$$= \frac{1 \cdot 5 \times 3 \cdot 5 \times 10^3}{84 \times 200}$$

$$= 0 \cdot 31 \text{ N/mm}^2 \text{ (Answer)}$$

6.14 Distribution of shearing stress in I-sections

The shearing stress in the web of the I-section shown in Fig. 6.31 is given when $b = t$ in equation [6.37]. In the flange, however, $b = B$ and consequently a much lower value of shearing stress results, giving the distribution diagram shown. The maximum value occurs at the $x-x$ axis, i.e. when $y_1 = 0$.

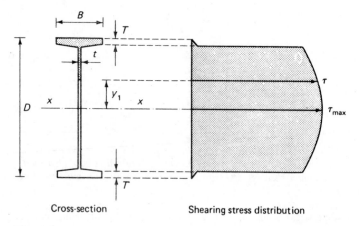

Fig. 6.31

EXAMPLE 6.18 Plot the shearing stress distribution diagram for the I-section in Fig. 6.32.

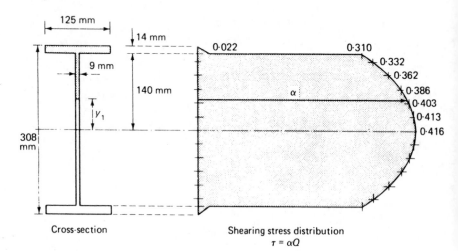

Cross-section

Shearing stress distribution
$\tau = \alpha Q$

Fig. 6.32

$$I_{xx} = \frac{125 \times 308^3}{12} - \frac{116 \times 280^3}{12} = (304 \cdot 36 - 212 \cdot 20)\ 10^6$$
$$= 92 \cdot 16 \times 10^6 \text{ mm}^4$$

When $y_1 \leq 140$ mm

$$A\bar{y} = 125 \times 14 \times 147 + 9\ (140 - y_1) \left(\frac{140 - y_1}{2} + y_1 \right)$$

$$= 257\ 250 + 4\cdot5\ (140 - y_1)\ (140 + y_1)$$
$$= 257\ 250 + 4\cdot5\ (140^2 - y_1^2)\ \text{mm}^3$$

Then at y_1

$$\tau = \frac{Q[257\ 250 + 4\cdot5\ (140^2 - y_1^2)]}{9 \times 92\cdot16 \times 10^6}$$

For Q in kN and $\tau = \text{N/mm}^2$

let
$$\alpha = \frac{\tau}{Q} = \frac{257\ 250 + 4\cdot5\ (140^2 - y_1^2)}{9 \times 92\cdot16 \times 10^3}$$

$$= \frac{57\ 167 + 19\ 600 - y_1^2}{184\cdot32 \times 10^3}$$

$$= \frac{76\ 767 - y_1^2}{184\cdot32 \times 10^3}\ \text{N/mm}^2\ \text{per kN}$$

y_1	0	25	50	75	100	125	140
y_1^2	0	625	2 500	5 625	10 000	15 625	10 000
$76\ 767 - y_1^2$	76 767	76 142	74 267	71 142	66 767	61 142	57 167
α	0·416	0·413	0·403	0·386	0·362	0·332	0·310

When $y_1 \geq 140$ and $b = 125$ mm

$$A\bar{y} = 125\ (154 - y_1)\ \frac{154 + y_1}{2} = 62\cdot5\ (154^2 - y_1^2)\ \text{mm}^3$$

Then at y_1

$$\tau = \frac{Q[62\cdot5\ (154^2 - y_1^2)]}{125 \times 92\cdot16 \times 10^6}$$

and
$$\alpha = \frac{154^2 - y_1^2}{2 \times 92\cdot16 \times 10^3}$$

$$= \frac{23\ 716 - y_1^2}{184\cdot32 \times 10^3}\ \text{N/mm}^2\ \text{per kN}$$

When $y_1 = 140$ mm

$$\alpha = \frac{23\ 716 - 140^2}{184\cdot32 \times 10^3}$$

$$= 0\cdot0223\ \text{N/mm}^2\ \text{per kN}$$

When $y_1 = 154$, $\alpha = 0$

The distribution diagram is plotted in terms of α against y_1, and the shearing stress at a particular position in the section is then given by

$$\tau = \alpha \times Q\ (\text{kN})\ \text{N/mm}^2$$

For example, if $Q = 100$ kN, the maximum shearing stress occurring at the neutral axis will be

$$\tau_{max} = 0 \cdot 416 \times 100 = 41 \cdot 6 \text{ N/mm}^2$$

As a convenient approximation, the shearing stress is often calculated by assuming that all the shearing is afforded by the web, i.e. ignoring the resistance of the flanges. In which case

$$\tau = \frac{100 \times 1\,000}{9 \times 280} = 39 \cdot 7 \text{ N/mm}^2$$

The approximate value thus underestimates the true value by about 4–5%, at the neutral axis, but overestimates the values nearer the flanges.

EXERCISES CHAPTER 6

1. Calculate the stress set up in a steel strip of thickness 3 mm when it is bent to form a circular hoop of diameter $1 \cdot 5$ m. Young's modulus = 210 kN/mm^2.

2. A steel wire of diameter $1 \cdot 6$ mm is pulled over a pulley of 1 m diameter with a tension of 250 N. Calculate the maximum tensile stress induced in the wire. Young's modulus = 210 kN/mm^2.

3. A timber joist of cross-section 75 mm $\times$ 150 mm is simply supported over a span of 4 m and carries a concentrated load of 5 kN at the mid-span point. Calculate the maximum bending stress in the joist:

(a) when the 150 mm side is parallel to the plane of bending, and
(b) when the 75 mm side is parallel to the plane of bending.

4. A timber joist of thickness 100 mm and depth 375 mm cantilevers from an encastré support and is $2 \cdot 5$ m long. Calculate the safe load that the joist may carry uniformly distributed over its whole length if the allowable bending stress is 7 N/mm^2.

5. A universal beam section of depth 348 mm has a second moment of area about its major axis of 82×10^6 mm^4. The beam spans $3 \cdot 6$ m and is simply supported at each end. If the permissible bending stress is 165 N/mm^2, calculate the magnitude of the centrally placed concentrated load that the beam may safely carry in addition to a uniform load of 20 kN/m extending over the whole length.

6. When a straight steel bar of rectangular section is subject to a uniform bending moment about its x–x axis, it bends in the form of a circular arc whose radius is 7 m. When the same moment is applied about the y–y axis, the radius of curvature is 2 m. Given that the cross-sectional dimension perpendicular to the x–x axis is 25 mm, determine the other dimension. Also calculate the magnitude of the bending moment and the maximum stress in the bar in both cases. Young's modulus = 210 kN/mm^2.

7. Compare, in the form of a ratio, the load-carrying capacity of a timber plank which has a cross-section of 250 mm $\times$ 40 mm. Assume that the load may be applied either perpendicular to the x–x axis or perpendicular to the y–y axis.

8. A universal beam section is 406 mm deep and has a second moment of area of 215×10^6 mm^4. Calculate the maximum span over which the beam, when simply supported, will carry a uniform load of 20 kN/m if the permissible stress is 165 N/mm^2.

9. Show that the bending stress due to self-weight in a solid circular shaft which is horizontal and simply supported over a span L is given by the following expression:

$$f_b = \frac{\gamma g L^2}{D}$$

Where: D = diameter of the shaft; γ = density of the material; g = 9·81 m/s^2.

Using this expression, calculate the maximum bending stress due to self-weight occurring in a steel shaft of diameter 150 mm and span 4 m. Density of steel = 7 850 kg/m^3.

10. A brass tube of external diameter 75 mm and internal diameter 60 mm is simply supported horizontally over a span of 4 m. Given that the density of the brass is 8 200 kg/m^3 and that the maximum allowable stress is 85 N/mm^2, calculate the magnitude of the centrally placed concentrated load that may be carried by the shaft in addition to its own weight.

11. A metal beam of 75 mm square section is to be replaced by a tube of the same material. The new beam must sustain the same maximum bending moment at the same maximum bending stress. If the external diameter of the tube is to be 120 mm, calculate the internal diameter required. What percentage saving in the mass of material used will result from this replacement?

12. An I-section beam having the cross-section shown in Fig. 4.38 is simply supported over a span of 5 m. Determine the maximum uniform load that the beam may carry:

(a) when its $x-x$ axis is horizontal, and
(b) when its $y-y$ axis is horizontal.

The permissible bending stress in both cases is 150 N/mm^2.

13. A welded steel I-section beam having a cross-section such as that shown in Fig. 4.39 is to carry a uniform load of 75 kN/mm when simply supported over a span of 10 m. Calculate the maximum bending stress in the beam.

14. A box-section beam having a cross-section such as that shown in Fig. 4.40 is simply supported over a span of 8 m and carries a uniform load of 25 kN/m perpendicular to its $x-x$ axis together with a centrally placed concentrated load of 55 kN perpendicular to its $y-y$ axis. Calculate the maximum bending stress in the beam.

15. Fig. 4.44 shows a universal beam section to which additional flange plates are welded. Calculate the percentage increase in the moment of resistance of the section that results from adding the flange plates.

16. The compound girder whose cross-section is shown in Fig. 4.49 is simply supported and carries a uniform load of 45 kNm over its entire span of 8·2 m. Calculate the bending stresses acting at the flange/flange-plate interface.

17. Fig. 4.50 shows the cross-section of a gantry girder which spans 4 m between simple supports and carries two wheel loads spaced $1 \cdot 25$ m apart. If the permissible bending stress is 165 N/mm^2, calculate the maximum wheel load that the girder may carry. What will be the bending stress in the top fibres of the girder under these loading conditions?

18. Fig. 6.33 shows a timber cantilever carrying a concentrated load at the free end. The beam is of constant breadth 150 mm, but the depth varies uniformly from 300 mm at A to 200 mm at B. Calculate the maximum bending stress in the beam at (a) the support and (b) at a point $0 \cdot 6$ m from the support.

19. Fig. 6.34 shows a box-section cantilever which carries a uniform load perpendicular to the $x-x$ axis and a concentrated load at the free end perpendicular to the $y-y$ axis. Calculate the maximum compressive and tensile bending stresses in the beam.

20. A timber beam of cross-section 150 mm $\times$ 80 mm is strengthened by the addition of steel flange plates 12 mm thick as shown in Fig. 6.35. The beam is simply supported over a span of 4 m and carries a uniform load of 20 kN over its entire length. Calculate the maximum bending stresses in the steel and timber. Young's moduli: steel = 210 kN/mm^2; timber = 11 kN/mm^2.

21. A flitched beam comprising a steel plate sandwiched between two timber joists is shown in Fig. 6.36. The three components are secured together along their full length. The stress in the timber must not exceed 8 N/mm^2 and that in the steel must not exceed 150 N/mm^2. Calculate the magnitude of the uniform load that the beam may safely carry when the span is 5 m. Young's moduli: steel = 210 kN/mm^2; timber = $10 \cdot 5$ kN/mm^2.

22. An existing timber beam is to be strengthened by the addition of two steel plates as shown in Fig. 6.37 so that the beam may safely carry a uniform load of 120 kN when simply supported over a span of 6 m. if the stress in the timber is not to exceed 7 N/mm^2 and that in the steel is not to exceed 150 N/mm^2, calculate the thickness of the steel plates required. ($E_s/E_t = 20$.)

23. A flitched timber beam comprises a steel plate 18 mm thick $\times$ 200 mm deep sanwiched between two timber joists each 100 mm wide $\times$ 300 mm deep. The three components are symmetrically arranged and secured firmly to each other along their full length. The beam is simply supported over a span of 8 m and carries a uniform load of 60 kN. Calculate the stresses in the timber and steel. ($E_s/E_t = 20$.)

24–27. Calculate the elastic design factors required for the design of reinforced concrete beams using the following permissible stresses and assuming $m = 15$.

24	p_{cb} =	$8 \cdot 5$ N/mm^2	p_{st} =	140 N/mm^2
25	p_{cb} =	$8 \cdot 5$ N/mm^2	p_{st} =	230 N/mm^2
26	p_{cb} =	$10 \cdot 0$ N/mm^2	p_{st} =	230 N/mm^2
27	p_{cb} =	$11 \cdot 0$ N/mm^2	p_{st} =	230 N/mm^2

28. A reinforced concrete beam of rectangular cross-section has an effective depth of 400 mm and a breadth of 300 mm. Calculate the depth from the top fibre to

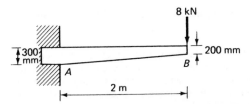

Fig. 6.33

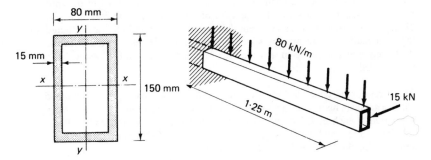

Fig. 6.34

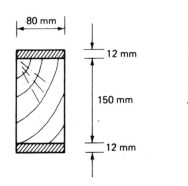

Fig. 6.35

Fig. 6.36

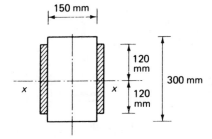

Fig. 6.37

234

the neutral axis if the permissible stresses are $7 \cdot 0$ N/mm² for the concrete and 140 N/mm² for the steel and $m = 15$.

29. A reinforced concrete beam of rectangular section has an effective depth of 500 mm and a breadth of 250 mm. Determine the plastic moment of resistance of the beam with respect to the strength of the concrete which has a 28-day cube strength of 25 N/mm². For a balanced section, what area of tensile steel reinforcement will be required? ($f_y = 250$ N/mm².)

30. A reinforced concrete beam of rectangular section is required to transmit a bending moment of 65 kNm. The breadth of the beam is to be 250 mm. The 28-day cube strength of the concrete will be 20 N/mm² and the yield strength of the steel reinforcement may be taken as 250 N/mm². Determine the effective depth required for the beam and the area of tensile steel reinforcement that must be provided.

31. A reinforced concrete lintel is required to have a breadth of 225 mm and to carry a total inclusive uniform load of 40 kN over a span of 3 m when simply supported at each end. The permissible stresses are $7 \cdot 0$ N/mm² and 140 N/mm² for the concrete and steel respectively and $m = 15$. Determine the effective depth required for the lintel and also the area of steel reinforcement required.

32. A timber beam of rectangular section 300 mm × 120 mm is simply supported over a span of 5 m and carries a uniform load of 8 kN/m along its entire length. Determine the maximum shearing stress in the beam.

33. Plot the shearing stress distribution over the cross-section of the I-beam shown in Fig. 6.38. All variations must be shown on the diagram.

34. Plot the shearing stress distribution over the cross-section shown in Fig. 6.39 at a point where the vertical shearing force is 20 kN.

35. By plotting the appropriate diagrams, determine for the section shown in Fig. 6.40:

(a) the percentage of the shearing force taken in the web, and
(b) the percentage of the bending moment taken in the flanges.

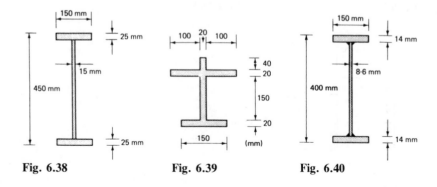

Fig. 6.38   **Fig. 6.39**   **Fig. 6.40**

7 Deflection of beams

Symbols

A	=	constant of integration
a	=	distance to a force or load
B	=	constant of integration
b	=	breadth of section
d	=	depth of section
E	=	Young's modulus of elasticity
EI	=	$E \times I$ = flexural stiffness of beam
FBM	=	'free' bending moment
I	=	second moment of area (about a given axis)
L	=	length, span
M	=	bending moment
P	=	force
R	=	radius of curvature
R	=	reaction force at a given point
S	=	distance along an arc
W	=	concentrated load, total uniform load
w	=	uniform load per unit length
x	=	distance measured in the x-direction
y	=	vertical displacement or deflection of the longitudinal axis
y	=	distance measured in the y-direction
Δ	=	displacement relative to a tangent
θ	=	angle of slope of the bent longitudinal axis
σ	=	stress, normal stress

7.1 Bending and stiffness

As stated previously, the application of lateral loads to a beam will cause it to deform in a mode that is called *bending*. The implications of this deformation in terms of the stresses induced are discussed in Chapter 6. In addition to considering the material strength of a beam the designer has also to think of the amount of deformation that may take place as a result of it bending.

The STIFFNESS of a beam is a measure of the resistance it has to bending deformation: the stiffer the beam the less it will bend. The relationship between stiffness and bending is important from two points of view. Firstly, it is necessary to limit the amount of deflection that takes place in a structural member, since excessive deflection movement can be both unsightly and inconvenient; some damage to or impairment of use of fittings and finishes may also occur. Secondly, in structures where the bending effects are distributed through rigid joints to other members, such as in the case of continuous beams and portal frames, the proportionate bending moments that are distributed are directly related to the relative stiffnesses of the components involved.

7.2 Curvature and bending — definitions

When a beam of constant section is subjected to *pure bending* (see Section 6.2), it is assumed to bend in the form of a circular arc having a radius of curvature R such that:

$$\frac{E}{R} = \frac{M}{I} \quad \text{(from equation [6.3])}$$

or
$$\frac{1}{R} = \frac{M}{EI} \qquad [7.1]$$

At a given point P on the longitudinal axis of a beam thus deformed, having been initially straight and horizontal, the following definitions apply (Fig. 7.1).

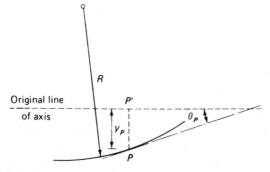

Fig. 7.1

CURVATURE at $P = R$ = the radius of curvature of the bent longitudinal axis of the beam

SLOPE at P = θ_p = the angle made between the bent longitudinal axis of the beam at P and the original line of the unbent longitudinal axis

DEFLECTION at $P = y_p$ = the vertical displacement of point $P = PP'$

7.3 Mathematical relationship between bending moment, slope and deflection

Consider points P and Q which are close together on the longitudinal axis of a beam subject to a bending moment M (Fig. 7.2). The deflected shape is a circular arc of radius R about a centre of curvature O.

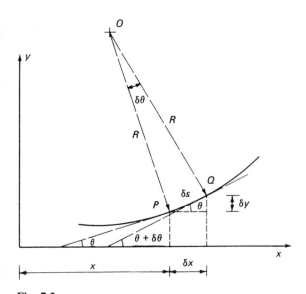

Fig. 7.2

Let the positions of P and Q be x and $x + \delta x$ from the origin respectively. Let δs = distance along the arc from P to Q

Now $$\delta s = R\delta\theta$$

Then $$\frac{1}{R} = \frac{\delta\theta}{\delta s}$$

or as $\delta s \to 0$ and $\delta x \to 0$, $$\frac{1}{R} = \frac{d\theta}{ds} = \frac{d\theta}{dx} \qquad [7.2]$$

Also as $\delta s \to 0$, $$\theta_p = \theta_Q = \theta$$

The angle θ will also be very small giving $\theta = \tan\theta = \dfrac{\delta y}{\delta x} = \dfrac{\mathrm{d}y}{\mathrm{d}x}$

Then substituting in equation [7.2]:

$$\frac{1}{R} = \frac{\mathrm{d}\left(\dfrac{\mathrm{d}y}{\mathrm{d}x}\right)}{\mathrm{d}x} = \frac{\mathrm{d}^2 y}{\mathrm{d}x^2}$$

and from equation [7.1]

$$\frac{1}{R} = \frac{M}{EI}$$

Hence $\qquad\qquad \dfrac{M}{EI} = \dfrac{\mathrm{d}^2 y}{\mathrm{d}x^2}$ [7.3]

which is known as the GENERAL DIFFERENTIAL EQUATION FOR DEFLECTION.

From this general equation the following relationships may be stated:

	Differential coefficients	Integrals
BENDING MOMENT (M)	$= EI\,\dfrac{\mathrm{d}^2 y}{\mathrm{d}x^2}$	$= M$
SLOPE (θ)	$= \dfrac{\mathrm{d}y}{\mathrm{d}x}$	$= \displaystyle\int \frac{M}{EI}\,\mathrm{d}x$
DEFLECTION (y)	$= y$	$= \displaystyle\int \theta\,\mathrm{d}x = \int\int \frac{M}{EI}\,\mathrm{d}x\,\mathrm{d}x$

Thus, mathematical analyses are carried out by either writing down the expression for y in terms of x and then differentiating the resulting expression with respect to x, or by writing down the expression for M in terms of x and then integrating with respect to x. The latter procedure is the most usual and is illustrated in the examples which follow.

7.4 Sign convention and units

It is important that a conventional system of signs and units be used in all numerical exercises. The correct sign and correct units should be ascribed to all numerical answers, as well as being stated throughout the intermediate stages of a calculation. Table 7.1 gives the convention of signs and units that will be adopted in this book, this system being compatible with that given in Table 5.1 p. 142 for shearing force and bending moment.

Table 7.1 Sign convention and units for bending moment, slope and deflection

Effect	Symbol	Sign Convention		Units
		Positive (+)	Negative (−)	
Bending moment	M	‿ sagging	⌢ hogging	N mm N m kNm
Slope	θ	→ xx ← ╱ or ╲ slope upwards in direction of x positive	→ xx ← ╲ or ╱ slope downwards in direction of x positive	radians
Deflection	y	↑ translation upwards	↓ translation downwards	mm m

EXAMPLE 7.1 Obtain expressions for the slope and deflection at the free end of the cantilever shown in Fig. 7.3.

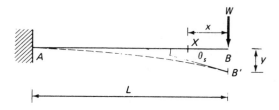

Fig. 7.3

Consider a section X which is at distance x from the free end.

Then at X

$$M = -Wx = EI \frac{d^2y}{dx^2}$$

Integrating
$$EI\theta = \int - Wx \, dx + A$$
$$= - \frac{Wx^2}{2} + A$$

Now when $x = L$, $\quad \theta = 0 \quad \therefore A = \frac{WL^2}{2}$

Giving
$$EI\theta = - \frac{Wx^2}{2} + \frac{WL^2}{2}$$

Integrating again,
$$EIy = \int \theta \, dx$$
$$= \int \left(- \frac{Wx^2}{2} + \frac{WL^2}{2} \right) dx + B$$
$$= - \frac{Wx^3}{6} + \frac{WL^2 x}{2} + B$$

Now when $x = L$, $\quad y = 0 \quad\quad \therefore B = - \frac{WL^3}{2} + \frac{WL^3}{6}$
$$= - \frac{WL^3}{3}$$

Giving
$$EIy = - = \frac{Wx^3}{6} + \frac{WL^2 x}{2} - \frac{WL^2}{3}$$

Thus at B (when $x = 0$):

$$\theta_B = \frac{WL^2}{2EI} \text{ (Answer)}$$

$$\text{and } y_B = - \frac{WL^3}{3EI} \text{ (Answer)}$$

EXAMPLE 7.2 Obtain expressions for the maximum slope and maximum deflection occurring in the uniformly loaded and simply supported beam shown in Fig. 7.4.

θ_{max} occurs at the support: $\theta_{max} = \theta_A = \theta_B$

y_{max} occurs at mid-span

Consider a section X which is at distance x from support A.

Then at X:
$$EI \frac{d^2 y}{dx^2} = M = \frac{wL}{2} \times x - wx \times \frac{x}{2}$$
$$= \frac{wLx}{2} - \frac{wx^2}{2}$$

Integrating
$$EI\theta = \int \left(\frac{wLx}{2} - \frac{wx^2}{2} \right) dx + A$$
$$= \frac{wLx^2}{4} - \frac{wx^3}{6} + A$$

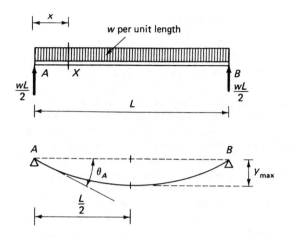

Fig. 7.4

Now when $x = \dfrac{L}{2}$, $\theta = 0$ $\quad \therefore \quad A = -\dfrac{wL^3}{16} + \dfrac{wL^3}{48}$

$$= -\dfrac{wL^3}{24}$$

Giving $\quad EI\theta = \dfrac{wLx^2}{4} - \dfrac{wx^3}{6} - \dfrac{wL^3}{24}$

Integrating again $\quad EIy = \displaystyle\int \left(\dfrac{wLx^2}{4} - \dfrac{wx^3}{6} - \dfrac{wL^3}{24} \right) \mathrm{d}x + B$

$$= \dfrac{wLx^3}{12} - \dfrac{wx^4}{24} - \dfrac{wL^3x}{24} + B$$

Now when $x = 0$, $y = 0$ $\quad \therefore \quad B = 0$

Giving $\quad EIy = \dfrac{wLx^3}{12} - \dfrac{wx^4}{24} - \dfrac{wL^3x}{24}$

Now $\theta_{max} = \theta_A$ (when $x = 0$)

Hence $\quad \theta_{max} = -\dfrac{wL^3}{24EI}$ (Answer)

And $y_{max} = y\left(\text{when } x = \dfrac{L}{2} \right)$

Hence $\quad y_{max} = \left(\dfrac{wL^4}{12 \times 8} - \dfrac{wL^4}{24 \times 16} - \dfrac{wL^4}{24 \times 2} \right) \dfrac{1}{EI}$

$$= \dfrac{wL^4}{EI} \left(\dfrac{4 - 1 - 8}{384} \right)$$

$$= -\dfrac{5\,wL^4}{384\,EI} \text{ (Answer)}$$

7.5 Mohr's area moment theorems

In Section 7.3 a mathematical relationship was shown to exist between the bending moment and the deflection in a loaded beam. Now it is usual to show the bending moment distribution along the beam in the form of a graph — the *bending moment diagram* — which is in general made up of a series of geometric curves. A similar diagram may be drawn to show the distribution of M/EI along a beam — for beams of constant cross-section this will be identical to the bending moment diagram except for the scale. Mohr's theorems relate the areas of the M/EI diagram to the slope and deflection in the beam at a given point. Where the diagrams consist of comparatively simple geometric shapes, a quick and easy method of analysis is possible. A large part of structural analysis, both simple and advanced, depends upon the application in one form or another of these two theorems. Their function as an analytical tool is of fundamental importance, particularly to civil and structural engineering students.

Mohr's 1st Theorem: THE CHANGE IN SLOPE BETWEEN ANY TWO POINTS ON A LOADED BEAM IS EQUAL TO THE AREA OF THE M/EI DIAGRAM BETWEEN THOSE POINTS

$$\theta_B - \theta_A = \int_{\theta_A}^{\theta_B} d\theta = \int_A^B \frac{M}{EI} dx \qquad [7.4]$$

Mohr's 2nd Theorem: FOR AN ORIGINALLY STRAIGHT BEAM, THE VERTICAL DISPLACEMENT OF A POINT ON THE BEAM RELATIVE TO THE TANGENT AT A SECOND POINT IS EQUAL TO THE FIRST MOMENT OF AREA OF THE M/EI DIAGRAM BETWEEN THE TWO POINTS, TAKEN ABOUT THE FIRST POINT (i.e. THE POINT AT WHICH THE DISPLACEMENT IS MEASURED)

$$\Delta = \int_A^B \frac{Mx}{EI} dx \qquad [7.5]$$

These statements may be proved as follows. Consider an originally straight length of beam AB which, when loaded, takes up a deformed shape $A'B'$, as shown in Fig. 7.5. As in the general theory of bending, $A'B'$ is assumed to be a circular arc of radius R.

For a short length of beam PQ situated at a distance x from B, let: M = the average bending moment over this length, δs = the length of arc PQ, δx = the distance between P and Q measured parallel to the x-axis, $\delta\theta$ = the angle subtended by PQ at the centre of curvature.

From equation [7.1]: $\qquad \dfrac{M}{EI} = \dfrac{1}{R} = \dfrac{\delta\theta}{\delta s}$

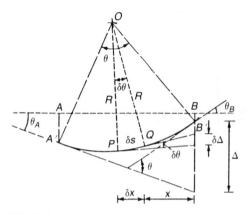

Fig. 7.5

Then as $\delta s \to 0$ and $\delta x \to 0$

$$\frac{M}{EI} = \frac{\mathrm{d}\theta}{\mathrm{d}x}$$

(as equation [7.2])

Thus integrating between A and B:

$$\int_0^\theta \mathrm{d}\theta = \int_A^B \frac{M}{EI}\,\mathrm{d}x$$

Giving

$$\theta = \int_A^B \frac{M}{EI}\,\mathrm{d}x \qquad [7.6]$$

i.e. AREA OF THE M/EI DIAGRAM BETWEEN A AND B.

Now let Δ = the intercept of the tangent at the vertical through B and $\delta\Delta$ = the intercept difference of the tangents at P and Q at the vertical through B.

When $\delta\theta$ is small; $\delta\Delta = x\,\delta\theta$, and as $\delta\theta \to 0$, $\mathrm{d}\Delta = x\,\mathrm{d}\theta$. Thus, integrating between A and B

$$\int_0^\Delta \mathrm{d}\Delta = x \int_0^\theta \mathrm{d}\theta = \int_A^B \frac{Mx}{EI}\,\mathrm{d}x$$

Giving

$$\Delta = \int_A^B \frac{Mx}{EI}\,\mathrm{d}x \qquad [7.7]$$

i.e. THE FIRST MOMENT OF AREA OF THE M/EI DIAGRAM BETWEEN A AND B TAKEN ABOUT B.

Note: If EI is constant along the length of beam under consideration, the M/EI diagram area becomes the BM diagram area/EI.

244

EXAMPLE 7.3   Obtain the expressions required in Example 7.1 using Mohr's area moment theorems.

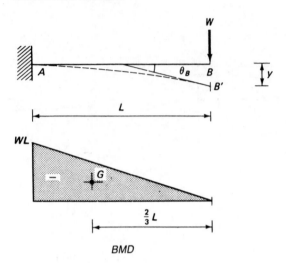

Fig. 7.6

First draw the bending moment diagram (Fig. 7.6).

Applying Mohr's 1st theorem

The change in slope between A and $B = 0 - \theta_B$

$$= \frac{\text{Area of the BMD between A and B}}{EI}$$

Hence

$$\theta_B = \frac{WL}{EI} \times \frac{L}{2}$$

$$= \frac{WL^2}{2EI} \text{ (Answer)}$$

Applying Mohr's 2nd theorem

The tangent at A is horizontal and therefore:

y = displacement of B below the tangent at A

$$= \frac{\text{first moment of the } BMD \text{ area about } B}{EI}$$

Hence

$$y_B = -\frac{WL^2}{2EI} \times \frac{2}{3}L$$

$$= -\frac{WL^3}{3EI} \text{ (Answer)}$$

EXAMPLE 7.4 Obtain expressions for the slope and deflection at the free end of a cantilever carrying a uniformly distributed load as shown in Fig. 7.7.

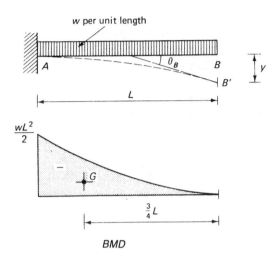

Fig. 7.7

See Fig. 5.11 (p. 148), for areas and centroids of parabolas.

Applying Mohr's 1st theorem

$$0 - \theta_B = \frac{\text{Area of the } BMD \text{ between } A \text{ and } B}{EI}$$

Hence
$$-\theta_B = \frac{wL^2}{2EI} \times \frac{L}{3}$$

Giving
$$\theta_B = \frac{wL^2}{6EI} \text{ (Answer)}$$

Applying Mohr's 2nd theorem

$$y = \text{displacement of } B \text{ below the tangent at } A$$
$$= \frac{\text{first moment of } BMD \text{ area about } B}{EI}$$

Hence
$$y_B = -\frac{wL^3}{6EI} \times \frac{3}{4}L$$

$$= -\frac{wL^4}{8EI} \text{ (Answer)}$$

*EXAMPLE 7.5   Obtain expressions for the slope at the supports and the
deflection at mid-span for a simply supported beam carrying a centrally placed
concentrated load as shown in Fig. 7.8.*

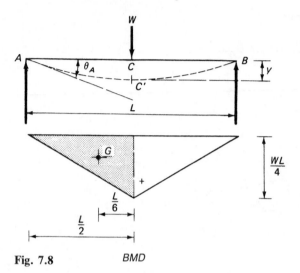

Fig. 7.8   BMD

Applying Mohr's 1st theorem

The maximum deflection will occur at the mid-span point C and therefore the
slope at C is zero:

$$\frac{dy}{dx} = \theta_C = 0$$

Therefore $-\theta_A + 0 = \dfrac{\text{Area of the } BMD \text{ between } A \text{ and } C \text{ (shaded)}}{EI}$

Hence $-\theta_A = \dfrac{WL}{4EI} \times \dfrac{L}{2} \times \dfrac{1}{2}$

Giving $\theta_A = -\dfrac{WL^2}{16EI}$ (Answer)

Applying Mohr's 2nd theorem

$$y_C = \Delta_{AC} = \text{displacement of } A \text{ above the tangent at } C$$
$$= \text{first moment of } M/EI \text{ diagram area}$$
$$\text{between } A \text{ and } C \text{ taken about } A$$

Hence $-y_C = \dfrac{WL^2}{16EI} \times \dfrac{L}{2} \times \dfrac{2}{3}$

Giving $y_C = -\dfrac{WL^3}{48EI}$ (Answer)

EXAMPLE 7.6   Obtain expressions for the slope at the supports and the deflection at mid-span for a simply supported beam carrying a uniformly distributed load as shown in Fig. 7.9.

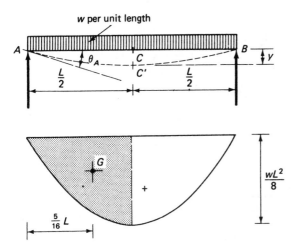

Fig. 7.9

Applying Mohr's 1st theorem

The maximum deflection will occur at the mid-span point C and therefore the slope at C is zero.

$$\frac{dy}{dx} = \theta_C = 0$$

Therefore $- \theta_A + 0 = \dfrac{\text{Area of the } BMD \text{ between } A \text{ and } C \text{ (shaded)}}{EI}$

Hence $\qquad - \theta_A = \dfrac{WL^2}{8EI} \times \dfrac{L}{2} \times \dfrac{2}{3}$

Giving $\qquad \theta_A = - \dfrac{WL^3}{24EI}$ (Answer)

Applying Mohr's 2nd theorem

$$y_C = \Delta_{AC} = \text{displacement of } A \text{ above the tangent at } C$$
$$= \text{first moment of } M/EI \text{ diagram area}$$
$$\text{between } A \text{ and } C \text{ taken about } A$$

Hence $\qquad - y_C = \dfrac{wL^3}{24EI} \times \dfrac{5L}{16}$

Giving $\qquad y_C = - \dfrac{5wL^4}{384EI}$ (Answer)

7.6   Deflection formulae — standard cases

There are several standard cases of loaded beams which occur frequently, either singly in simple situations, or in combination in more complex arrangements. Many apparently complicated systems can be rationalized into a series of standard cases and the principle of superposition used to provide solutions.

In Table 7.2 the deflection formulae for some of the more common cases are given, together with other relevant expressions.

FURTHER WORKED EXAMPLES

EXAMPLE 7.7   Figure 7.10 shows a propped cantilever. Prove that the force in the prop at B is equal to $\frac{3}{8}$ wL when the deflection at B is zero.

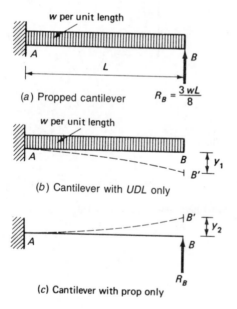

(a) Propped cantilever $R_B = \dfrac{3\,wL}{8}$

(b) Cantilever with *UDL* only

(c) Cantilever with prop only

Fig. 7.10

If the prop were to be removed, the deflection at B would be:

$$y_1 = -\frac{wL^4}{8EI} \qquad \text{(Table 7.2)}$$

Since the net deflection at B is zero, the replacement of the prop will cause, as it were, an upward deflection of y_2.

where
$$y_2 = +\frac{R_B L^3}{3EI} \qquad \text{(Table 7.2)}$$

Table 7.2 Common deflection formulae

Loading case	Maximum deflection (occurring at C)	Other properties
Simply supported beam A–B, central point load W at C, spans $\frac{L}{2}$, $\frac{L}{2}$	$y_{max} = -\dfrac{WL^3}{48\,EI}$	
Simply supported beam A–B, uniform load w per unit length, C at centre	$y_{max} = -\dfrac{5wL^4}{384EI}$	
Fixed-ended beam A–B, central point load W at C, moments M_A, M_B	$y_{max} = -\dfrac{WL^3}{192EI}$	$M_A = M_A = \dfrac{WL}{8}$
Fixed-ended beam A–B, uniform load w per unit length, C at centre, moments M_A, M_B	$y_{max} = -\dfrac{wL^4}{384EI}$	$M_A = M_B = -\dfrac{wL^2}{12}$
Propped cantilever, fixed at A (moment M_A), supported at B (R_B), uniform load w per unit length, C at $0.5785L$	$y_{max} = -\dfrac{27wL^4}{5\,000EI}$	$M_A = \dfrac{9wL^2}{128}$ $R_B = \dfrac{3wL}{8}$
Cantilever fixed at A (moment M_A), uniform load w per unit length over length L to C	$y_{max} = -\dfrac{wL^4}{8EI}$	$M_A = \dfrac{wL^2}{2}$
Cantilever fixed at A (moment M_A), point load W at free end C, length L	$y_{max} = -\dfrac{WL^3}{3EI}$	$M_A = WL$

But

$$y_1 + y_2 = 0$$

Therefore

$$\frac{R_B L^3}{3EI} = \frac{wL^4}{8EI}$$

Hence

$$R_B = \frac{wL^4}{8EI} \times \frac{3EI}{L^3}$$

$$= \frac{3}{8} wL \text{ (Answer)}$$

EXAMPLE 7.8 *Figure 7.11 shows a simply supported beam carrying a concentrated load at mid-span. Determine (a) the maximum deflection, and (b) the slope and deflection at the quarter-span point D. EI = 80 × 10³ kNm²*

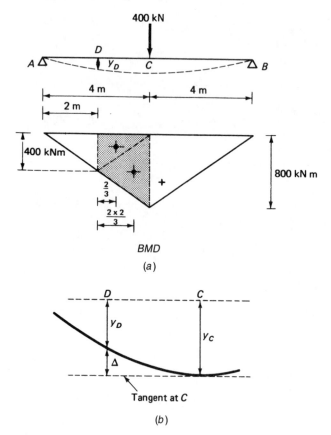

Fig. 7.11

$$BM \text{ at mid-span} = \frac{WL}{4} = \frac{400 \times 8}{4}$$

$$= 800 \text{ kNm}$$

(a) For the deflection at mid-span, use the standard-case formula given in Table 7.2:

$$y_c = -\frac{WL^3}{48EI} = -\frac{400 \times 8^3 \times 10^3}{48 \times 80 \times 10^3}$$

$$= -56 \cdot 7 \text{ mm (Answer)}$$

(b) *Applying Mohr's 1st theorem*

Since $\quad \theta_C = 0$

$$-\theta_D + 0 = \frac{Area \text{ of BMD between } D \text{ and } C \text{ (shaded)}}{EI}$$

Therefore
$$\theta_D = -\frac{(800 + 400)2}{2EI}$$

$$= -\frac{1\,200}{80 \times 10^3} = -0 \cdot 015 \text{ radians (Answer)}$$

Applying Mohr's 2nd theorem

Let $\quad \Delta_D = y_C - y_D$

Since the tangent at C is horizontal: (see Figure 7.11b)

$$\Delta_D = \text{displacement of } D \text{ above the tangent at } C$$

$$= \frac{\text{first moment of the } BMD \text{ area between } D \text{ and } C \text{ taken about } D}{EI}$$

$$= -\frac{400 \times 2}{2EI} \times \frac{2}{3} - \frac{800 \times 2}{2EI} \times \frac{2 \times 2}{3}$$

$$= -\frac{800}{3EI} - \frac{3\,200}{3EI}$$

$$= \frac{-4\,000}{3 \times 80 \times 10^3} = -0 \cdot 0167 \text{ m} = -16 \cdot 7 \text{ mm}$$

Therefore
$$y_D = y_C - \Delta_D$$

$$= -56 \cdot 7 + 16 \cdot 7 = -40 \cdot 0 \text{ mm (Answer)}$$

EXAMPLE 7.9 A timber joist of rectangular section is to carry a uniformly distributed load of 15 kN over the whole of its span of 4 m. Determine the dimensions of a suitable section that will not deflect more than 20 mm. E for timber = 10 kN/mm².

The maximum deflection will occur at mid-span and therefore from Table 7.2:

$$y_{max} = -\frac{5wL^4}{384EI} = -\frac{5WL^3}{384EI} \qquad \text{(since } W = wL)$$

Therefore

$$-20 \text{ mm} = \frac{-5 \times 15 \times 4^3 \times 10^9}{384 \times 10I}$$

Hence

$$I_{reqd} = \frac{5 \times 15 \times 4^3 \times 10^9}{384 \times 10 \times 20}$$

$$= \frac{1\,500 \times 10^6}{24} = 62 \cdot 5 \times 10^6 \text{ mm}^4$$

The proportions of the depth to the breadth of the joist should be about 3:1, then putting $b = d/3$:

$$I = \frac{bd^3}{12} = \frac{d^4}{36}$$

Then

$$d = (36 \times 62 \cdot 5 \times 10^6)^{\frac{1}{4}} = 218 \text{ mm}$$

and

$$b = \frac{218}{3} \qquad\qquad = 73 \text{ mm}$$

Choosing practical dimensions, say 220 mm × 75 mm or 225 mm × 75 mm

(Answer)

EXAMPLE 7.10 Draw the shearing force and bending moment diagrams for the beam shown in Fig. 7.12(a) which is carried on simple supports at B and C. EI = constant.

The two-span system shown in sketch (a) may be considered as the combination of the two one-span systems shown in (b) and (c) respectively, (b) is a *UDL* of 60 kNm carried over 8 m, giving a deflection *downwards* at B of y_B; and (c) is a concentrated load acting upwards of R_B at the centre of the 8 m span giving a deflection upwards at B of y_C.

Then

$$y_B + y_C = 0$$

Substituting, using standard formulae from Table 7.2:

$$\frac{-5 \times 60 \times 8^4}{384EI} + \frac{R_B \times 8^3}{48EI} = 0$$

Therefore

$$R_B = \frac{5 \times 60 \times 8 \times 48}{384} = 300 \text{ kN}$$

and

$$R_A = R_C = \frac{60 \times 8 - 300}{2} = 90 \text{ kN}$$

Bending moments

$$M_B = 90 \times 4 - 60 \times 4 \times 2 = 360 - 480 = -120 \text{ kNm}$$

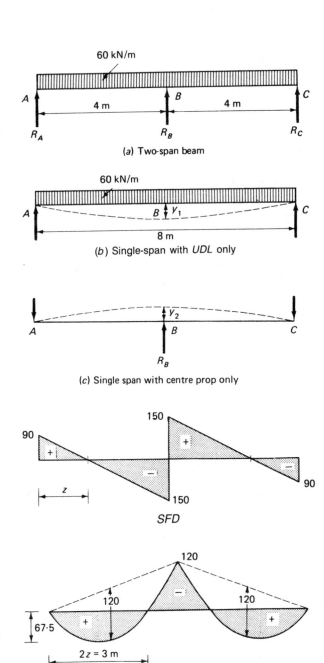

(a) Two-span beam

(b) Single-span with *UDL* only

(c) Single span with centre prop only

SFD

BMD

Fig. 7.12

Position of maximum sagging moment (z):

At Z: $Q_z = 0 = 90 - 60z$ $\therefore z = 1 \cdot 5$ m

Then maximum sagging moment:

$$M_z = 90 \times 1 \cdot 5 - \frac{60 \times 1 \cdot 5^2}{2} = 135 - 67 \cdot 5 = 67 \cdot 5 \text{ kNm}$$

or $$M_z = \frac{w(2z)^2}{8} = \frac{60 \times 3^2}{8} = 67 \cdot 5 \text{ kNm}$$

Also $$FBM_{AB} = FBM_{BC} = \frac{60 \times 4^2}{8} = 120 \text{ kNm}$$

EXAMPLE 7.11   The cantilever shown in Fig. 7.13 is built in at A and supported at the lower end of an 8 mm diameter rod at B. The upper end of the rod is secured to a rigid support at C. If the cantilever is horizontal and the rod

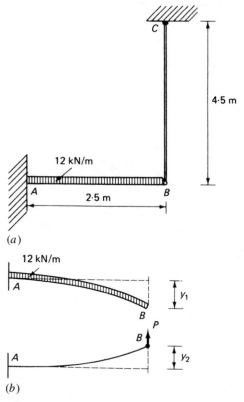

Fig. 7.13

unstrained before the application of the load, determine the stress in the rod and the bending moment in the beam at A due to the uniform load shown. The second moment of area of the beam is 15×10^6 mm^4 and Young's modulus is the same for both the rod and the beam.

Referring to Fig. 7.13(*b*).

Let y_1 = the deflection at the free end *B*, when unsupported by the rod, due to the 12 kN load.

y_2 = the deflection at the free end *B* due to a force *P* acting upwards

P = the tension force induced in the rod

ΔL = the extension of the rod due to the force *P*.

Now the deflection at *B* will be equal to the extension of the rod.

Then $\quad y_1 - y_2 = \Delta L$

The following substitutions may be made using standard-case formulae:

$$y_1 = \frac{wL^4}{8EI} = \frac{12 \times 2 \cdot 5^4 \times 10^9}{8E \times 15 \times 10^6} = \frac{3 \cdot 91 \times 10^3}{E}$$

$$y_2 = \frac{PL^3}{3EI} = \frac{P \times 2 \cdot 5^3 \times 10^9}{3E \times 15 \times 10^6} = \frac{0 \cdot 3472P \times 10^3}{E}$$

$$\Delta L = \frac{PL}{AE} = \frac{P \times 4 \cdot 5 \times 10^3 \times 4}{8^2 \times \pi \times E} = \frac{0 \cdot 0895P \times 10^3}{E}$$

So $\quad \dfrac{3 \cdot 91 \times 10^3}{E} - \dfrac{0 \cdot 3472P \times 10^3}{E} = \dfrac{0 \cdot 0895P \times 10^3}{E}$

giving $\qquad\qquad 3 \cdot 91 = (0 \cdot 3472 + 0 \cdot 0895)P$

Therefore $P = \dfrac{3 \cdot 91}{0 \cdot 4367} = 8 \cdot 95$ kN

The stress in the rod, $\qquad \sigma = \dfrac{8 \cdot 95 \times 10^3 \times 4}{8^3 \times \pi} = 178$ N/mm^2 (Answer)

$$\text{The bending moment at } A = 8 \cdot 95 \times 2 \cdot 5 - 12 \times \frac{2 \cdot 5^2}{2}$$

$$= 22 \cdot 4 - 37 \cdot 5$$

$$= -15 \cdot 1 \text{ kNm (Answer)}$$

EXAMPLE 7.12   Draw the bending moment diagram for the fixed-ended beam shown in Fig. 7.14. EI = constant.

In a fixed-ended beam the ends are so restrained in direction that no rotation may take place, i.e. $\theta_A = \theta_B = 0$.

The supports therefore apply to the beam fixing moments, say of magnitude *M*.

The resulting bending moment diagram may be considered in two parts, one due to the load applied at *C*, and the other due to the fixing moments applied at the ends.

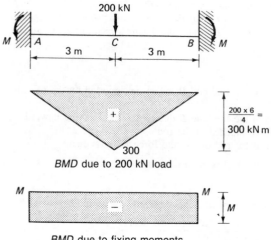

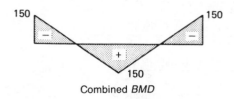

Fig. 7.14

Applying Mohr's 1st theorem

Change in slope $= \dfrac{\text{Area of } BMD}{EI}$

But the change in slope between A and B is zero!

Therefore $\theta_A - \theta_B = 0 = \dfrac{\text{area of } BMD \text{ between } A \text{ and } B}{EI}$

$$= \frac{200 \times 6}{4EI} \times \frac{6}{2} - \frac{M \times 6}{EI}$$

Giving $\qquad M = \dfrac{200 \times 6}{4 \times 2}$

$$= 150 \text{ kNm}$$

Then $\qquad M_B = \dfrac{200 \times 6}{4} - 150$

$$= 150 \text{ kNm}$$

7.7 Macaulay's method and superposition

From the foregoing sections it will have been seen that the bending moment expression changes at node points, i.e. at concentrated loads or at the beginning and end of uniform loads. In adapting the mathematical expressions for use with more complex multiple loading arrangements it would be convenient to have a single expression. This is particularly desirable when writing computer routines to perform calculations for bending moment, slope and deflection.

Such an expression may be established using a type of step function generally known as Macaulay's method. Consider the part-beam shown in Fig. 7.15. The bending moment at point K due to the loading shown may be written:

$$M_K = R_1 x - W_1[x - a_1] - W_2[x - a_2] - W_3[x - a_3] \quad [7.8]$$

This will serve as a general expression for values of x such that $0 < x < L$, provided that the quantities in the square brackets are eliminated (i.e. set to zero) when they are negative. Thus, the bending moment at point P is given by:

$$M_P = R_1 x - W_1[x - a_1]$$

since both $[x - a_2]$ and $[x - a_3]$ will be negative.

EXAMPLE 7.13 If, in Fig. 7.15, the following values apply, calculate the bending moments at points P, Q, and K.

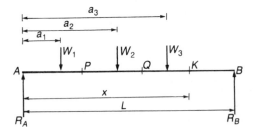

Fig. 7.15

$$
\begin{array}{lll}
W_1 = 20\ kN & W_2 = 30\ kN & W_3 = 25\ kN \\
a_1 = 1 \cdot 5\ m & a_2 = 3 \cdot 0\ m & a_3 = 5 \cdot 0\ m \\
AP = 2 \cdot 0\ m & AQ = 3 \cdot 5\ m & AK = 6 \cdot 0\ m \qquad L = 7 \cdot 0\ m
\end{array}
$$

Reaction at A: $R_A = 40 \cdot 0\ kN$

$$
\begin{aligned}
M_x &= R_A x - W_1[x - a_1] - W_2[x - a_2] - W_3[x - a_3] \\
&= 40 \cdot 0\,x - 20[x - 1 \cdot 5] - 30[x - 3 \cdot 0] - 25[x - 5 \cdot 0]
\end{aligned}
$$

At P, $x = 2 \cdot 0$ m:

$$\therefore\ M_p = 40 \times 2 - 20[0 \cdot 5] = 70\ kNm \text{ (Answer)}$$

At Q, $x = 3.5$ m:
$$\therefore\ M_Q = 40 \times 3.5 - 20[2.0] - 30[0.5] = 85 \text{ kNm (Answer)}$$
At K, $x = 6.0$ m:
$$\therefore\ M_K = 40 \times 6.0 - 20[4.5] - 30[3.0] - 25[1.0]$$
$$= 25 \text{ kNmm (Answer)}$$

The Macaulay expression can be utilized to good effect in setting up a simple computer *spreadsheet* analysis (see also Chapter 5 and Section 1.14). By using the *principle of superposition*, i.e. that the resultant effect of several identical effects acting simultaneously is equal to the algebraic sum of those effects, the bending moment can be calculated at given points due to any number of loads. This method can also be extended to calculate values for the slope and deflection at a point along a beam by integrating the basic Macaulay expression, i.e. eqn [7.12].

(a) Slope and deflection in a simply supported beam carrying concentrated loads

Consider first a single concentrated load on a simply supported beam (Fig. 7.16).

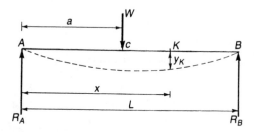

Fig. 7.16

Reaction at A:
$$R_A = \frac{W(L-a)}{L}$$

Bending moment at K: $\quad M_K = R_A x - W[x - a]$

Thus, $\quad EI\dfrac{d^2y}{dx^2} = -M_K = -R_A x + W[x - a]$ $\qquad$ [7.9]

$$EI\frac{dy}{dx} = -\frac{R_A x^2}{2} + \frac{W}{2}[x - a]^2 + C \qquad [7.10]$$

$$EIy = -\frac{R_A x^3}{6} + \frac{W}{6}[x - a]^3 + Cx + D \qquad [7.11]$$

when $\qquad x = 0$, $y = 0$ and so $D = 0$

When $\qquad x = L$, $y = 0$ and $C = \dfrac{R_A L^2}{6} - \dfrac{W}{6L}[L - a]^3$

Substituting for C and then R_A:

$$EI\theta = -\frac{R_A x^2}{2} - \frac{W}{2}[x - a]^2 + \frac{R_A L^2}{6} - \frac{W}{6L}(L - a)^3$$

$$\therefore \quad \theta = \frac{W}{6EIL}\{(L - a)(L^2 - 3x^2) - (L - a)^3 + 3L[x - a]^2\}$$

$$[7.12]$$

Similarly,

$$EIy = -\frac{R_A x^3}{6} + \frac{W}{6}[x - a]^3 + \frac{R_A Lx}{6} - \frac{Wx}{6L}(L - a)^3$$

$$\therefore \quad y = \frac{W}{6EIL}\{x(L - a)(2aL - x^2 - a^2) + L[x - a]^3\} \quad [7.13]$$

Remember that when the value of the square bracket is negative it is set to zero. The special cases of these expressions are:

(i) When $x = a$ $\qquad\qquad\qquad \theta = \frac{W}{3EIL}(L - a)(aL - 2a^2)$ [7.14]

$$y = \frac{Wa}{3EIL}(L - a)(aL - a^2)$$

(ii) When $x = a = L/2$ $\qquad \theta = 0$ and $y = \frac{WL^3}{48EI}$

(iii) When $x = 0$ and $a = L/2$ $\quad \theta = \frac{WL^3}{16EI}$ and $y = 0$

(b) Simply supported beam with a partial uniform load

Fig. 7.17(a) shows a uniformly distributed load of length b on a simply supported beam of span L, where $b < L$ and $a > 0$. The method explained in Section 7.7(a) above may be used, but the termination of the load at point D produces an additional boundary condition for the equation. This problem is overcome by replacing load $C-D$ with a load $C-B$ acting downward, together with a load $D-B$ acting upward, both of an equal intensity of w per metre (Fig. 7.17(b)). Loads $C-B$ and $D-B$ acting together are equivalent to the original load $C-D$.

Reaction at A: $\qquad R_A = \frac{wb}{L}\left(L - a - \frac{b}{2}\right)$

Bending moment at any point K:

$$M_K = R_A x - \frac{w}{2}[x - a]^2 + \frac{w}{2}[x - (a + b)]^2 \quad [7.15]$$

and since $\qquad M_K = -EI\frac{d^2y}{dx^2}$

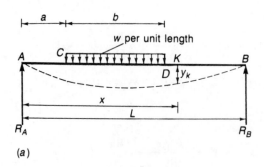

(a)

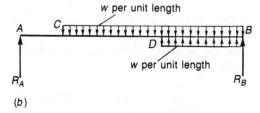

(b)

Fig. 7.17

$$EI \frac{dy}{dx} = -\frac{R_A x^2}{2} + \frac{w}{6} [x - a]^3 - \frac{w}{6} [x - (a + b)]^3 + C$$

$$EIy = -\frac{R_A x^3}{6} + \frac{w}{24} [x - a]^4 - \frac{w}{24} [x - (a + b)]^4 + Cx + D$$

When $\qquad x = 0, y = 0$ and so $D = 0$

When $\qquad x = L, y = 0$ and

$$C = \frac{R_A L^2}{6} - \frac{w}{24L} (L - a)^4 + \frac{w}{24L} (L - a - b)^4$$

Substituting for C and then R_A:

$$EI\theta = -\frac{R_A x^2}{2} + \frac{w}{6} [x - a]^3 - \frac{w}{6} [x - (a + b)]^3$$

$$+ \frac{R_A L^2}{6} - \frac{w}{24L} \{(L - a)^4 - (L - a - b)^4\}$$

$$\therefore \theta = \frac{w}{24EIL} \{4b \left(L - a - \frac{b}{2}\right)(L^2 - 3x^2) - (L - a)^4$$

$$+ (L - a - b)^4 + 4L[x - a]^3 - 4L[x - (a + b)]^3\} \quad [7.16]$$

Similarly,
$$EIy = -\frac{R_A x^3}{6} + \frac{w}{24} [x - a]^4 + \frac{w}{24} [x - (a + b)]^4$$

$$+ \frac{R_A L^2 x}{6} - \frac{wx}{24L} \{(L - a)^4 - (L - a - b)^4\}$$

$$\therefore \quad y = \frac{wx}{24EIL} \left\{ \left(L - a - \frac{b}{2} \right)(L^2 - x^2) - (L - a)^4 \right.$$

$$\left. + (L - a - b)^4 + L[x - a]^4 + L[x - (a + b)]^4 \right\} \quad [7.17]$$

```
DEFSSB1   DEFLECTION OF A SIMPLE BEAM WITH POINT LOADS
------------------------------------------------------------
     E =      65   kN/mm^2
     I =  200000   mm^4                    | W
                                           |
         <---------- a ------------->|
                                     V
     _____
         <----- al ----->| RL
         <-------------------------- ar ---------->| RR
------------------------------------------------------------
------------------------------------------------------------
                   RL        RR        LOADS
------------------------------------------------------------
R or W value:     14.5      15.5      10      12       8
Distance al/ar/a:   2        10       4        6       9
------------------------------------------------------------
------------------------------------------------------------
       x       y         Component deflection due to each load
------------------------------------------------------------
      0  -.019974   -1740        0     160     24      -2
      .5 -.015469  -1343.06      0     120     18    -1.5
      1  -.010545   -913.5       0      80     12      -1
     1.5 -.005342  -462.188      0      40      6     -.5
      2      0         0         0       0      0       0
     2.5  .0053421  462.1875     0     -40     -6      .5
      3   .0105449   913.5       0     -80    -12       1
     3.5  .0154687  1343.063     0    -120    -18     1.5
      4   .0199744   1740        0    -160    -24       2
     4.5  .0234896  2093.438     0  -233.75   -30     2.5
      5   .0270449  2392.5       0    -250    -36       3
     5.5  .0295713  2626.313     0  -281.25   -42     3.5
      6   .0310256   2784        0    -320    -48       4
     6.5  .0308454  2854.688     0  -358.75  -94.5    4.5
      7   .030391   2827.5       0    -390    -72       5
     7.5  .028504   2691.563     0  -406.25  -67.5    5.5
      8   .0252564   2436        0    -400    -72       6
     8.5  .0207204  2049.938     0  -363.75  -76.5    6.5
      9   .0149679  1522.5       0    -290    -72       7
     9.5  .0077252  842.8125     0  -171.25  -49.5   -19.5
     10      0         0         0       0      0       0
```

Fig. 7.18

262

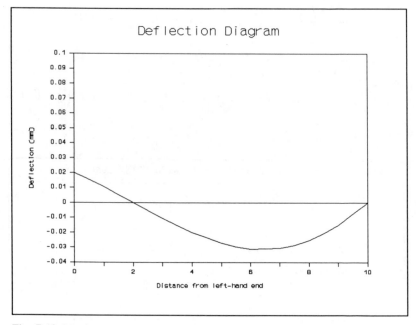

Fig. 7.18 (cont)

Remember that when the value of the square bracket is negative it is set to zero. The special cases of these expressions are:

(i) When a = 0, b = L and x = L/2 $\theta = 0$ and $y = \dfrac{5wL^4}{48EI}$

(ii) When a = 0, b = L and x = 0 $\theta = \dfrac{wL^3}{24EI}$ and $y = 0$

Now, using the principle of superposition, the slope and deflection at a given point on a beam carrying several concentrated or uniform loads can be obtained by adding together values calculated for each individual load. This technique is illustrated in the computer spreadsheet printout shown in Fig. 7.18.

EXERCISES   CHAPTER 7

1. Obtain expressions for the slope and deflection at the free end of the cantilever shown in Fig. 7.19.

2. Obtain expressions for the maximum slope and deflection occurring in the simply supported beam shown in Fig. 7.20.

3. A universal beam is simply supported and carries a uniform load of 400 kN over its entire span of 8 m. The second moment of area of the beam section is

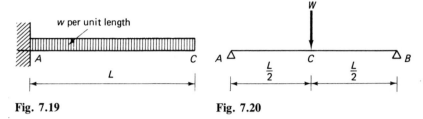

Fig. 7.19   **Fig. 7.20**

404×10^6 mm^4 and Young's modulus is 210 kN/mm^2. Calculate the maximum deflection.

4. A universal beam is simply supported and carries a single concentrated load of 105 kN at the mid-point of its span of 7·5 m. The second moment of area of the section is 100×10^6 mm^4 and Young's modulus is 210 kN/mm^2. Calculate the maximum deflection due to this load.

5. A steel cantilever beam is 3·5 m long and has a second moment of area of 180×10^6 mm^4. If the maximum deflection must not exceed 10 mm, calculate the magnitude of the concentrated load that may be carried at the free end. Young's modulus = 210 kN/mm^2.

6. A universal beam is simply supported over a span of 10 m and carries a uniform load of 8 kN/m along its entire length. The second moment of area of the section is 640×10^6 mm^4 and Young's modulus is 210 kN/mm^2. If the maximum deflection must not exceed 1/350 of the span, calculate the magnitude of the concentrated load that may be safely carried at mid-span in addition to the uniform load which is inclusive of the load due to self-weight.

7. A timber beam of breadth 75 mm and depth 225 mm is simply supported and carries an inclusive uniform load of 5 kNm along its entire span of 2·2 m. Calculate the maximum deflection. Young's modulus = 10 kN/mm^2.

8. A timber beam of breadth 100 mm and depth 300 mm is simply supported and carries a concentrated load of 15 kN at the centre of its span of 4·2 m. Calculate the maximum deflection due to this load. Young's modulus = 10 kN/mm^2.

9. A universal beam is built in at each end and carries an inclusive uniform load of 90 kN along its entire length of 8 m. The second moment of area of the section is 212×10^6 mm^4. Calculate the maximum deflection due to this load. Young's modulus = 210 kN/mm^2.

10. A timber beam is built in at each end of a 5 m span and carries a concentrated load of 5 kN at mid-span. Determine a suitable size for the beam in order to satisfy the condition that the maximum deflection must not exceed 1/333 of the span. Young's modulus = 11 kN/mm^2.

11−14. Figs 7.21−7.24 show a steel cantilever loaded in various ways. In each case the second moment of area is constant between A and C and has a value of 20×10^6 mm^4. Using the area moment theorems, determine the slope and deflection at both B and C. Young's modulus = 210 kN/mm^2.

264

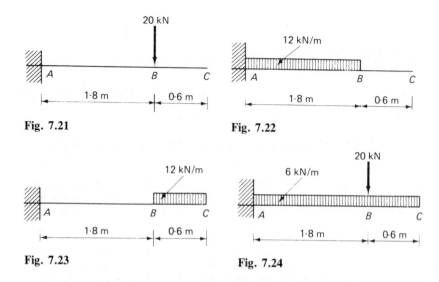

Fig. 7.21

Fig. 7.22

Fig. 7.23

Fig. 7.24

15–16. Figs 7.25 and 7.26 show a steel cantilever loaded in different ways, but in both cases propped at C such that the deflection at C is zero. Also in both cases, the second moment of area has a constant value between A and C of 20×10^6 mm^4. Using the area moment theorems, determine the slope and deflection at point B. Young's modulus $= 210$ kN/mm^2.

17–20. Figs 7.27–7.30 show a steel cantilever loaded in various ways, but in each case propped at B so that the deflection at B is zero. Also in each case, the second moment of area has a constant value between A and C of 20×10^6 mm^4. Using the area moment theorems, determine the slope at point B and the deflection at point C. Young's modulus $= 210$ kN/mm^2.

21. Fig. 7.31 shows a steel cantilever, the second moment of area of which has a constant value between A and D of 120×10^6 mm^4. Determine the deflection at points B, C and D. Young's modulus $= 210$ kN/mm^2.

22. Fig. 7.32 shows the plan view of an arrangement of two steel beams in which beam CBD is supported under its mid-span point by the free end of cantilever AB. The beam is also simply supported at C and D and carries a uniform load between these points of 20 kN/m. The second moments of area at 60×10^6 mm^4 and 120×10^6 mm^4 for CBD and AB respectively and Young's modulus is 210 kN/mm^2 for both. Calculate the deflection at B and draw the bending moment diagrams for both the beam and the cantilever.

23. Fig. 7.33 shows a cantilever AB fixed to a rigid support at A and pinned to the lower end of a steel rod BC at B. The upper end of the steel rod is secured to a rigid support C. The diameter of the rod is 8 mm and the second moment of area of the beam is 150×10^6 mm^4. If the beam is horizontal and the rod unstrained before the application of the load, determine the force acting in the rod and the bending moment in the beam at A due to the 40 kN load. Young's modulus is the same for both the beam and the rod.

265

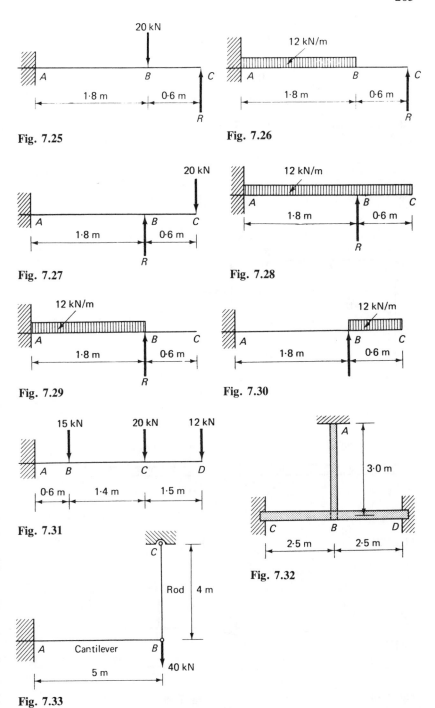

Fig. 7.25

Fig. 7.26

Fig. 7.27

Fig. 7.28

Fig. 7.29

Fig. 7.30

Fig. 7.31

Fig. 7.32

Fig. 7.33

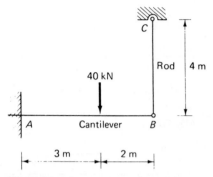

Fig. 7.34

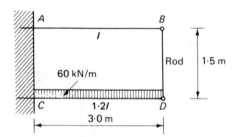

Fig. 7.35

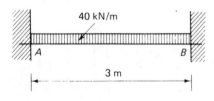

Fig. 7.36

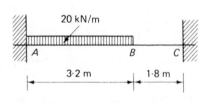

Fig. 7.37

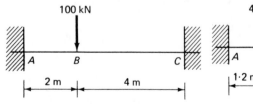

Fig. 7.38

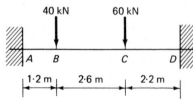

Fig. 7.39

24. As Exercise 23, but for a different position of the 40 kN load. (Fig. 7.34).

25. Figure 7.35 shows an arrangement of two cantilever beams and a steel rod, in which the ends of the cantilevers are built in at A and at C and the rod pinned to the free ends of the cantilevers at B and at D. The rod has a diameter of 15 mm. If the cantilevers are horizontal and the rod unstrained before the application of the load, determine the stress in the rod and the bending moments in the beams at A and at C due to the uniform loading shown. Young's modulus is the same for all three components. For both beams: $I = 100 \times 10^6$ mm^4.

26−29. Figs 7.36−7.39 show a series of fixed-ended beams each having constant second moment of area. In each case, determine the magnitude of the fixing moments and sketch the bending moment diagram.

8 Further work on stress and strain. Combined stresses

Symbols

B	=	breadth
D, d	=	diameter, distance
E	=	Young's modulus of elasticity
G	=	modulus of rigidity or shear modulus
K	=	bulk modulus
L	=	length, span
p	=	pressure
t	=	thickness
V	=	volume
α	=	angle of an oblique plane
α_p	=	angle of a principal plane
Δ	=	linear displacement
ϵ	=	strain
θ	=	angle
ν	=	Poisson's ratio
σ	=	stress (generally), oblique applied stress
σ_n	=	normal component stress
σ_R	=	resultant stress
σ_1	=	major principal stress
σ_2	=	minor principal stress
τ	=	shearing stress
ϕ	=	shearing strain (angle)
ϵ_s	=	shearing strain

8.1 Introduction

Elsewhere in this book, where stress and strain have been considered, attention has been concentrated mainly on the effects of a single stress. There are, however, many situations in which a number of stresses may be acting simultaneously. Sometimes this will be due to the simultaneous application of more than one force-effect, such as combined direct tension and torsion or combined bending and torsion. In some cases several stresses may be induced by the application of a single force-effect, as for example in the case of lateral loading on a beam which produces both bending and shearing stresses. After a few preliminary definitions, the main work in this chapter will be concerned with *compound stress* states or systems. A *compound stress* system is set up when two or more direct and/or shearing stresses are induced simultaneously at a given section or point in a member.

8.2 Shearing stress and shearing strain

Just as axial stresses cause shortening or lengthening, i.e. axial strain, in a member, shearing stresses cause deformation which may be termed shearing strain. The measure of this shearing deformation, and thus the shearing strain, is given by the angular displacement that results. When the rectangular prism shown in Fig. 8.1 is subjected to a shearing stress τ it is deformed to become the parallelogram prism whose corners are now $A'B'CD$. The face AB is displaced by an amount Δ relative to face CD. The proportional deformation may be expressed thus:

$$\epsilon_s = \frac{\Delta}{h} = \tan \phi$$

Since only small strains occur in structural components, ϕ will be small, so that:

$$\tan \phi = \phi \text{ radians}$$

The shearing strain is therefore:

$$\epsilon_s = \phi \text{ radians} \qquad [8.1]$$

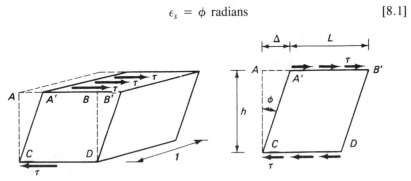

Fig. 8.1

270

Up to the limit of proportionality, as in the case of direct stress and strain, Hooke's law is obeyed whereby:

$$\frac{\tau}{\epsilon_s} = \text{constant}$$

This constant is termed the **Modulus of rigidity** or the **Shear modulus** (**G**)

Hence $\quad\dfrac{\tau}{\epsilon_s} = G \qquad$ (SHEAR MODULUS) $\qquad$ [8.2]

8.3 Lateral strain. Poisson's ratio

When a bar is subjected to a longitudinal stress a direct strain takes place in the direction of the stress. At the same time a strain will take place at right angles to the direction of the applied stress; this is termed the LATERAL STRAIN (Fig. 8.2). Where the deformations remain elastic,

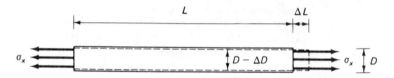

Fig. 8.2

the ratio of the lateral strain to the longitudinal strain will be constant for a given material. This ratio is called POISSON'S RATIO.

$$\text{Poisson's ratio,} \quad \nu = \frac{\text{lateral strain}}{\text{longitudinal strain}}$$

$$\nu = \frac{\epsilon_y}{\epsilon_x} \qquad [8.3]$$

Typical values of ν are given in Table 3.2, but for the majority of structural materials the value lies between 0·25 and 0·35. Referring to Fig. 8.2:

$$\text{Longitudinal strain, } \epsilon_x = \frac{\Delta L}{L}$$

$$\text{Lateral strain, } \epsilon_y = -\frac{\Delta D}{D}$$

$$= -\nu\epsilon_x$$

Note that the lateral strain is of opposite sign, so that if the longitudinal dimension is increased the lateral dimension will be decreased and vice versa.

EXAMPLE 8.1   A steel bar of rectangular cross-section 100 mm × 40 mm is subjected to an axial tension of 240 kN. Determine the changes that result in the cross-sectional dimensions. $E = 200\ kN/mm^2$, Poisson's ratio $\nu = 0\cdot3$.

Area of section $= 100 \times 40 = 4\,000$ mm^2

Longitudinal tensile stress, $\sigma_x = \dfrac{240 \times 1\,000}{4\,000}$

$$= 60\ \text{N/mm}^2$$

Longitudinal strain, $\epsilon_x = \dfrac{\sigma_x}{E}$

$$= \dfrac{60}{200 \times 1\,000}$$

$$= 0\cdot3 \times 10^{-3}$$

Lateral strain, $\epsilon_y = -\nu\epsilon_x$

$$= -0\cdot3 \times 0\cdot3 \times 10^{-3}$$

$$= -0\cdot09 \times 10^{-3}$$

Hence the 100 mm side contracts by $0\cdot09 \times 10^{-3} \times 100 = 0\cdot009$ mm
and the 40 mm side contracts by $0\cdot09 \times 10^{-3} \times 40 \qquad = 0\cdot0036$ mm

(Answer)

8.4   Biaxial stresses and strains

Components such as plates and shells are often subject to a system of two mutually perpendicular or *biaxial* stresses. Fig. 8.3(*a*) shows a rectangular element subject to stresses σ_x and σ_y; Fig. 8.3(*b*) shows the

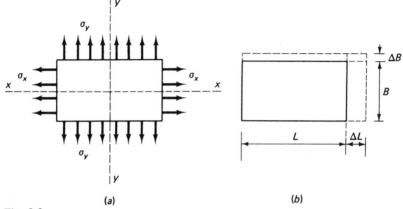

(*a*)   (*b*)

Fig. 8.3

resulting deformation of the element. Each stress will produce both longitudinal and lateral strain components. The table below shows how the strain components are related to the stresses, assuming that the material is isotropic, i.e. $E_x = E_y$.

| | Strain caused by | |
	σ_x	σ_y
Strain in the direction of σ_x	$\dfrac{\sigma_x}{E}$	$-\dfrac{\nu\sigma_y}{E}$
Strain in the direction of σ_y	$-\dfrac{\nu\sigma_x}{E}$	$\dfrac{\sigma_y}{E}$

Hence the total strain in the direction of σ_x, $\epsilon_x = (\sigma_x - \nu\sigma_y)/E$ [8.4]

and the total strain in the direction of σ_y, $\epsilon_y = (\sigma_y - \nu\sigma_x)/E$ [8.5]

By inspection of Fig. 8.3(b), the change in area is:

$$\Delta A = B\,\Delta L + L\,\Delta B + \Delta L\,\Delta B$$

of which $\Delta L\,\Delta B$ is very small and may be neglected.

The areal strain is therefore:

$$\frac{\Delta A}{A} = \frac{B\,\Delta L + L\,\Delta B}{BL}$$

$$= \frac{\Delta L}{L} + \frac{\Delta B}{B}$$

Hence $\epsilon_A = \epsilon_x + \epsilon_y$ [8.6]

i.e. for small strains, the areal strain is equal to the sum of the biaxial linear strains.

A special cases exists when $\sigma_x = \sigma_y$ and therefore $\epsilon_x = \epsilon_y$

giving $\epsilon_A = 2\epsilon_{x,y}$

EXAMPLE 8.2   *Figure 8.4 shows a flat steel panel which is subject to biaxial tensile stresses. Determine the value of σ_y at which the strain in this direction (ϵ_y) will be zero. What increase will have taken place in the 1·80 m dimension? Young's modulus = 210 kN/mm², Poisson's ratio = 0·3.*

Strain in the direction of σ_y, ϵ_y $= \dfrac{\sigma_y - \nu\sigma_x}{E}$

$\therefore \quad 0 = \sigma_y - 0\cdot3 \times 40$

Hence $\sigma_y = 12$ N/mm² (Answer)

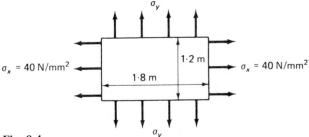

Fig. 8.4

Strain in the direction of σ_x, ϵ_x $= \dfrac{\sigma_x - \nu\sigma_y}{E}$

$$= \frac{40 - 0\cdot03 \times 12}{210 \times 1\ 000}$$

$$= \frac{36\cdot4}{210 \times 1\ 000}$$

Then the increase in length
in the direction of σ_x = ΔL_x = $\epsilon_x \times 1\cdot8 \times 1\ 000$

$$= \frac{36\cdot4 \times 1\cdot8 \times 10^3}{210 \times 1\ 000}$$

$$= 0\cdot312 \text{ mm (Answer)}$$

8.5 Triaxial stress and volumetric strain

There are several instances of stress systems consisting of three triaxially perpendicular direct stresses that may occur in engineering situations connected with building and civil engineering. For example, in the material stored in bunkers and silos and in soils beneath foundations. Another example may be found in the change in volume of a fluid-storage vessel resulting from changes in internal pressure. Figure 8.5

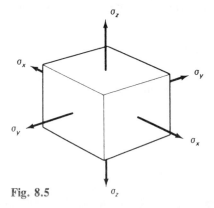

Fig. 8.5

shows a cuboidal element subject to direct triaxial stresses σ_x, σ_y, σ_z. Each of the three stresses will produce a longitudinal strain component together with two perpendicular lateral strain components, the inter-relationships of which are given in the table below:

	Strain caused by		
	σ_x	σ_y	σ_z
Strain in the direction of σ_x	$\dfrac{\sigma_x}{E}$	$-\dfrac{\nu\sigma_y}{E}$	$-\dfrac{\nu\sigma_z}{E}$
Strain in the direction of σ_y	$-\dfrac{\nu\sigma_x}{E}$	$\dfrac{\sigma_y}{E}$	$-\dfrac{\nu\sigma_z}{E}$
Strain in the direction of σ_z	$-\dfrac{\nu\sigma_x}{E}$	$-\dfrac{\nu\sigma_y}{E}$	$\dfrac{\sigma_z}{E}$

Hence the total strains are:

$$\text{in the direction of } \sigma_x,\ \epsilon_x = \frac{\sigma_x - \nu(\sigma_y + \sigma_z)}{E} \tag{8.7}$$

$$\text{in the direction of } \sigma_y,\ \epsilon_y = \frac{\sigma_y - \nu(\sigma_x + \sigma_z)}{E} \tag{8.8}$$

$$\text{in the direction of } \sigma_z,\ \epsilon_z = \frac{\sigma_z - \nu(\sigma_x + \sigma_y)}{E} \tag{8.9}$$

Using a proof similar to that adopted to give equation [8.6], the following volumetric strain relationship may be deduced:

Volumetric strain, $\qquad \dfrac{\Delta V}{V} = \epsilon_x + \epsilon_y + \epsilon_z$ $\qquad$ [8.10]

In the special case of an element subject to a constant all-round stress, such as the hydrostatic pressure p illustrated in Fig. 8.6, it may be shown that, for small strains:

$$\frac{\Delta V}{V} = \epsilon_v = 3\,\epsilon_{x,y,z} \tag{8.11}$$

From Section 3.8, $\qquad \dfrac{\text{pressure}}{\text{volumetric strain}} = \text{bulk modulus}$

i.e. $\qquad\qquad\qquad \dfrac{p}{\epsilon_V} = K$

Therefore $\qquad\qquad\qquad \epsilon_V = 3\,\epsilon_{x,y,z} = \dfrac{p}{K}$ $\qquad$ [8.12]

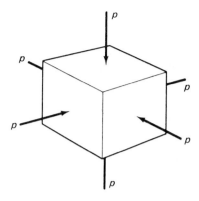

Fig. 8.6

EXAMPLE 8.3   A spherical vessel of diameter 3 m and plate thickness 25 mm is full of water under a pressure of 3 N/mm² above atmospheric. Calculate the volume of water expelled from the vessel when a valve is opened and the pressure allowed to equalize with that of the atmosphere. Assume that no gravity drainage can take place. Bulk modulus for water, K = 2·25 kN/mm²; Young's modulus for the vessel material, E = 210 kN/mm²; Poisson's ratio for the vessel material, ν = 0·3.

The maximum stress in the shell of a spherical vessel is given by $\sigma = pd/4t$

where
$$p = \text{the internal pressure}$$
$$d = \text{the diameter of the vessel}$$
$$t = \text{the thickness of the shell}$$

In this case
$$\sigma = \frac{3 \times 1\cdot5 \times 10^3}{4 \times 25} = 45 \text{ N/mm}^2$$

This stress acts normal to the diametral planes giving rise to a two-dimensional system in which the diametral strain will be

$$\epsilon_D = \frac{\sigma(1 - \nu)}{E} \qquad \text{(from eqn [8.4])}$$

Then
$$\epsilon_D = \frac{45(1\cdot0 - 0\cdot3)}{210 \times 10^3}$$

The reduction in volume of the vessel itself caused by the decrease in pressure is (eqn [8.11]):

$$\Delta V = 3\,\epsilon_D V = \frac{3 \times 45 \times 0\cdot7}{210 \times 1\,000} \times \frac{4\pi}{3} \times 1\cdot5^3 \times 10^9$$

$$= \frac{45}{10^5} \times \frac{4\pi}{3} \times 1\cdot5^3 \times 10^9$$

The volumetric strain on the water due to the change in pressure is given by equation [8.12]:

$$\epsilon_V = \frac{p}{K} = \frac{3}{2 \cdot 25 \times 1\,000}$$

Therefore the increase in the volume of the water as the pressure is reduced is:

$$\Delta V_W = \epsilon_V \times V$$

$$= \frac{3}{2 \cdot 25 \times 1\,000} \times \frac{4\pi}{3} \times 1 \cdot 5^3 \times 10^9 \text{ mm}^3$$

Hence the volume of water expelled when the pressure is reduced to that of the atmosphere will be:

$$V_E = \Delta V + \Delta V_W$$

$$= \left(\frac{45}{10^5} + \frac{3}{2 \cdot 25 \times 1\,000} \right) \frac{4\pi}{3} \times 1 \cdot 5^3 \times 10^9$$

$$= (0 \cdot 015 + 0 \cdot 444)\, 4\pi \times 1 \cdot 5^3 \times 10^6$$

$$= 25 \cdot 2 \times 10^6 \text{ mm}^3$$

$$= 0 \cdot 0252 \text{ m}^3 \text{ (Answer)}$$

8.6 Oblique stresses and resultants

It is usual to refer most stresses and strains to planes which include two of the three spatial axes (x, y, z). Any stress that is obliquely applied may be resolved into component stresses which are referred to as the xyz planes (Fig. 8.7). In a similar way, any two 'pure' stresses may be resolved into a single oblique resultant as shown in Fig. 8.8.

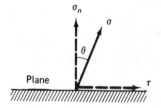

Fig. 8.7

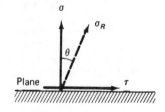

Fig. 8.8

In Fig. 8.7: σ = oblique applied stress

σ_n = normal component stress ('pure')

τ = tangential component stress ('pure')

Where $\sigma_n = \sigma \cos \theta$

$\tau = \sigma \sin \theta$

and $\sigma = \sqrt{(\sigma_n^2 + \tau^2)}$ [8.13]

In Fig. 8.8: σ = applied normal stress ('pure')

τ = applied tangential stress ('pure')

σ_R = oblique resultant stress

Where $\sigma_R = \sqrt{(\sigma^2 + \tau^2)}$

and $\theta = \arctan \dfrac{\tau}{\sigma}$ [8.14]

Thus, a principle may be established that, for a given plane, subject to one or more stresses, the corresponding pure normal and pure tangential stresses acting on that plane are simply calculable; and that, for any system of 'pure' stresses, the oblique resultant stresses are calculable.

8.7 Normal and tangential stresses on an oblique plane

Consider the bar of rectangular section shown in Fig. 8.9 which is subject to a uniaxial tensile stress σ. The normal and tangential stresses are required on the oblique plane AC. The prism ABC (which has a thickness b) is subject to the stresses shown in Fig. 8.10. Since the prism is in equilibrium, the forces acting on it (Fig. 8.11) may be resolved as follows:

Applied force on AB = $\sigma \times$ area of plane AB = σbd

Normal force on AC = $\sigma_n \times$ area of plane AC = $\sigma_n \dfrac{bd}{\cos \alpha}$

Tangential force on AC = $\tau \times$ area of plane AC = $\tau \dfrac{bd}{\cos \alpha}$

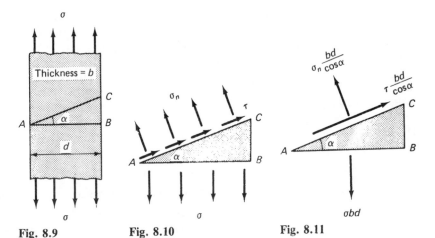

Fig. 8.9   **Fig. 8.10**   **Fig. 8.11**

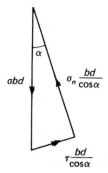

Fig. 8.12

The triangle of forces (Fig. 8.12) then yields the following relationships:

(i) $\qquad \sigma_n \dfrac{bd}{\cos\alpha} = \sigma bd \, \cos\alpha$

Hence the normal stress, $\sigma_n = \sigma \cos^2\alpha$ $\qquad\qquad$ [8.15]

(ii) $\qquad \tau \dfrac{bd}{\cos\alpha} = \sigma bd \, \sin\alpha$

Hence the tangential stress, $\tau = \sigma \sin\alpha \, \cos\alpha$

$$= \tfrac{1}{2}\sigma \, \sin2\alpha \qquad\qquad [8.16]$$

Equation [8.16] is of some importance; consider the range of values for the tangential stress (τ) which correspond to the possible values of angle α.

$$\text{when } \alpha = 0, \quad \sin2\alpha = 0 \; \therefore \; \tau = 0$$
$$\text{when } \alpha = 90°, \quad \sin2\alpha = 0 \; \therefore \; \tau = 0$$
$$\text{when } \alpha = 45°, \quad \sin2\alpha = 1 \; \therefore \; \tau = \tfrac{1}{2}\sigma$$

THUS IN A MEMBER SUBJECT TO UNIAXIAL TENSION OR COMPRESSION THE MAXIMUM TANGENTIAL (SHEARING) STRESS IS EQUAL TO ONE-HALF OF THE APPLIED STRESS AND ACTS ON A PLANE INCLINED AT 45° TO THAT OF THE APPLIED STRESS.

The normal and tangential stresses acting on plane AC may be resolved into a single resultant stress.

Resultant stress on plane AC, $\sigma_R = \sqrt{(\sigma_n{}^2 + \tau^2)}$

$$= \sigma\sqrt{(\cos^4\alpha + \sin^2\alpha \, \cos^2\alpha)}$$
$$= \sigma\sqrt{\{\cos^2\alpha(\cos^2\alpha + \sin^2\alpha)\}}$$
$$= \sigma \cos\alpha \qquad\qquad [8.17]$$

8.8 Oblique planes and biaxial normal stresses

The prism ABC shown in Fig. 8.13 is of unit thickness and AC is also unity.

So that area of plane $AB = \cos\alpha$

and area of plane $BC = \sin\alpha$

The prism is subject to biaxial normal stresses σ_x and σ_y and it is required to find the normal and tangential (shearing) stresses acting on the plane AC. The equilibrium of the prism results from the forces shown in Fig. 8.14. Resolving in the direction of σ_n:

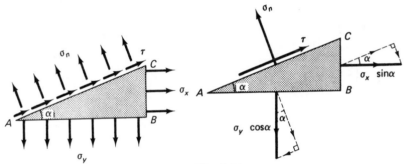

Fig. 8.13   **Fig. 8.14**

$$\sigma_n = (\sigma_y \cos\alpha)\ \cos\alpha + (\sigma_x \sin\alpha)\ \sin\alpha$$
$$= \sigma_y \cos^2\alpha + \sigma_x \sin^2\alpha$$
$$= \tfrac{1}{2}(\sigma_y + \sigma_x) + \tfrac{1}{2}(\sigma_y - \sigma_x) \cos2\alpha \qquad [8.18]$$

Resolving in the direction of τ:

$$\tau = (\sigma_y \cos\alpha)\ \sin\alpha - (\sigma_x \sin\alpha)\ \cos\alpha$$
$$= (\sigma_y - \sigma_x)\ \sin\alpha\ \cos\alpha$$
$$\tfrac{1}{2}(\sigma_y - \sigma_x)\ \sin2\alpha \qquad [8.19]$$

From eqn [8.19]:

when $\alpha = 0$, $\sin2\alpha = 0 \therefore \tau = 0$

when $\alpha = 90°$, $\sin2\alpha = 0 \therefore \tau = 0$

when $\alpha = 45°$, $\sin2\alpha = 1 \therefore \tau = \tfrac{1}{2}(\sigma_y - \sigma_x)$

THUS IN A MEMBER SUBJECT TO BIAXIAL TENSION OR COMPRESSION THE MAXIMUM TANGENTIAL (SHEARING) STRESS IS EQUAL TO ONE-HALF THE DIFFERENCE BETWEEN THE APPLIED STRESSES AND ACTS ON A PLANE INCLINED AT 45° TO THEIR DIRECTION

$$\tau_{max} = \tfrac{1}{2}(\sigma_y - \sigma_x) \qquad [8.20]$$

8.9 Oblique planes and a general two-dimensional stress system

The prism ABC shown in Fig. 8.15 is of unit thickness and AC is also unity.

So that $\qquad$ area of plane $AB = \cos\alpha$

and $\qquad$ area of plane $BC = \sin\alpha$

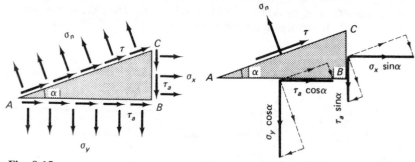

Fig. 8.15   **Fig. 8.16**

The prism is subject to a general two-dimensional stress system of applied normal and tangential (shearing) stresses and it is required to find the normal and tangential (shearing) stresses acting on plane AC. The equilibrium of the prism results from the forces shown in Fig. 8.16. Resolving in the direction of σ_n:

$$
\begin{aligned}
\sigma_n &= (\sigma_y \cos\alpha)\,\cos\alpha + (\sigma_x \sin\alpha)\,\sin\alpha + (\tau_a \cos\alpha)\,\sin\alpha \\
&\qquad + (\tau_a \sin\alpha)\,\cos\alpha \\
&= \sigma_y \cos^2\alpha + \sigma_x \sin^2\alpha + 2\tau_a \sin\alpha \cos\alpha \\
&= \tfrac{1}{2}\sigma_y (1 + \cos2\alpha) + \tfrac{1}{2}\sigma_x (1 - \cos2\alpha) + \tau_a \sin2\alpha \\
&= \tfrac{1}{2}(\sigma_y + \sigma_x) + \tfrac{1}{2}(\sigma_y - \sigma_x) \cos2\alpha + \tau_a \sin2\alpha \qquad [8.21]
\end{aligned}
$$

Resolving in the direction of τ:

$$
\begin{aligned}
\tau &= (\sigma_y \cos\alpha)\,\sin\alpha - (\sigma_x \sin\alpha)\,\cos\alpha + (\tau_a \sin\alpha)\,\sin\alpha \\
&\qquad - (\tau_a \cos\alpha)\,\cos\alpha \\
&= (\sigma_y - \sigma_x) \sin\alpha \cos\alpha + \tau_a (\sin^2\alpha - \cos^2\alpha) \\
&= \tfrac{1}{2}(\sigma_y - \sigma_x) \sin2\alpha + \tfrac{1}{2}\tau_a (1 - \cos2\alpha - 1 - \cos2\alpha) \\
&= \tfrac{1}{2}(\sigma_y - \sigma_x) \sin2\alpha - \tau_a \cos2\alpha \qquad [8.22]
\end{aligned}
$$

From eqn [8.22]:

$$
\begin{aligned}
&\text{when } \alpha = 0, && \tau = -\tau_a \\
&\text{when } \alpha = 90°, && \tau = \tau_a \\
&\text{when } \alpha = 45°, && \tau = \tfrac{1}{2}(\sigma_y - \sigma_x)
\end{aligned}
$$

Then the maximum tangential (shearing) stress will be:

$$\tau_{max} = \tfrac{1}{2}(\sigma_y - \sigma_x) \qquad [\text{or} \pm \tau_a] \qquad [8.23]$$

8.10 Principal planes and principal stresses

It will be seen from the last few lines of the preceding section that, as the angle of the plane (α) varies from 0 to 90°, so the tangential (shearing) stress acting on the plane varies from $-\tau$ to $+\tau$. It follows, therefore, that there must be a value of α at which the tangential stress on the plane is zero. Such a plane, a plane upon which the tangential (shearing) stress is zero, is called a *PRINCIPAL PLANE*. The normal stress acting on a principal plane is called a *PRINCIPAL STRESS*.

> PRINCIPAL PLANE = plane upon which the shearing stress
>   is zero
>
> PRINCIPAL STRESS = the normal stress acting on a
>   principal plane

Consider again the prism *ABC* subject to a general two-dimensional stress system (Fig. 8.15); but let *AC* be a *principal plane*. The normal stress acting on plane *AC* will therefore be a *principal stress*. The forces acting on the prism are shown in Fig. 8.17.

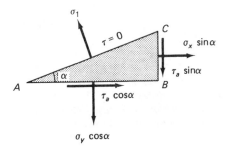

Fig. 8.17

Resolving in the direction of σ_y:

$$\sigma_1 \cos\alpha_p = \sigma_y \cos\alpha_p + \tau_a \sin\alpha_p$$

Therefore $\qquad \sigma_1 = \sigma_y + \tau_a \tan\alpha_p \qquad\qquad\qquad [8.24]$

Resolving in the direction of σ_x:

$$\sigma_1 \sin\alpha_p = \sigma_x \sin\alpha_p + \tau_a \cos\alpha_p$$

Therefore $\qquad \sigma_1 = \sigma_x + \tau_a \cot\sigma_p \qquad\qquad\qquad [8.25]$

From eqns [8.24] and [8.25]:

$$\tan \alpha_p = \frac{\sigma_1 - \sigma_y}{\tau_a} = \frac{\tau_a}{\sigma_1 - \sigma_x}$$

Thus
$$(\sigma_1 - \sigma_y)(\sigma_1 - \sigma_x) = \tau_a^2 \qquad [8.26]$$

The roots of eqn [8.26] are:

$$\sigma_1 = \tfrac{1}{2}(\sigma_x + \sigma_y) + \tfrac{1}{2}\sqrt{\{(\sigma_x - \sigma_y)^2 + 4\,\tau_a^2\}} \qquad [8.27]$$

the **Major Principal Stress** or Maximum Direct Stress

and
$$\sigma_2 = \tfrac{1}{2}(\sigma_x + \sigma_y) - \tfrac{1}{2}\sqrt{\{(\sigma_x - \sigma_y)^2 + 4\,\tau_a^2\}} \qquad [8.28]$$

the **Minor Principal Stress** or Minimum Direct Stress

The major and minor principal stresses occur on mutually perpendicular planes, the value of α being found by substituting the appropriate values in either eqn [8.24] or eqn [8.25], or alternatively by equating these two equations thus:

$$\sigma_y + \tau_a \tan \alpha = \sigma_x + \tau_a \cot \alpha$$

Therefore
$$0 = (\sigma_x - \sigma_y) + \tau_a (\cot \alpha - \tan \alpha)$$

$$= -(\sigma_y - \sigma_x) + \tau_a \left(\frac{2}{\tan 2\alpha} \right)$$

Hence
$$\tan 2\alpha = \frac{2\tau_a}{\sigma_y - \sigma_x} \qquad [8.29]$$

EXAMPLE 8.4 At a point in a member subject to load the stress system induced is defined by the stresses acting on the elemental prism shown in Fig. 8.18. Determine the principal stresses and the orientation of the principal planes. Also determine the magnitude and direction of the maximum shearing stress.

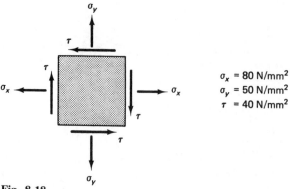

$\sigma_x = 80$ N/mm²
$\sigma_y = 50$ N/mm²
$\tau = 40$ N/mm²

Fig. 8.18

From eqn [8.27]:

the major principal stress, $\sigma_1 = \frac{1}{2}(80 + 50) + \frac{1}{2}\sqrt{\{(80 - 50)^2 + 4 \times 40^2\}}$
$= \frac{1}{2}(130 + \sqrt{\{(900 + 6\,400)\}})$
$= \frac{1}{2}(130 + 85 \cdot 4)$
$= 107 \cdot 7$ N/mm^2 (Answer)

From eqn [8.28]:

the minor principal stress, $\sigma_2 = \frac{1}{2}(130 - 85 \cdot 4)$
$= 22 \cdot 3$ N/mm^2 (Answer)

From eqn [8.29]:

$$\tan 2\alpha = \frac{2 \times 40}{80 - 50} = \frac{80}{30}$$

$$\therefore \quad 2\alpha = \arctan 2 \cdot 667$$
$$= 69 \cdot 4° \text{ or } 180 \cdot 0° + 69 \cdot 4°$$

Therefore the angle of inclination of the principal planes,

$$\alpha = 34 \cdot 7° \text{ or } 124 \cdot 7° \text{ (Answer)}$$

Thus the principal planes are at right-angles to each other and are orientated as shown in Fig. 8.19.

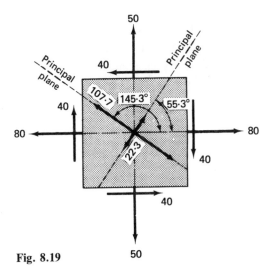

Fig. 8.19

For the maximum shearing stress use eqn [8.20]:

$$\tau_{max} = \frac{1}{2}(107 \cdot 7 - 22 \cdot 3) = 42 \cdot 7 \text{ N/mm}^2 \text{ (Answer)}$$

This maximum value will occur on planes inclined at 45° to the principal planes (or the direction of the principal stresses).

8.11 Mohr's stress circle

By obtaining a geometrical solution for the equations derived in Section 8.10 a reasonably simple graphical method may be devised for calculations involving principal stresses. Figure 8.20 shows a prismatic element subject to a general two-dimensional direct and tangential stress combination. The resulting major and minor principal stresses are σ_1 and σ_2 respectively, acting on principal planes inclined at angle α to the x- and y-axes respectively.

The graphical procedure to obtain the principal stresses is as follows — refer to Fig. 8.21:

1. Draw the perpendicular axes $x-x$ and $y-y$ and mark off the following intercepts on the x-axis, plotting tensile stresses to the right of y-axis and compressive stresses to the left.

$$OP = \sigma_y, \quad OR = \sigma_x$$

2. Erect the perpendiculars PQ and RS such that

$$PQ = RS = \tau \text{ (the applied tangential or shearing stress)}$$

3. Draw a circle of diameter SQ whose centre lies on the x-axis at B.

4. The circle construction now has the following properties:

$$
\begin{aligned}
\text{length } OC &= \sigma_1 && \text{(the major principal stress)} \\
\text{length } OA &= \sigma_2 && \text{(the minor principal stress)} \\
\text{angle } Q\hat{B}P &= 2\alpha && \text{(twice the principal plane angle)} \\
\text{angle } Q\hat{A}P &= \alpha && \text{(the principal plane angle)} \\
\text{length } BD &= \tau_{\max} && \text{(the maximum shearing stress)}
\end{aligned}
$$

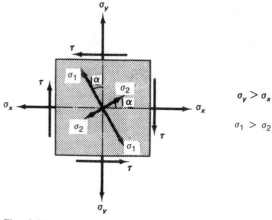

$$\sigma_y > \sigma_x$$

$$\sigma_1 > \sigma_2$$

Fig. 8.20

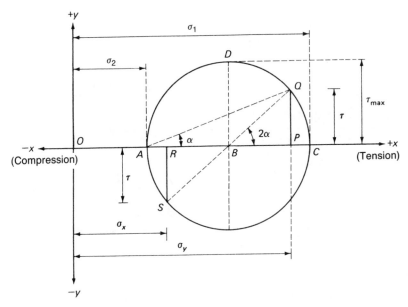

Fig. 8.21

The graphical procedure to obtain the normal and tangential (shearing) stresses on any plane inclined at angle α to a principal plane is similar. The intercepts equal to the principal stresses are marked off (OA and OC) and AC bisected to give the centre of the circle at B. The circle is now drawn and AQ added, inclined at angle α to the x-axis.

Then   length $PQ = \tau_\alpha$ (the tangential or shearing stress on the plane)

and   length $OP = \sigma_n$ (the normal stress on the plane)

To obtain the maximum shearing stress, put $\alpha = 45°$.

Then   angle $D\hat{A}B = 45°$ and length $BD = \tau_{max}$.

EXAMPLE 8.5   Obtain the solution to Example 8.4 using Mohr's stress circle.

Refer to Fig. 8.22:

1. To an appropriate scale, plot $OR = 50$ and $OP = 80$ on the right-hand side of the y-axis along the x-axis.
2. Erect the perpendiculars $PQ = 40$ and $RS = -40$ and then bisect RP in B.
3. Draw the circle about centre B and scale off the required values.

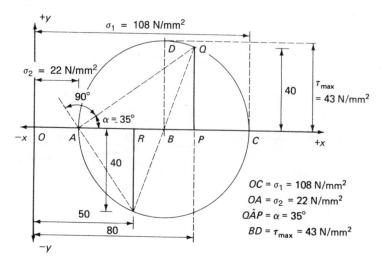

Fig. 8.22

EXAMPLE 8.6 At a point in a member subject to load the principal stresses are 60 N/mm² tensile and 30 N/mm² compressive. Determine the maximum shearing stress acting at the point and also the normal and tangential stresses acting on plane AB (Fig. 8.23).

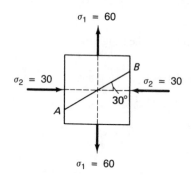

Fig. 8.23

Refer to Fig. 8.24:
1. To an appropriate scale, plot the principal stresses:

$$OA = -30 \text{ N/mm}^2 \text{ on the left-hand (compression) side}$$
$$OC = 60 \text{ N/mm}^2 \text{ on the right-hand (tension) side}$$

2. Bisect AC in B and draw the circle about centre B.

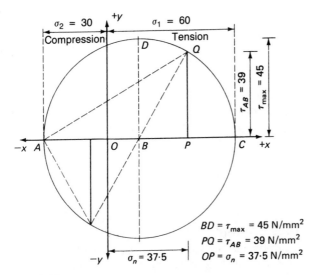

$\sigma_2 = 30$ Compression +y $\sigma_1 = 60$ Tension

$BD = \tau_{max} = 45 \text{ N/mm}^2$
$PQ = \tau_{AB} = 39 \text{ N/mm}^2$
$OP = \sigma_n = 37.5 \text{ N/mm}^2$

Fig. 8.24

3. Plot AQ, where $Q\hat{A}C = 30°$ and erect the perpendicular PQ.
4. Scale off the required values.

Proof for Mohr's stress circle construction

The validity of Mohr's stress circle construction may be demonstrated as follows.

Refer to Fig. 8.21:

(a)
$$\sigma_1 = OB + BC$$
$$= OR + \tfrac{1}{2}RP + BQ$$
$$= \sigma_x + \tfrac{1}{2}(\sigma_y - \sigma_x) + \sqrt{\{(BP)^2 + (PQ)^2\}}$$
$$= \tfrac{1}{2}(\sigma_y + \sigma_x) + \sqrt{\{\tfrac{1}{4}(\sigma_y - \sigma_x)^2 + \tau^2\}}$$
$$= \tfrac{1}{2}(\sigma_y + \sigma_x) + \tfrac{1}{2}\sqrt{\{(\sigma_y - \sigma_x)^2 + 4\tau^2\}}$$

which agrees with eqn [8.27].

(b)
$$\sigma_2 = OB - AB$$
$$= OB - BC$$
$$= \tfrac{1}{2}(\sigma_y + \sigma_x) - \tfrac{1}{2}\sqrt{\{(\sigma_y - \sigma_x)^2 + 4\tau^2\}}$$

which agrees with eqn [8.28].

(c)
$$\tan 2\alpha = \frac{PQ}{BP}$$

$$= \frac{\tau}{\frac{1}{2}(\sigma_y - \sigma_x)}$$

$$= \frac{2\tau}{\sigma_y - \sigma_x)}$$

which agrees with eqn [8.29].

(d) $\qquad \tau_{max} = BD$
$$= BC$$
$$= \tfrac{1}{2}(\sigma_y - \sigma_x)$$

which agrees with eqn [8.20].

8.12 Triaxial Stress Systems

In a triaxial system *three* principal stresses (σ_1, σ_2 and σ_3) will occur aligned in mutually perpendicular directions. For the conventional arrangement where $\sigma_1 > \sigma_2 > \sigma_3$, the Mohr circle diagram is shown in Fig. 8.25. The three circles represent two-dimensional stress states in planes parallel to each of the three axes:

> Circle 1: parallel to axis 1; has a diameter $= \sigma_2 - \sigma_3$
> Circle 2: parallel to axis 2; has a diameter $= \sigma_1 - \sigma_3$
> Circle 3: parallel to axis 3; has a diameter $= \sigma_1 - \sigma_2$

The normal and tangential stresses on a plane inclined to all three axes are given by point P. The maximum shearing stress is given by half the difference between the major and minor principal stresses:

$$\tau_{max} = \tfrac{1}{2}(\sigma_1 - \sigma_3)$$

Certain ostensibly triaxial problems reduce to a biaxial system, which is then easier to analyse. For example, the effect of loading soil masses beneath foundations and roads, and in embankments, produces compressive stress systems.

(a) Biaxially symmetrical system

For example, the stresses occurring at a point under a circular loaded area (Fig. 8.26(a)). The horizontal stresses and strains will be equal.

$$\sigma_1 > (\sigma_2 = \sigma_3) \quad \text{and} \quad \epsilon_2 = \epsilon_3$$

(b) Plane strain systems

For example, the stresses occurring at a point under a long strip footing (Fig. 8.26(b)). In this case, the strain in the direction along the length of the wall will be zero, $\epsilon_2 = 0$.

From eqns [8.7] and [8.9]:

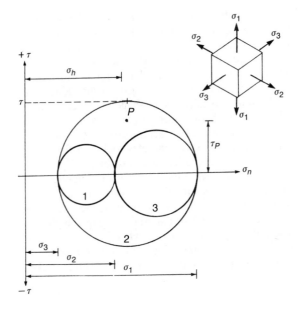

Fig. 8.25

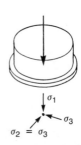 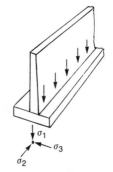

(a) Biaxially symmetrical system   (b) Plane strain system

Fig. 8.26

$$\epsilon_2 = 0 = \frac{\sigma_2 - \nu(\sigma_1 - \sigma_3)}{E}$$

Therefore $\qquad \sigma_2 = \nu(\sigma_1 - \sigma_3)$

If the soil is saturated and no volume change takes place, Poisson's ratio is $0 \cdot 5$ and therefore $\sigma_2 = \frac{1}{2}(\sigma_1 - \sigma_3)$

8.13 Failure of brittle materials in compression

When a piece or a component of brittle material fails under compressive loading it is natural to describe the occurrence as a *compressive failure*.

However, it should be realized that this description refers to the *applied* stress condition which led to the failure and not to the stress condition which caused the actual disruption of the fibres along the plane of failure. The mode of deformation in which the longitudinal dimension decreases and the lateral dimension increases produces a tendency for the material to flow; inwards from the top and bottom and outwards at the sides (Fig. 8.27). The specimen may become typically barrel-shaped, especially if the material is soft and ductile (low E value) or soft and plastic. This tendency to flow is resisted by the internal shearing resistance of the material. In brittle materials, such as brick, concrete, rock and firm-to-hard soils, two modes of brittle fracture failure are possible.

Tension crack failure. When a vertical compressive load is applied, both shortening and lateral expansion occur. Thus, a lateral tensile stress is induced. Vertical cracks appear when the brittle limit of the material is exceeded. This type of failure is common in materials having a high cohesive strength and where the aspect ratio h/b is very low.

Diagonal shear failure. In the case of materials with low-to-medium cohesive strength, e.g. soils, soft rocks, some concrete and some bricks,

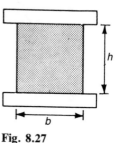
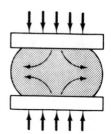

Fig. 8.27

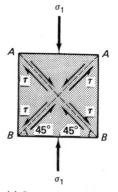

(a) Concrete cube under applied uniaxial compressive stress σ_l

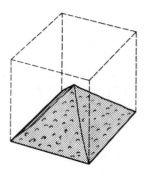

(b) Typical pyramidal shape after failure

Fig. 8.28

the induced maximum shear stress may exceed the shear strength of the material before the brittle tensile strength is reached. The maximum shear stress occurs on a plane inclined at an angle of 45° to the direction of applied stress (Section 8.7); in some highly frictional materials the shear plane angle may be greater than 45°. Observations made on specimens of concrete, brick, rock and soils which have been tested in uniaxial compression show this notion to be fundamentally true. The observed failure planes are not always at exactly 45° owing to a number of factors which are ignored in the theoretical analysis, such as the differences in grain size and density, the presence of air voids and the misalignment of the applied loads. Nevertheless, the student should find the examination of such broken specimens enlightening. Some typical examples are shown in Figs. 8.28 to 8.30.

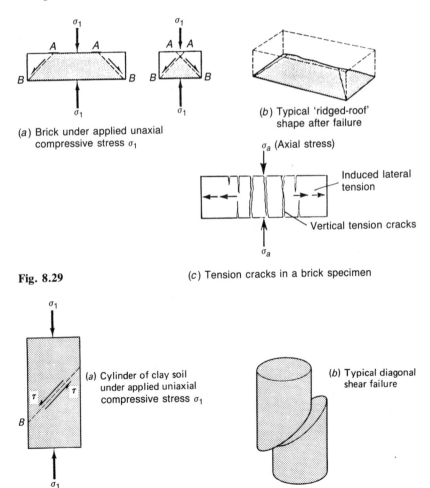

(a) Brick under applied unaxial compressive stress σ_1

(b) Typical 'ridged-roof' shape after failure

(c) Tension cracks in a brick specimen

Fig. 8.29

(a) Cylinder of clay soil under applied uniaxial compressive stress σ_1

(b) Typical diagonal shear failure

Fig. 8.30

292

8.14 Cracking in concrete beams and the principle of shear reinforcement

Figure 8.31 shows a portion of a reinforced concrete beam under load. The small shaded element will be subject to vertical and horizontal (complementary) shearing stresses as shown in Fig. 8.32; the horizontal bending stress (tensile) is assumed to be taken entirely by the reinforcing steel. If the shearing stresses are resolved at the top-left and bottom-right corners of the element, a diagonal direct tensile stress results. This diagonal tension will tend to produce cracks inclined at 45° to the neutral axis of the beam. Since the concrete has virtually no tensile strength the diagonal tension has to be resisted by the provision of shear reinforcement, either in the form of bars bent up at 45° (Fig. 8.33(a)) or in the form of hoops of steel called stirrups (Fig. 8.33(b)).

LOAD

Neutral axis

Reinforcing steel

Fig. 8.31

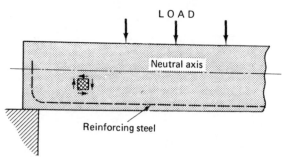

Fig. 8.32

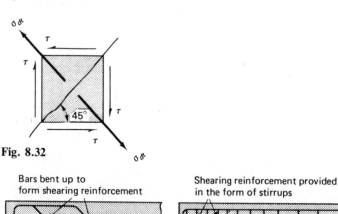

Bars bent up to form shearing reinforcement

Tension steel

45° 45°

(a)

Shearing reinforcement provided in the form of stirrups

(b)

Fig. 8.33

EXERCISES CHAPTER 8

1. A steel bar of diameter 50 mm is subject to an axial tension of 204 kN. Determine the decrease in the diameter of the bar.
Young's modulus = 210 kN/mm²; Poisson's ratio = 0·3.

2. In a tensile test on a steel rod the following data was recorded:

Applied load   = 16·8 kN
Original diameter = 12·00 mm   Reduction in diameter   = 0·00246 mm
Gauge length   = 200·0 mm   Reduction in gauge length = 0·141 mm

Determine Poisson's ratio and Young's modulus for the material.

3. Prove that if Poisson's ratio for a substance is 0·5 there will be no change in volume when a specimen of the substance is stretched or compressed without lateral restraint.
  Determine the change in volume that takes place in a concrete cylinder of diameter 250 mm and length 1 m when it is subject to a longitudinal axial load of 200 kN.
Young's modulus = 14 kN/mm²; Poisson's ratio = 0·3.

4. A steel tube of external diameter 25 mm and internal diameter 15 mm is subject to a longitudinal tension load. Determine the change in internal volume per metre length that will take place when the axial load is increased by 20 kN.
Young's modulus = 210 kN/mm²; Poisson's ratio = 0·3.

5. Fig. 8.34 shows a steel plate of thickness 12 mm which is subject to tensile loads along its edges. Determine the changes that take place in the length and breadth of the plate from the unstressed condition.
Young's modulus = 210 kN/mm²; Poisson's ratio = 0·3.

6. Find the change in volume of a sheet of steel 2 m × 2 m × 25 mm thick when tensile stresses of 90 N/mm² and 60 N/mm² are applied at right-angles to each other in the plane of the sheet and perpendicular to its edges.
Young's modulus = 200 kN/mm²; Poisson's ratio = 0·28.

7. The principal stresses at a point in a piece of material are 40 N/mm² tension and 15 N/mm² compression. Determine the normal and shearing stresses acting

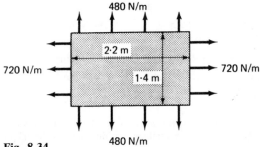

480 N/m

2·2 m

720 N/m   720 N/m

1·4 m

Fig. 8.34   480 N/m

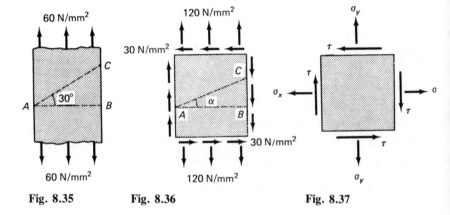

Fig. 8.35   **Fig. 8.36**   **Fig. 8.37**

on a plane inclined at 30° to the direction of the major principal stress. Also determine the magnitude and direction of the maximum shearing stress.

8. At a certain point in a piece of material the principal stresses are 60 N/mm² tension and 30 N/mm² compression. Determine the normal and shearing stresses acting on a plane inclined at 40° to the direction of the major principal stress. Determine also the magnitude and direction of the maximum shearing stress.

9. Fig. 8.35 shows part of a steel rod of uniform section subject to a longitudinal tensile stress of 60 N/mm². Determine the normal and tangential stresses acting on the plane AC.

10. Fig. 8.36 shows a square element of material which is subject to a uniaxial tension and a shearing stress simultaneously applied. Determine the normal and shearing stresses acting on the plane AC when:

$$\text{(a) } \alpha = 30°, \text{ (b) } \alpha = 45°, \text{ (c) } \alpha = 60°$$

11. Fig. 8.37 shows a square element of material subject to biaxial direct stresses and shearing stresses. For each of the cases given below determine the direction of the principal planes and the magnitude of the principal stresses, together with the magnitude and direction of the maximum shearing stress.

(a) $\sigma_x = 60 \text{ N/mm}^2,$ $\sigma_y = 80 \text{ N/mm}^2,$ $\tau = 30 \text{ N/mm}^2$
(b) $\sigma_x = -50 \text{ N/mm}^2,$ $\sigma_y = 70 \text{ N/mm}^2,$ $\tau = 40 \text{ N/mm}^2$
(c) $\sigma_x = 100 \text{ N/mm}^2,$ $\sigma_y = -60 \text{ N/mm}^2,$ $\tau = 40 \text{ N/mm}^2$
(d) $\sigma_x = -80 \text{ N/mm}^2,$ $\sigma_y = -55 \text{ N/mm}^2,$ $\tau = 30 \text{ N/mm}^2$

(negative sign indicates compression)

12. At a point in a beam there is a longitudinal tensile stress of 85 N/mm² and a shearing stress of 45 N/mm². Determine the magnitude and direction of the major principal stress acting at this point.

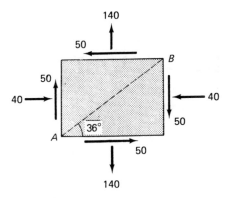

Fig. 8.38

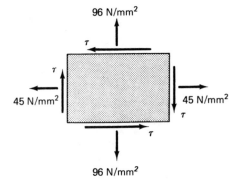

Fig. 8.39

13. At a point in a transmission shaft there is a shearing stress of 60 N/mm² due to the driving torque and a longitudinal compressive stress of 20 N/mm² due to an end thrust. Determine the magnitude of the principal stresses and the direction of the principal planes.

14. Fig. 8.38 shows a rectangular piece of material subject to the applied stresses indicated in N/mm². Determine the following:

(a) the magnitude and direction of the principal stresses,
(b) the magnitude and direction of the maximum shearing stress,
(c) the normal and tangential stresses acting along the plane of the diagonal *AB*.

15. Fig. 8.39 shows a rectangular element subject to biaxial direct stresses and shearing stresses. Determine the maximum allowable value for the shearing stress (τ):

(a) when the maximum permissible tensile stress for the material is 120 N/mm²,
(b) when the maximum permissible shearing stress for the material is 40 N/mm²

9 Torsion in circular shafts

Symbols

A	=	area of cross-section
D, d	=	diameter
G	=	modulus of rigidity or shear modulus
J	=	polar second moment of area
k	=	torsional stiffness
N	=	number of revolutions per minute
R, r	=	radius
T	=	torque or twisting moment
y	=	radius of elemental ring
θ	=	angle of twist over a given length
τ	=	shearing stress
ϕ	=	longitudinal strain

9.1 Introduction

In engineering practice torsion may occur in many situations, such as in shafts connected to motors and pulleys and in other rotating mechanisms: aircraft wings, fuselages and control systems are also subject to complex torques. In building structures torsion may be induced in beams and columns as a result of eccentric transverse loading. It will be sufficient in this book to examine only the effects of torsion on circular solid and hollow shafts. The reader will be acquainted with the basic principles and should subsequently be able to refer, with some confidence, to more advanced texts if the need arises.

9.2 Solid circular shafts

Consider the solid circular bar of metal shown in Fig. 9.1(a) which is rigidly held at one end and upon which is painted a longitudinal stripe. When a couple is applied to the free end the bar will twist, as indicated by the distortion of the stripe in Fig. 9.1(b). If a transverse cut were now to be made through the bar at point X the effect of the couple would be to cause the unimpeded rotation of the free portion. A typical view of the bar at this stage is shown in Fig. 9.1(c).

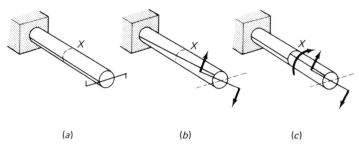

(a)   (b)   (c)

Fig. 9.1

In order that the (uncut) bar may resist the tendency to rotate a shearing stress must be developed between the two portions. The bar may be considered as consisting of a large number of thin discs placed end to end. In order that the applied torque may be transmitted along the length of the bar to the support a shearing action must be set up between adjacent discs. The shearing stress is therefore developed on all transverse sections of bar between the point of application of the couple and the support.

When a circular-section bar or shaft is subject only to a couple centred on its longitudinal axis PURE TORSION is induced. This condition will now be studied in the case of the solid circular shaft shown in Fig. 9.2. Let the twisting moment or torque of the couple $= T$.

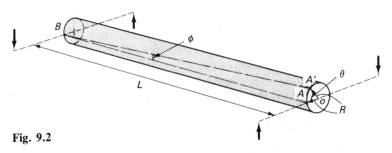

Fig. 9.2

The diagram shows that, under the action of this torque, the shaft is twisted so that point A moves to A'. (AB being initially straight and parallel to the axis of the shaft).

Then the relative angle of twist at the ends $= \theta$ radians
and the longitudinal strain over length $L \quad = \phi$.
Since the angles θ and ϕ will be small: arc $AA' = L\phi = R\theta$.

Now from the work done in Chapter 8, it will be remembered that the shear strain is proportional to the shearing stress (up to the limit of proportionality)

i.e. $\qquad\qquad \tau = \phi G \qquad$ where $G =$ modulus of rigidity
$\qquad\qquad\qquad\qquad\qquad\qquad$ (shear modulus)

or $\qquad\qquad \phi = \dfrac{\tau}{G}$

Then $\qquad\qquad \phi = \dfrac{\tau}{G} = \dfrac{R\theta}{L}$

or, rewriting, $\dfrac{\tau}{G} = \dfrac{R\theta}{L}$ $\qquad\qquad\qquad\qquad\qquad\qquad$ [9.1]

EXAMPLE 9.1 *When a solid circular shaft of diameter 200 mm is subjected to a pure torque the relative rotation between cross-sections 2 m apart is 0·6 degrees. Calculate the maximum shearing stress induced in the shaft. Shear moudlus = 82 kN/mm².*

Rewriting eqn [9.1]:

$$\tau = \frac{G\theta R}{L}$$

$$\text{Angle of twist } \theta = 0\cdot6° \times \frac{2\pi}{360} = \frac{\pi}{300} \text{ radians}$$

The maximum shearing stress occurs at the perimeter, i.e. when $R = 100$ mm

Maximum shearing stress: $\qquad \tau = 82 \times 1\,000 \times \dfrac{\pi}{300} \times \dfrac{100}{2 \times 1\,000}$

$$= \frac{82\pi}{6}$$

$$= 42\cdot9 \text{ N/mm}^2 \text{ (Answer)}$$

Now consider the distribution of shearing stress over transverse sections of the shaft. A cross-section of the shaft is shown in Fig. 9.3. Consider an elemental annulus or ring of radius y:

The area of this ring may be given as: $\qquad\qquad\qquad\qquad 2\pi y\, dy$

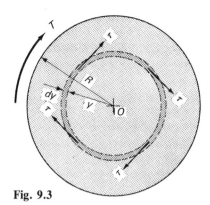

Fig. 9.3

The total shearing force acting on the ring is therefore: $\tau 2\pi y \, dy$

and the moment of this force about the longitudinal axis is: $\tau 2\pi y^2 \, dy$

However from eqn [9.1]: $\tau = \dfrac{G\theta y}{L}$

Therefore the moment of the shearing force on the ring is: $\dfrac{G\theta}{L} 2\pi y^3 \, dy$

If the shaft is considered to be made up of an infinitely large number of elemental rings of radii varying from O to R, the *total torque* on the shaft is given by the sum of the moments of shearing force on all of these rings:

$$T = \int_O^R \left(\frac{G\theta}{L} 2\pi y^3 \, dy \right)$$

or taking out the constant factors:

$$T = \frac{G\theta}{L} 2\pi \int_O^R y^3 \, dy$$

Now from Chapter 4 it will be seen that $2\pi \int_O^R y^3 \, dy$ is the *polar second moment of area* of the cross-section.

Putting $2\pi \int_O^R y^3 \, dy = J$

$$T = \frac{G\theta}{L} J$$

or, rewriting, $\dfrac{T}{J} = \dfrac{G\theta}{L}$ [9.2]

The combination of eqns [9.1] and [9.2] produces what is termed the **GENERAL TORSION EXPRESSION**

$$\frac{T}{J} = \frac{\tau}{R} = \frac{G\theta}{L}$$ [9.3]

which provides a summary of the fundamental relationships involved in the condition of PURE TORSION.

A useful comparison may be made with the GENERAL BENDING EXPRESSION derived in Chapter 6, i.e.

$$\frac{M}{I} = \frac{\sigma}{y} = \frac{E}{R}$$

By rearranging equation [9.3]

$$\tau = \frac{TR}{J} = \frac{G\theta R}{L} \qquad [9.4]$$

It can be seen that the shearing stress is proportional to the radius. Hence at the centre of the shaft, i.e. $R = 0$, $\tau = 0$

and at an intermediate radius, r, i.e. $R = r$, $\tau = \dfrac{Tr}{J}$

Therefore the *maximum shearing stress* occurs at the perimeter of the shaft.

Also from eqn [9.3] it can be seen that the maximum shearing stress is proportional to the applied torque, i.e.

$$\frac{T}{\tau} = \frac{J}{R} = \text{constant for a given shaft} \qquad [9.5]$$

EXAMPLE 9.2   Calculate the maximum torque that may be transmitted by a solid circular steel shaft of diameter 120 mm and length 2·2 m if the shearing stress is not to exceed 75 N/mm². Shear modulus = 82 kN/mm².

First determine the polar second moment of area of the section.

$$J = \frac{\pi D^4}{32} = \frac{\pi \times 120^4}{32} = 20 \cdot 36 \times 10^6 \text{ mm}^4$$

Now from eqn [9.3]:
$$\frac{T}{J} = \frac{\tau}{R}$$

Rearranging:
$$= \frac{J\tau}{R}$$

$$= \frac{20 \cdot 4 \times 10^6 \times 75}{60}$$

$$= 25 \cdot 4 \times 10^6 \text{ Nmm}$$

$$= 25 \cdot 4 \text{ kNm (Answer)}$$

9.3   Hollow circular shafts

In the case of a hollow circular shaft the limiting values of the radius will be R and r, *instead of R and O* (Fig. 9.4). The minimum shearing stress

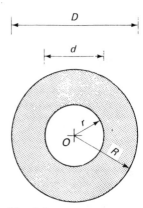

Fig. 9.4

will not therefore be zero as it is in a solid shaft, but will correspond to the stress acting at radius r. The expression for the polar second moment of area also differs from that for a solid shaft:

for a solid shaft $\qquad J = \dfrac{\pi D^4}{32}$ [9.6]

for a hollow shaft $\qquad J = \dfrac{\pi}{32} (D^4 - d^4)$ [9.7]

Providing the appropriate value for J is used, however, eqn [9.3] applies equally to both solid and hollow shafts of circular cross-section.

9.4 Torsional stiffness

As in the case of beams subject to bending (Ch. 6) the value of the applied moment, in this case the torque, required to cause unit displacement gives a useful indication of the flexibility of the shaft. The torque required to cause unit rotational displacement is termed the **Torsional stiffness (k)**

$$k = \frac{T}{\theta} = \frac{GJ}{L} \text{ in units of torque per radian of twist} \qquad [9.8]$$

EXAMPLE 9.3 A hollow circular shaft is required to transmit a torque of 6 000 Nm and is 4 m long. If the maximum permissible angle of twist is 2 degrees over the whole length, determine the diameters required when in a ratio of 2:1. What will be the maximum and minimum shearing stresses in the shaft at the time when this torque is developed? G = 70 kN/mm².

From eqn [9.3]: $\qquad\qquad J = \dfrac{TL}{C\theta}$ (1)

Now for a hollow shaft: $\qquad J = \dfrac{\pi}{32}(D^4 - d^4)$

but $\qquad\qquad\qquad\qquad D = 2d$, then $J = -[(2d)^4 - d^4]$

$$= \dfrac{\pi}{32}(15d^4)$$

$$= \dfrac{15\pi d^4}{32}$$

The limiting angle of twist over 4 m $= 2 \times \dfrac{2\pi}{360} = \dfrac{\pi}{90}$ radians

Substituting in (1) above: $\qquad \dfrac{15\pi d^4}{32} = \dfrac{6\,000 \times 1\,000 \times 4 \times 1\,000 \times 90}{70 \times 1\,000 \times \pi}$

Therefore: $\qquad\qquad\qquad d^4 = \dfrac{24\,000 \times 1\,000 \times 90 \times 32}{70 \times \pi^2 \times 15}$

$$= 6 \cdot 67 \times 10^6 \text{ mm}^4$$

Hence $\qquad\qquad\qquad\qquad d = 50 \cdot 8 \text{ mm}$

and $\qquad\qquad\qquad\qquad D = 101 \cdot 6 \text{ mm} \quad \text{(Answer)}$

EXAMPLE 9.4 A solid circular shaft of diameter 225 mm is to be replaced by a hollow shaft in which the ratio of diameters is 1·5:1. Determine the size of the hollow shaft required to transmit the same torque as the solid shaft at the same maximum shearing stress. What percentage economy in mass of material used will this change bring about?

From eqn [9.5]:

$$\frac{T}{\tau} = \frac{J}{R} = \text{constant}$$

The condition specified in the question is that $\dfrac{T}{\tau} = \text{constant}$

Therefore

$$\left(\frac{J}{R}\right)_{\text{solid}} = \left(\frac{J}{R}\right)_{\text{hollow}} = \text{constant}$$

Substituting

$$\frac{225^4 \times \pi \times 2}{32 \times 225} = \frac{\pi}{32}[(1 \cdot 5d)^4 - d^4] \times \frac{2}{1 \cdot 5d}$$

$$225^3 = \frac{\left(\dfrac{3d}{2}\right)^4 - d^4}{\dfrac{3d}{2}}$$

$$= \frac{65d^3}{24}$$

Thus $\qquad d^3 = \dfrac{225^3 \times 24}{65}$

$$= 0 \cdot 420 \times 10^6$$

Then the required inside diameter $d = 162$ mm
and the required outside diameter $D = 162 \times 1 \cdot 5$ (Answer)

$$= 243 \text{ mm}$$

The mass of material contained in each shaft is proportional to its cross-sectional area.

$$A_{\text{solid}} = \frac{\pi}{4} \times 225^2 \qquad = \frac{\pi}{4} \times 50\ 625$$

$$A_{\text{hollow}} = \frac{\pi}{4} (243^2 - 162^2) \qquad = \frac{\pi}{4} \times 32\ 805$$

Then the percentage saving in mass of material

$$= \frac{50\ 625 - 32\ 805}{50\ 625} \times 100$$

$$= 35\% \text{ (Answer)}$$

9.5 Work done and power transmitted

When a force moves in a straight line with constant velocity the *work done* is given by the product of the magnitude of the force and the distance through which it has moved. The *power* transmitted by this action is defined as the *rate* at which this work is done, i.e. the work done in unit time.

$$\text{Work done} = \text{force} \times \text{distance}$$

$$\text{Power} = \frac{\text{work done}}{\text{time}}$$

The 'distance' travelled by a rotating body is measured by the number of radians through which it rotates. The work done by a torque acting on a shaft is therefore given by the product of the magnitude of the torque and the amount of rotation in radians.

For one revolution of the shaft: the work done $= T \times 2\pi$
(since the shaft turns through 2π radians in one revolution)
If the shaft is rotating at N revolutions per minute, then
work done $= T \times 2\pi N$ (units of work per minute)

Usually the torque will be measured in Newton metres (Nm) and therefore the units of work will also be Nm. However, it is more usual to give work in joules (J) which are equal numerically to Newton metres.

$$1 \text{ joule} = 1 \text{ Newton metre}$$

Power is measured in watts (W),

$$1 \text{ W} = 1 \text{ J/s}$$
$$= 1 \text{ Nm/s}$$
$$= \frac{1}{60} \text{ Nm/min}$$

Hence, the power transmitted by a shaft rotating at N revolutions per minute and subject to torque of T Nm will be given by:

$$\text{Power} = \frac{T2\pi N}{60} \text{ Watts} \qquad [9.9]$$

EXAMPLE 9.5 *The specification for a solid steel shaft states that the maximum shearing stress must not exceed 40 N/mm^2 and that the angle of twist must not exceed one degree in twenty diameters of length along the shaft. Calculate the diameter of shaft required to transmit 750 kW when rotating at 250 rpm. G = 82 kN/mm^2.*

First consider the torque required to deliver the specified power, 750 kW at 250 rpm.

From eqn [9.9]:
$$T = \frac{\text{power} \times 60}{2\pi N}$$
$$= \frac{750 \times 1\,000 \times 60}{2\pi \times 250}$$
$$= 28\,650 \text{ Nm}$$
$$= 28 \cdot 65 \times 10^6 \text{ Nmm}$$

There are two design criteria: $\hat{\tau} \not> 40 \text{ N/mm}^2$

and $\theta \not> 1 \text{ deg/20 diameters}$

If the maximum shearing stress is the more critical:

$$\frac{T}{J} = \frac{40}{R} \qquad \text{(from eqn [9.3])}$$

Then
$$\frac{J}{R} = \frac{T}{40}$$

Therefore
$$\frac{\pi D^4 \times 2}{32 \times D} = \frac{28 \cdot 65 \times 10^6}{40}$$

Giving
$$D^3 = \frac{28 \cdot 65 \times 16 \times 10^6}{40\pi}$$
$$= 3 \cdot 64 \times 10^6 \text{ mm}^3$$

Hence $D = 154$ mm (Answer)

If the maximum angle of twist is the more critical:

$$\frac{T}{J} = \frac{G\pi}{20D \times 180}$$ (from eqn [9.3])

since $\theta = 1° = \dfrac{\pi}{180}$ radians and $L = 20D$

Then $$\frac{J}{D} = \frac{T \times 20 \times 180}{G \times \pi}$$

Therefore $$\frac{\pi D^3}{32} = \frac{28 \cdot 65 \times 10^6 \times 3\,600}{82 \times 1\,000 \times \pi}$$

Giving $$D^3 = \frac{28 \cdot 64 \times 3 \cdot 6 \times 32 \times 10^6}{82 \times \pi^2}$$

$$= 4 \cdot 078 \times 10^6 \text{ mm}^3$$

Hence $D = 160$ mm

Therefore the maximum angle of twist is the more critical in this case and so the shaft must have a diameter of at least 160 mm.

EXAMPLE 9.6 A solid circular shaft is to transmit 1·15 MW when rotating at 240 rpm. The shaft will be in two sections, connected together with a flange coupling which has six equally spaced bolts on a pitch circle having a diameter of 1·5 times that of the shaft. The allowable shearing stresses are 75 N/mm² for the shaft and 100 N/mm² for the bolts. Determine (a) the diameter of shaft required, and (b) the diameter of bolts required.

First, determine the torque transmitted at full power.

From eqn [9.9]: $$T = \frac{\text{power} \times 60}{2\pi N}$$

$$= \frac{1 \cdot 15 \times 10^6 \times 60}{2\pi \times 240}$$

$$= 45\,757 \text{ Nm}$$

To determine the diameter of shaft required to transmit this torque rearrange eqn [9.3]:

$$J = \frac{TR}{\tau}$$

$$= \frac{45\,757 \times 10^3 \times R}{75}$$

$$= 610\,100\ R \text{ mm}^4$$

But $$J = \frac{\pi D^4}{32}$$

Therefore

$$\frac{\pi D^4}{32} = 610\ 100 \times \frac{D}{2}$$

Giving

$$D^3 = \frac{610\ 100 \times 32}{2\pi}$$

$$= 3 \cdot 107 \times 10^6 \text{ mm}^3$$

Hence

$$D = 146 \text{ mm (Answer)}$$

The coupling bolts will therefore lie on a pitch circle of diameter

$146 \times 1 \cdot 5 = 219$ mm   (say 220 mm)   (see Fig. 9.5)

The whole of the torque is transmitted through the coupling via the six bolts, so that the shear moment for each bolt will amount to one-sixth of the total torque. Let the shearing force developed in each bolt $= P$ newtons

Then the total torque transmitted $= 6 \times P \times 110$ Nmm

$$= 0 \cdot 66 \ P \text{ Nm}$$

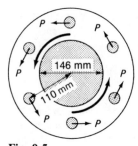

Fig. 9.5

Hence

$$P = \frac{45\ 757}{0 \cdot 66}$$

$$= 69\ 329 \text{ N}$$

Since the allowable shearing stress in the bolts is 100 N/mm²:

$$\text{area of bolt required} = \frac{69\ 329}{100}$$

$$= 693 \text{ mm}^2$$

$$\text{Hence diameter of bolt required} = \sqrt{\frac{693 \times 4}{\pi}}$$

$$= 30 \text{ mm (Answer)}$$

EXERCISES   CHAPTER 9

1. Calculate the torque exerted by a shaft of diameter 50 mm when the maximum shearing stress is 20 N/mm².

2. What torque, when applied to a solid circular shaft of diameter 250 mm, will produce a maximum shearing stress of 75 N/mm^2?

3. A circular shaft of length 2 m and diameter 75 mm is subject to a torque which causes it to twist through an angle of one degree. Calculate the maximum shearing stress induced in the shaft and also the magnitude of the torque. $G = 82$ kN/mm^2.

4. Determine the maximum shearing stress in a solid steel shaft of diameter 250 mm if the angle of twist is one degree over a length of 1·8 m and the modulus of rigidity is 80 kN/mm^2.

5. A solid steel shaft is subject to a torque of 730 Nm and the allowable shearing stress for the material is 70 N/mm^2. Determine the diameter required for the shaft.

6. A hollow shaft is to have an inside/outside diameter ratio of 0·6 and will be required to transmit a torque of 185 kNm. Calculate the diameter sizes required for the shaft if the allowable shearing stress is 70 N/mm^2.

7. A shaft of diameter 75 mm is subject to a torque of 4 kNm. Calculate the maximum shearing stress induced in the shaft when (a) it is solid, and (b) it is hollow with an internal diameter of 50 mm.

8. Calculate the shearing stress at the outer and inner surfaces of a hollow shaft in which the external diameter is 150 mm, the internal diameter is 100 mm and the applied torque is 30 kNm.

9. Calculate the diameter required for a solid shaft which is to transmit a torque of 15 kNm at a maximum shearing stress of 80 N/mm^2.

10. Calculate the torque that may be transmitted by a solid circular shaft of diameter 120 mm at a maximum shearing stress of 85 N/mm^2.

11. Design a hollow shaft in which the ratio of inside/outside diameters is 0·7 and the torque to be transmitted is 50 kNm. The allowable shearing stress is 80 N/mm^2.

12. Calculate the maximum permissible angle of twist that may be applied to a solid circular shaft of diameter 25 mm and length 1 m if the allowable shearing stress is 70 N/mm^2 and the shear modulus is 80 kN/mm^2.

13. Calculate the angle of twist in a solid circular shaft of diameter 80 mm and length 5 m when the torque transmitted is 5 kNm at a maximum shearing stress of 80 N/mm^3. $G = 82$ kN/mm^2.

14. Calculate the angle of twist in a propeller shaft subject to a torque of 160 kNm and having the following dimensions:

$$\text{outside diameter} = 600 \text{ mm} \quad \text{length} = 30 \text{ m}$$
$$\text{inside diameter} = 400 \text{ mm} \quad G = 82 \text{ kN/mm}^2$$

15. In a torsion test a brass rod of diameter 25 mm was subject to a torque of 220 Nm when the angle of twist was observed to be 5° over a length of 1 m. Calculate the modulus of rigidity for the material. If the shaft is 1·8 m long, determine the angle of twist induced by a torque which produces a maximum shearing stress of 120 N/mm².

16. Calculate the percentage economy in the mass of material that results in replacing a solid shaft with a hollow shaft in which the internal diameter is 2/3 of the external diameter and which will transmit an equal torque at the same maximum shearing stress.

17. Calculate the maximum shearing stress in a solid circular shaft of diameter 75 mm when it is transmitting 20 kW at 30 rpm.

18. Design a solid circular shaft required to transmit 50 kW at 120 rpm, if the permissible shearing stress for the material is 60 N/mm².

19. The engine of a ship delivers 4·5 MW through a propeller shaft rotating at 350 rpm. Design a hollow shaft in which the ratio of diameters is 0·6 and the maximum shearing stress is not to exceed 80 N/mm². If the shaft is 15 m long, calculate the angle of twist from end to end, when full power is being transmitted. $G = 80$ kN/mm².

20. What maximum power may be transmitted through a solid circular shaft of 50 mm diameter rotating at 800 rpm if the permissible shearing stress is 85 N/mm²?

21. A hollow steel shaft of external diameter 200 mm and internal diameter 120 mm transmits 800 kW when rotating at 150 rpm. Calculate the shearing stress induced at the outer and inner surfaces of the shaft.

22. A hollow shaft of diameter ratio 0·6 is required to transmit 600 kW when rotating at 120 rpm. The maximum shearing stress must not exceed 60 N/mm² and the angle of twist must not exceed 1° in every 3 m. Determine the dimensions of the lightest shaft that will meet these requirements. $G = 75$ kN/mm².

23. Two sections of a shaft are connected together by a flange coupling containing six bolts on a 220 mm pitch circle. The shaft transmits a torque of 45 kNm and the allowable shearing stress in the bolts is 115 N/mm³. Calculate the diameter of bolts required.

24. Two portions of a solid circular shaft of 150 mm diameter are coupled together with 8 bolts equally spaced on a 200 mm pitch circle. Determine the diameter of bolt required if the permissible shearing stress for the shaft is 70 N/mm² and that for the bolts is 85 N/mm².

10 Combined direct and bending stress

Symbols

A	=	area of cross-section
b	=	breadth of section
D	=	diameter
d	=	depth of section
E	=	Young's modulus of elasticity
e_x	=	eccentricity about $x-x$ axis
e_y	=	eccentricity about $y-y$ axis
e_0	=	eccentricity that produces zero stress at an extreme fibre
I	=	second moment of area (about a given axis)
L	=	length
L_c	=	critical buckling length
l	=	effective length of a strut
M	=	bending moment, moment of resistance
P	=	load, axial load
P_E	=	Euler buckling load
p_c	=	Compressive strength, critical strength
p_E	=	Euler buckling strength
p_y	=	yield strength
r	=	radius of gyration
y	=	distance from the neutral axis to any fibre
y_0	=	distance from the centroid to the neutral axis
$\bar{y}$	=	distance to the centroid

$$\alpha = \sqrt{\frac{P}{EI}}$$

σ = stress
σ_d = direct stress
σ_b = longitudinal bending stress
σ_c = buckling stress
λ = slenderness ratio

10.1 Bending in struts and ties

A number of situations arise in structural systems where the members become subject to a combination of direct (axial) and bending stresses acting parallel to the longitudinal axis. Such members may, for convenience, be considered as either struts or ties resisting direct stress primarily but which, in addition, must also resist longitudinal stresses resulting from bending moments. These bending moments may be induced in a number of different ways:

(a) due to an externally applied moment,
(b) due to the longitudinal loading being applied eccentrically with reference to the centroidal axis of the section,
(c) due to the member buckling.

10.2 Direct axial load and applied moment

Consider a member of depth d which is subject to an axial compressive load P together with a bending moment M externally applied about the $x-x$ axis (Fig. 10.1).

Then at distance y from the neutral axis the following stresses are induced:

due to the direct load P: $\sigma_d = \dfrac{P}{A}$ stress diagram (a)

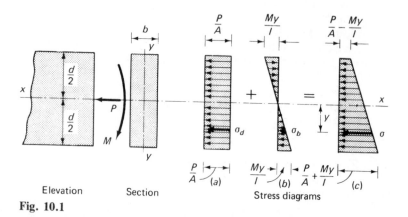

Elevation   Section   Stress diagrams

Fig. 10.1

due to the bending moment M: $\sigma_b = \dfrac{My}{I_{xx}}$ stress diagram (b)

For the combined application of the load and the moment together, the total stress at distance y from the neutral axis is given by the sum of these two stresses (i.e. applying the principle of superposition).

$$\sigma = \sigma_d \pm \sigma_b$$

$$= \dfrac{P}{A} \pm \dfrac{My}{I_{xx}} \quad \text{stress diagram } (c) \qquad [10.1]$$

Note that the bending stress will be additive on one side of the neutral axis and subtractive on the other. For the particular case of the rectangular section shown, the extreme fibre stress will be:

$$\sigma_{max} = \dfrac{P}{A} + \dfrac{Md}{2I_{xx}}$$

$$= \dfrac{P}{bd} + \dfrac{6M}{bd^2} \qquad [10.2]$$

$$\sigma_{min} = \dfrac{P}{A} - \dfrac{Md}{2I_{xx}} \qquad [10.3]$$

$$= \dfrac{P}{bd} - \dfrac{6M}{bd^2}$$

Since $I_{xx} = \dfrac{bd^3}{12}$

10.3 Position of neutral axis for combined direct and bending stress

From eqn [10.3] it will be seen that σ_{max} could be tensile if $Md/2I_{xx} > P/A$ or compressive if $Md/2I_{xx} < P/A$. The alternative stress distributions are shown in Fig. 10.2. The neutral axis, i.e. where the strain is zero, occurs where the combined stress is zero.

$$\sigma = 0 = \dfrac{P}{A} - \dfrac{My_0}{I_{xx}}$$

Therefore $y_0 = \dfrac{PI_{xx}}{AM}$ [10.4]

10.4 Eccentrically applied longitudinal load

Consider now the case of a member subject to a longitudinal load P which is applied on the $y-y$ axis, but which is displaced from the $x-x$ axis by an eccentricity of e_x (Fig. 10.3(a)).

By resolving parallel to the longitudinal axis and by taking moments about the $x-x$ axis it can be shown that the actual condition shown in

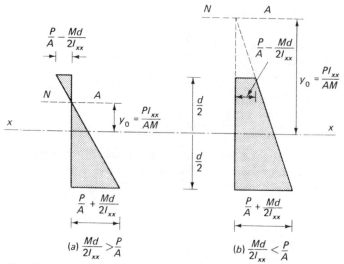

$\dfrac{P}{A} - \dfrac{Md}{2I_{xx}}$

$y_0 = \dfrac{PI_{xx}}{AM}$

$\dfrac{P}{A} - \dfrac{Md}{2I_{xx}}$

$y_0 = \dfrac{PI_{xx}}{AM}$

$\dfrac{d}{2}$

$\dfrac{d}{2}$

$\dfrac{P}{A} + \dfrac{Md}{2I_{xx}}$

$\dfrac{P}{A} + \dfrac{Md}{2I_{xx}}$

(a) $\dfrac{Md}{2I_{xx}} > \dfrac{P}{A}$

(b) $\dfrac{Md}{2I_{xx}} < \dfrac{P}{A}$

Fig. 10.2

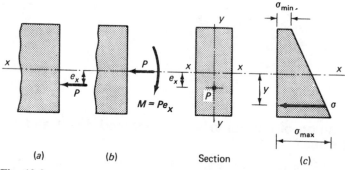

σ_{min}

$M = Pe_x$

σ_{max}

(a)   (b)   Section   (c)

Fig. 10.3

diagram (a) is equivalent to the combination shown in diagram (b) which consists of a thrust of the same magnitude, P, but axially applied, together with a moment M, where $M = Pe_x$.

The stress distribution resulting from the combined application of a direct axial load and a bending moment was considered in Section 10.2, from which:

$$\sigma = \frac{P}{A} \pm \frac{My}{I_{xx}}$$

$$= \frac{P}{A} \pm \frac{Pe_x y}{I_{xx}} \qquad [10.5]$$

Then $\qquad \sigma_{max} = \dfrac{P}{A} + \dfrac{Pe_x y}{I_{xx}} \qquad [10.6]$

and
$$\sigma_{\min} = \frac{P}{A} - \frac{Pe_x y}{I_{xx}}$$
[10.7]

10.5 Load applied eccentrically to both axes

Figure 10.4 shows a member of rectangular cross-section to a longitudinal load which is applied eccentrically to both axes. This load may be replaced by an equivalent system as follows:

$$\begin{array}{ccccc}
\text{Load } P \text{ eccentric} & = & \text{axial} & + & \text{moment about} & + & \text{moment about} \\
\text{about } x-x \text{ and } y-y & & \text{load } P & & x-x \text{ of } Pe_x & & y-y \text{ of } Pe_y
\end{array}$$

Applying the principle of superposition, the stress at any point (x, y) will be:

$$\sigma_{x,y} = \frac{P}{A} \pm \frac{M_x y}{I_{xx}} \pm \frac{M_y x}{I_{yy}}$$
[10.8]

This equation may be taken as the **General equation for combined stress** and is applicable to any combination of applied moments and/or eccentric loads. For the case shown in Fig. 10.4:

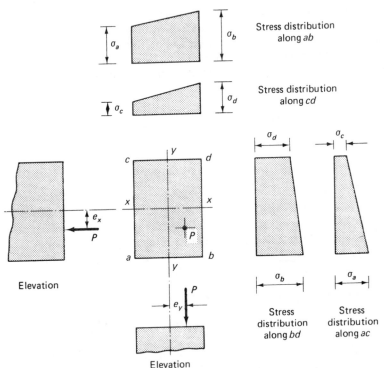

Fig. 10.4

314

$$M_x = Pe_x$$
$$M_y = Pe_y$$

Hence
$$\sigma_{x,y} = \frac{P}{A} \pm \frac{Pe_xy}{I_{xx}} \pm \frac{Pe_yx}{I_{yy}}$$

and the extreme fibre stresses at the corners will be:

$$\sigma_a = \frac{P}{A} + \frac{Pe_xy}{I_{xx}} - \frac{Pe_yx}{I_{yy}}$$

$$\sigma_b = \frac{P}{A} + \frac{Pe_xy}{I_{xx}} + \frac{Pe_yx}{I_{yy}}$$

$$\sigma_c = \frac{P}{A} - \frac{Pe_xy}{I_{xx}} - \frac{Pe_yx}{I_{yy}}$$

$$\sigma_d = \frac{P}{A} - \frac{Pe_xy}{I_{xx}} + \frac{Pe_yx}{I_{yy}}$$

10.6 The middle-third rule for rectangular sections

The position of load P which will locate the neutral axis (zero strain, zero stress) at the extreme fibre of the section is often an important consideration in the design of foundations and retaining walls and, in some cases, in the design of reinforced and prestressed concrete columns. Figure 10.5 illustrates this situation. The load P is applied with an eccentricity of e_0 such that the stress along the edge cd is zero. Then, from eqn [10.3]:

$$\sigma_{min} = 0 = \frac{P}{A} - \frac{Pe_0y}{I_{xx}}$$

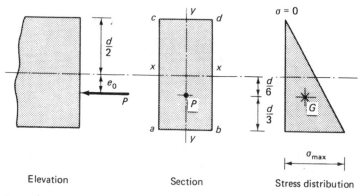

Elevation   Section   Stress distribution

Fig. 10.5

$$= \frac{P}{bd} - \frac{6Pe_0}{bd^2}$$

$$= \frac{P}{bd}\left(1 - \frac{6e_0}{d}\right)$$

Therefore $$e_0 = \frac{d}{6}$$

Alternatively, from the stress diagram given in Fig. 10.5, since P must pass through the centre of area of the triangle (G), then:

$$e_0 = \frac{d}{2} - \frac{d}{3} = \frac{d}{6}$$

The conclusion that may be drawn from this is that, so long as the eccentricity of the applied load does not exceed $d/6$, tensile stress will not occur in the section. Putting this another way, tension will not occur if the point of application of the load lies within the zone between $+d/6$ and $-d/6$, which is the middle-third of the section. A useful design guide is contained in what is known as the **Middle third rule**: 'TENSION WILL NOT OCCUR IF THE RESULTANT ECCENTRIC THRUST FALLS WITHIN THE MIDDLE-THIRD OF THE SECTION'.

In the case of eccentricity about both axes, tension will not occur if the resultant eccentric thrust falls within the **Middle-third kern**, i.e. the shaded area shown in Fig. 10.6.

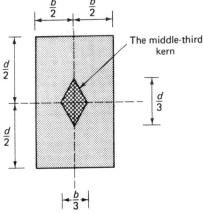

Fig. 10.6

10.7 The middle-quarter rule for circular sections

A design guide similar to the middle-third rule may be established for circular sections known (for obvious reasons) as the **middle-quarter rule** — see Fig. 10.7.

316

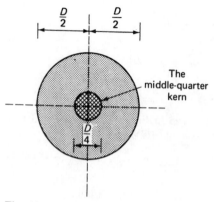

Fig. 10.7

Applying eqn [10.3]:

$$\sigma_{min} = 0 = \frac{P}{A} - \frac{Pe_0 y}{I_{xx}}$$

$$= \frac{P}{\frac{\pi}{4}D^2} - \frac{Pe_0 D/2}{\frac{\pi}{64}D^4}$$

$$= \frac{4P}{\pi D^2}\left(1 - \frac{8e_0}{D}\right)$$

Therefore

$$e_0 = \frac{D}{8}$$

EXAMPLE 10.1 A concrete foundation slab $2\cdot4$ m $\times$ $1\cdot8$ m is subject to a central concentrated load of $1\cdot20$ MN and a moment of 160 kNm which is applied parallel to the longer side (Fig. 10.8). Determine the maximum and minimum bearing pressures acting on the soil beneath the slab.

The maximum pressure will occur at $A-A$ and may be obtained using eqn [10.1]:

$$\sigma_{AA} = \frac{P}{A} + \frac{My}{I_{xx}}$$

$$= \frac{1\cdot20 \times 10^3}{2\cdot4 \times 1\cdot8} + \frac{160 \times 1\cdot2 \times 12}{1\cdot8 \times 2\cdot4^3}$$

$$= 278 + 93 = 371 \text{ kN/m}^2 \text{ (Answer)}$$

The minimum pressure will occur at $B-B$:

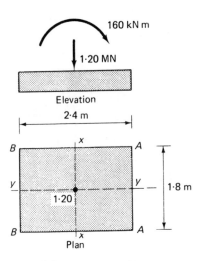

Fig. 10.8

$$\sigma_{BB} = \frac{P}{A} - \frac{My}{I_{xx}}$$

$$= 278 - 93 = 185 \text{ kN/m}^2 \text{ (Answer)}$$

EXAMPLE 10.2 A trapezoidal concrete footing having a mass of 16 Mg carries three column loads as shown in Fig. 10.9. Determine the maximum and minimum bearing pressure acting on the soil beneath the footing.

It is first necessary to calculate the geometrical properties of the footing slab using Table 4.1, p. 110.

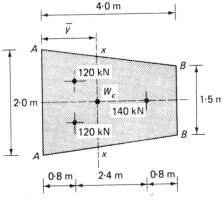

Fig. 10.9

Area,
$$A = (2 + 1\cdot 5)\frac{4}{2}$$
$$= 7\cdot 00 \text{ m}^2$$

Second moment of area about $x-x$,
$$I_{xx} = \frac{4^3}{36}\left[2\cdot 0 + 1\cdot 5 + \frac{2 \times 2\cdot 0 \times 1\cdot 5}{2\cdot 0 + 1\cdot 5}\right]$$
$$= \frac{16}{9}\left[3\cdot 5 + \frac{6\cdot 0}{3\cdot 5}\right] = 9\cdot 27 \text{ m}^4$$

Distance from $A-A$ to centroid,
$$\bar{y} = \frac{2\cdot 0 + 2 \times 1\cdot 5}{2\cdot 0 + 1\cdot 5} \times \frac{4}{3}$$
$$= 1\cdot 905 \text{ m}$$

Now the total load,
$$P = 120 + 120 + 140 + 16 \times 9\cdot 81$$
$$= 537 \text{ kN}$$

and the moment about $x-x$
$$M_{xx} = 2 \times 120\,(1\cdot 905 - 0\cdot 800)$$
$$\qquad\qquad - 140\,(3\cdot 200 \times 1\cdot 905)$$
$$= 265\cdot 2 - 181\cdot 3$$
$$= 83\cdot 9 \text{ kNm}$$

Then from eqn [10.1]:
$$\sigma_{max} = \sigma_{AA} = \frac{537}{7\cdot 00} + \frac{83\cdot 9 \times 1\cdot 905}{9\cdot 27}$$
$$= 77 + 17$$
$$= 94 \text{ kN/m}^2 \text{ (Answer)}$$
$$\sigma_{min} = \sigma_{BB} = \frac{537}{7\cdot 00} - \frac{83\cdot 9\,(4\cdot 000 - 1\cdot 905)}{9\cdot 27}$$
$$= 77 - 18$$
$$= 59 \text{ kN/m}^2 \text{ (Answer)}$$

EXAMPLE 10.3 A universal column section acting as a stanchion carries an axial load and two further loads from incoming beams supported on brackets. The arrangement of loads may be assumed to be as that shown in Fig. 10.10. Determine the extreme fibre stresses acting at the corners A, B, C and D. Properties of the universal column section: $A = 11\,400$ mm^2; $I_{xx} = 143 \times 10^6$ mm^4; $I_{yy} = 48 \times 10^6$ mm^4.

In this case, since there is bending about both the $x-x$ and $y-y$ axes, eqn [10.8] will be used:
$$\sigma_{x,y} = \frac{P}{A} \pm \frac{M_x y}{I_{xx}} \pm \frac{M_y x}{I_{yy}}$$

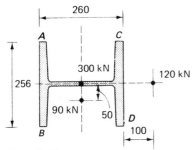

Fig. 10.10

Total load,	P	$= 300 + 120 + 90$	$=$	510 kN
Moment about $x-x$,	M_x	$= 120\,(130 + 100)$	$=$	27 600 kNmm
Moment about $y-y$,	M_y	$= 90 \times 50$	$=$	4 500 kNmm

Then the stresses at the corners are:

$$\sigma_A = \frac{P}{A} - \frac{M_x y}{I_{xx}} - \frac{M_y x}{I_{yy}}$$

$$= \frac{510 \times 10^3}{11\ 400} - \frac{27\ 600 \times 10^3 \times 130}{143 \times 10^6} - \frac{4\ 500 \times 10^3 \times 128}{48 \times 10^6}$$

$$= 44\cdot7 - 25\cdot1 - 12\cdot0 \qquad = 7\cdot6\ \text{N/mm}^2$$

$$\sigma_B = \frac{P}{A} - \frac{M_x y}{I_{xx}} + \frac{M_y x}{I_{yy}}$$

$$= 44\cdot7 - 25\cdot1 + 12\cdot0 \qquad = 31\cdot6\ \text{N/mm}^2$$

$$\sigma_C = \frac{P}{A} + \frac{M_x y}{I_{xx}} - \frac{M_y x}{I_{yy}}$$

$$= 44\cdot7 + 25\cdot1 - 12\cdot0 \qquad = 57\cdot8\ \text{N/mm}^2$$

$$\sigma_D = \frac{P}{A} + \frac{M_x y}{I_{xx}} + \frac{M_y x}{I_{yy}}$$

$$= 44\cdot7 + 25\cdot1 + 12\cdot0 \qquad = 81\cdot8\ \text{N/mm}^2 \quad (\text{Answer})$$

EXAMPLE 10.4 In a prestressed concrete beam the tendon is positioned as shown in Fig. 10.11(a) and applies a prestressing load of 340 kN. Determine the sagging bending moment which, when applied to the beam, will produce zero stress in the bottom fibres. Determine also the maximum compressive stress induced in the section due to the combined application of this moment and the prestressing force.

Properties of the section (ignoring the tendon, the effects of which are negligible):

Area		$A = 150 \times 450$	$= 67\ 500\ \text{mm}^2$

320

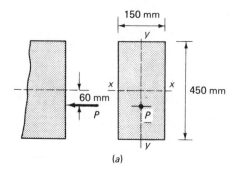

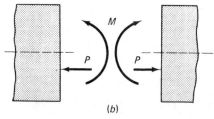

Fig. 10.11

Second moment of area, $I_{xx} = \dfrac{150 \times 450^3}{12} = 1\ 139 \times 10^6\ \text{mm}^4$

The combined system of the prestressing force and the moment acts as shown in Fig. 10.11(*b*)).

Then

$$\sigma_y = \frac{P}{A} \pm \frac{Pey}{I_{xx}} \mp \frac{M_x y}{I_{xx}}$$

$$= \frac{340 \times 10^3}{67\ 500} \pm \frac{340 \times 10^3 \times 60 \times 225}{1\ 139 \times 10^6} \mp \frac{M_x\ 225}{1\ 139 \times 10^6}$$

$$= 5 \cdot 04 \pm 4 \cdot 03 \mp \frac{225\ M_x}{1\ 139 \times 10^6} \quad (\text{N/mm}^2)$$

Therefore, if $\sigma_y = 0$ in the bottom fibres:

$$\frac{225\ M_x}{1\ 139 \times 10^6} = 5 \cdot 04 + 4 \cdot 03$$

Giving

$$M_x = \frac{9 \cdot 07 \times 1\ 139 \times 10^6}{225} = 45 \cdot 9 \times 10^6\ \text{N/mm}$$

$$= 46\ \text{kNm (Answer)}$$

Under these conditions, the maximum compressive stress will occur in the top fibres.

$$\sigma_{max} = 5 \cdot 04 - 4 \cdot 03 + 9 \cdot 07$$

$$= 10 \cdot 08\ \text{N/mm}^2\ \text{(Answer)}$$

EXAMPLE 10.5   Figure 10.12 shows a cast steel frame for a moulding press. The maximum tensile stress must not exceed 25 N/mm². Determine the maximum allowable magnitude for the pressing force P and also the maximum compressive stress that would be induced by the application of this force.

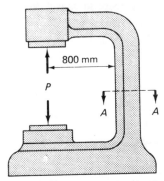

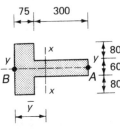

Section A–A

Fig. 10.12

Properties of the cross-section:

Area,
$$A = 220 \times 75 + 60 \times 300$$
$$= 16\ 500 + 18\ 000 \qquad = 34\ 500 \text{ mm}^2$$

Position of centroid,
$$\bar{y} = \frac{16\ 500 \times 37\cdot5 + 18\ 000 \times 225}{34\ 500}$$
$$= \frac{618\ 750 + 4\ 050\ 000}{34\ 500} \qquad = \qquad 135 \text{ mm}$$

Second moment of area,
$$I_{xx} = \frac{220 \times 75^3}{12} + 16\ 500(135 - 37\cdot5)^2 + \frac{60 \times 300^3}{12}$$
$$+ 18\ 000(225 - 135)^2$$
$$= (7\cdot7 + 156\cdot9 + 135\cdot0 + 145\cdot8)\ 10^6$$
$$= 445\cdot4 \times 10^6 \text{ mm}^4$$

From eqn [10.5]
$$\sigma_y = \frac{P}{A} \pm \frac{Pe\,y}{I_{xx}}$$
$$= P\left(\frac{1}{A} \pm \frac{ey}{I_{xx}}\right)$$

Now the maximum tensile stress will occur at *A*.
Then if $\sigma_y \not> 25$ N/mm² and $y = 135$ mm:
$$-25 = P\left[-\frac{1}{34\ 500} - \frac{(800 + 135)\ 135}{445\cdot4 \times 10^6}\right]$$
$$= P\ [-29\cdot0 - 283\cdot4]\ 10^{-6}$$

Therefore, the allowable pressing force,

$$P = \frac{25 \times 10^3}{312 \cdot 4}$$

$$= 80 \cdot 0 \text{ kN (Answer)}$$

When this pressing force is applied, the maximum compressive stress will occur at B, where $y = 375 - 135 = 240$ mm.

Therefore $\quad \sigma_{max} = 80 \cdot 0 \times 10^3 \left(-29 \cdot 0 + \frac{935 \times 240}{445 \cdot 4} \right) 10^{-6}$

$$= 80 \cdot 0 \, (-29 \cdot 0 + 503 \cdot 8) \, 10^{-3}$$

$$= 38 \cdot 0 \text{ N/mm}^2 \text{ (Answer)}$$

10.8 Buckling of slender columns and struts

If a long thin flexible rod is loaded longitudinally in compression, it is noticeable that it deflects readily near the centre of its length with a considerable amount of displacement. This phenomenon is called *BUCKLING* and occurs when the stresses in the rod are still well below those required to cause a shearing type failure. However, if the length of the rod is gradually reduced, whilst still applying the axial load, a length is eventually reached below which the tendency is for the rod not to buckle, but to fail internally. Columns and struts may therefore be described as either *SHORT* (or sometimes *STOCKY*) or *SLENDER* depending on the mode of the failure most likely to occur.

A *short column or strut* will fail internally by yielding (and distorting) in the case of ductile materials, such as mild steel; or by shearing in the case of brittle materials, such as concrete.

A *slender column or strut* will fail by buckling, that is to say a relatively large bending distortion will develop along its length; the member does not immediately collapse, but remains in bent equilibrium, unless the yield strength of the material is exceeded.

The buckling phenomenon is an example of *unstable equilibrium*, whereas the behaviour of a short strut is that of *stable equilibrium*. As further examples, consider the pin-ended vertical members shown in Fig. 10.13. Case (*a*) illustrates stable equilibrium, since the tensile load will tend to restore any forced displacement at the top end of the member. In case (*b*), it is clear that any displacement at the top end will precipitate collapse, and yet when the member is precisely vertical it will be in equilibrium, i.e. unstable equilibrium. In case (*c*), the initially vertical member is restrained at the top by two springs, each of the same length and stiffness (*k*). When a forced displacement (*u*) occurs at the top, the two springs react to restore equilibrium.

$$\text{Displacing moment} = Pu$$
$$\text{Restoring moment} = 2kuL$$

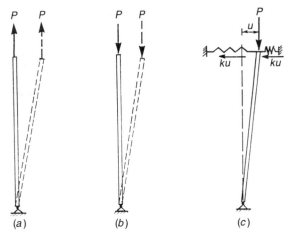

Fig. 10.13

The arrangement is *stable* when $Pu < 2kuL$ and *unstable* when $Pu > 2kuL$. The critical condition is therefore represented by $P_c = 2kL$. The value of P_c is known as the *critical load*. For a given load a *critical length* may also be deduced. In the case of slender structural columns or struts, the *critical buckling load* and the *critical buckling length* depend upon a number of factors, such as the shape and size of the cross-section, the relationship between the length of the column and its lateral dimensions, and the degree of fixity at the ends. For a strut of given length which is pinned at both ends, the minimum load at which buckling will occur may be determined using a mathematical analysis which produces what is known as the EULER FORMULA.

$$P_E = \frac{\pi^2 EI}{L^2} \qquad [10.9]$$

where: P_E = the **Euler buckling load**

E = Young's modulus for the material

I = the least second moment of area of the section

L = the length of the strut between the pinned ends

A proof of this formula is given in Section 10.11 for the benefit of those readers having some knowledge of differential equations. However, reference to this proof is not essential to the work of this chapter in general.

The magnitude of the buckling load given by eqn [10.9] is the appropriate value for initially straight struts which are PINNED at both ends and are subject to an axial load only. The EULER BUCKLING LOAD is also determinable for the following sets of cases:

1. Pin-ended struts subject to eccentric longitudinal load.

Table 10.1 Euler buckling loads and critical buckling lengths for axially loaded struts

End fixity conditions	Euler buckling load, P_0	Critical buckling length, L_c	Critical length ratio*
Both ends pinned	$\dfrac{\pi^2 EI}{L^2}$	$\pi\sqrt{\dfrac{EI}{P}}$	1
Both ends fixed in direction	$\dfrac{4\pi^2 EI}{L^2}$	$2\pi\sqrt{\dfrac{EI}{P}}$	2
One end pinned and the other fixed in direction	$\dfrac{2\cdot05\,\pi^2 EI}{L^2}$	$1\cdot43\,\pi\sqrt{\dfrac{EI}{P}}$	$1\cdot43$
One end fixed in direction and the other free	$\dfrac{\pi^2 EI}{4L^2}$	$\dfrac{\pi}{2}\sqrt{\dfrac{EI}{P}}$	$0\cdot5$

$$*\text{ratio} = \frac{\text{critical buckling length}}{\text{critical buckling length for both ends pinned}}$$

2. Pin-ended struts subject to both longitudinal and lateral loading.
3. Pin-ended struts with initial displacement or curvature.
4. Struts with one or both ends fixed in direction such that no rotation takes place.

The cases referred to in set 4 are of particular interest and therefore the solutions for the Euler buckling loads are given for a number of these in Table 10.1. A useful comparison of the effect of the end-fixity conditions may be made by considering for each case the CRITICAL BUCKLING LENGTH, that is to say the length at which the column just begins to buckle at a given load P. Refer to the last two columns of Table 10.1.

10.9 Practical buckling criteria

The Euler theory provides an elegant mathematical solution for the buckling of struts, but the assumptions embodied in it with regard to the initial straightness of the strut and the eccentricity of the loading are impossible to reproduce in practice with any reasonable degree of accuracy. The situation is further complicated by the fact that the departure of the practical conditions from those represented by the mathematical ideal is also impossible to assess accurately. Furthermore, the structural designer requires a quick and easy reference system rather than a series of involved equations, the solution of which may be time-consuming and costly. For these reasons a number of empirical formulae have been devised for use in the practical design of struts.

In most design procedures it is convenient to consider the *allowable* or *permissible* levels of stress which the actually induced stresses must not exceed. From the Euler formula (eqn [10.9]) the *critical buckling strength* is:

$$p_E = \frac{P_E}{A} = \frac{\pi^2 EI}{AL^2}$$

But from eqn [4.8] the ratio I/A may be expressed as r^2, where $r =$ the radius of gyration of the section.

Therefore $\qquad p_E = \dfrac{\pi^2 E}{(L/r)^2} \qquad$ [for pin-ended struts]

or, generally

$$p_E \propto \frac{1}{(L/r)^2} \qquad [10.10]$$

The empirical formulae are based on a similar relationship and give either the critical buckling stress, or the allowable stress, in terms of two measurable factors:

1. The yield stress (or other critical value) of the material.
2. The slenderness ratio of the strut (see below).

$$\text{SLENDERNESS RATIO} = \frac{\text{EFFECTIVE LENGTH OF THE STRUT}}{\text{RADIUS OF GYRATION}}$$

$$\lambda = \frac{l}{r}$$

where $l =$ the effective length of the strut (see Section 10.10).

Since the degree of fixity may not be the same about both axes, two slenderness ratio values may be determined for a given strut:

$$\lambda_{xx} = \frac{l_{xx}}{r_{xx}} \qquad [10.11]$$

$$\lambda_{yy} = \frac{l_{yy}}{r_{yy}} \qquad [10.12]$$

The greater value usually occurs about the $y-y$ axis and will be the more critical of the two.

For struts having a rectangular section, e.g. timber sections, the slenderness ratio is usually taken as $\lambda = l/b$

where $\qquad l$ = effective length

$\qquad\qquad\qquad b$ = the least lateral dimension

The relationship between the allowable stress and the slenderness ratio is usually presented in the form of a graph or table. The designer therefore calculates the value of the slenderness ratio and obtains the allowable stress from the appropriate table. When the actual stresses have been evaluated they are then compared with the allowable stresses. In good design, in order to satisfy the dual interests of safety and economy, the actual stress value should be equal to or just less than the allowable value.

10.10 Effective length of struts

Recommended values for the effective lengths of struts are given in various British Standards and Codes of Practice relating to design. The examples shown in Fig. 10.14 are given as a general guide. It will be seen that the effective length of any strut is related to the length between connections and may be considered to be approximately equal to the length between the points of contraflexure which develop as buckling occurs. The empirical values given here should now be compared with the mathematically derived CRITICAL LENGTH RATIOS given in Table 10.1.

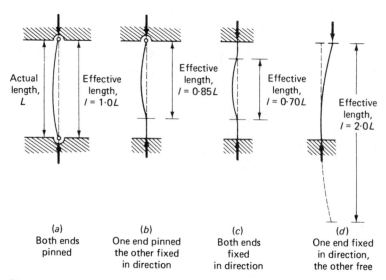

(a)
Both ends pinned

(b)
One end pinned the other fixed in direction

(c)
Both ends fixed in direction

(d)
One end fixed in direction, the other free

Fig. 10.14

10.11 Proof of the Euler formula

Consider a strut of uniform section (as shown in Fig. 10.15(a), of constant flexural rigidity (EI), of length L, and which is pinned at either end A and B. The strut deflects due to buckling when a load P is applied; the deflected mode is shown in Fig. 10.15(b). Since the flexural rigidity has a lower value about the y−y axis, it will be about this axis that buckling will occur.

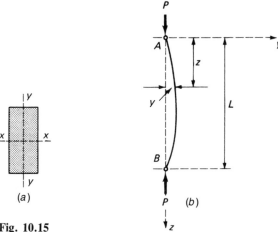

Fig. 10.15

At a position z from end A, the bending moment in the strut will be:

$$M = -Py$$

From eqn [7.3]: $$M = EI\frac{d^2y}{dz^2}$$

Therefore $$EI\frac{d^2y}{dz^2} + Py = 0$$

or, putting $$\frac{P}{EI} = \alpha^2, \quad \frac{d^2y}{dz^2} + \alpha^2y = 0$$

This is a differential equation for which the general solution is:

$$y = A\cos\alpha z + B\sin\alpha z$$

in which the constants A and B must be found to suit the particular boundary conditions which, in this case, are:

when $z = 0, y = 0$
and when $z = L, y = 0$

Then either $A = 0$ or $B\sin\alpha L = 0$.
But if $A = 0$, buckling does not occur! Therefore, $B\sin\alpha L = 0$

This equation is satisfied when $\alpha L = \pi, 2\pi, 3\pi$, etc.

The lowest value is π and hence $\alpha = \dfrac{\pi}{L}$

and since

$$\frac{P}{EI} = \alpha^2$$

the Euler buckling load is given by: $P_E = \dfrac{\pi^2 EI}{L^2}$ [10.13]

EXAMPLE 10.6 *Calculate the Euler buckling load for a mild steel rod of length 4 m and diameter 25 mm when pinned at each end and axially loaded. If this rod is to be used again with an initially straight axis, determine the maximum permissible deflection that may be allowed during testing procedures. Young's modulus = 210 kN/mm^2; Yield stress of the steel = 240 N/mm^2.*

Second moment of area of the section, $I = \dfrac{\pi \times 25^4}{64} = 19\ 175$ mm^4

The Euler buckling load, $P_E = \dfrac{\pi^2 EI}{L^2}$

$$= \frac{\pi^2 \times 210 \times 10^3 \times 19\ 175 \times 10^{-3}}{4^2 \times 10^6}$$

$$= 2 \cdot 48 \text{ kN}$$

Permanent deformation will take place when the maximum compressive stress reaches the yield stress. The maximum stress will be the combined direct and bending stress.

$$\sigma = \frac{P_E}{A} + \frac{My}{I}$$

Let the deflection at which this maximum stress equals the yield stress be equal to Δ mm.

The bending moment $= P\Delta$

Then $240 = \dfrac{2 \cdot 48 \times 10^3 \times 4}{\pi \times 25^2} + \dfrac{2 \cdot 48\Delta \times 12 \cdot 5}{19\ 175}$

Giving $\Delta = (240 - 5)\dfrac{19\ 175}{2 \cdot 48 \times 10^3 \times 12 \cdot 5}$

$$= 145 \text{ mm}$$

Therefore the deflection must not be permitted to exceed 145 mm, otherwise the strut will become permanently deformed.

10.12 Other buckling formulae

The Euler buckling theory has severe limitations from a practical point of view since the ideal circumstances upon which it is based can never be achieved in real conditions. In practice, there will always be some initial curvature of the strut. In addition, the direct stress in the strut is neglected. In some cases, as deflection increases with load and with it the bending moment, a bending stress failure may occur before the Euler buckling load is reached.

Consider the expression for the stress at buckling given in Section 10.9:

$$p_E = \frac{\pi^2 E}{(L/r)^2}$$

For structural mild steel, typical values are: $p_y = 275$ N/mm^2, $E = 205$ N/mm^2.

Then the value of L/r corresponding to $p_E = p_y$ will be:

$$\lambda_0 = \sqrt{\left(\frac{\pi^2 \times 205 \times 10^3}{275}\right)} = 86$$

Thus, for values of $(L/r) < 86$ the indicated Euler buckling strength is greater than the yield strength. Various empirical formulae have been proposed to overcome this difficulty and to establish a rational design expression which is valid over a practical range of (L/r) values.

Rankine-Gordon formula

When direct stress and buckling stress are considered to be acting simultaneously the critical strength (p_c) of a strut of any length will be given by:

$$\frac{1}{p_c} = \frac{1}{p_y} + \frac{1}{p_E} \tag{10.14}$$

where p_y = yield strength of the material

p_E = Euler buckling stress corresponding to appropriate conditions of end fixity

For a pin-ended strut, $p_E = \pi_2 E/(L/r)^2$
Substituting into eqn [10.14] and rearranging,

Critical strength, $$p_c = \frac{p_y}{1 + a(L/r)^2} \tag{10.15}$$

which is known as the *Rankine* or *Rankine-Gordon* formula.

Where a = a constant dependent on end-fixity conditions and the yield strength of the material, i.e. for a pinned-end strut of mild steel having $p_y = 275$ N/mm^2 and $E = 205$ kN/mm^2:

$$a = \frac{275}{\pi^2 \times 205 \times 10^6} = 0.000136$$

The parabolic formula

Sometimes called Johnson's formula, this has been widely recommended in the USA, e.g. by the American Railway Engineers Association, and is intended to follow the Euler curve for slender columns.

$$p_c = p_y[1 - b(L/r)^2] \qquad [10.16]$$

Where b = a constant devised to make this curve fit tangentially with the Euler curve; for a pinned-end strut when $L/r < 140$, b is usually taken as $0 \cdot 000023$.

The straight-line formula

This type of formula is also common in the USA and is intended to give sufficiently accurate estimated values over a limited range of L/r.

$$p_c = p_y[1 - c(L/r)] \qquad [10.17]$$

where c = a constant dependent on end fixity conditions and the yield strength of the material, typical values of c would be $0 \cdot 005$ with both ends pinned and $0 \cdot 004$ with both ends fixed.

The Perry strut formula

The Perry strut formula is recommended for construction steelwork design in BS 5950: Part 1: 1985 *Structural use of steelwork in building*.

Compressive strength, $\quad p_c = \dfrac{p_E p_y}{\phi + \sqrt{(\phi^2 - p_E p_y)}} \qquad$ [10.18]

where
$\quad \phi = \frac{1}{2}(p_y + (\eta + 1)p_E)$

$\quad p_E = \pi^2 E / \lambda^2 \quad$ (the Euler buckling strength)

$\quad p_y$ = yield strength of the steel

$\quad \eta$ = Perry factor

$\quad\quad = 0 \cdot 001 a(\lambda - \lambda_0) \quad$ (but not less than zero)

$\quad \lambda$ = slenderness ratio (l/r)

$\quad \lambda_0$ = limiting slenderness ratio = $0 \cdot 2 \sqrt{(\pi^2 E / p_y)}$

$\quad a$ = Robertson factor, for which values are given between $2 \cdot 0$ and $8 \cdot 0$ according to the shape and thickness of the steel section and the axis of buckling ($x-x$ or $y-y$); see the examples given below.

	a
Hot-rolled hollow section; $y-y$ buckling	$2 \cdot 0$
Rolled I-sections; $\ngtr$ 40 mm thick; $y-y$ buckling	$3 \cdot 5$
Rolled H-sections; $\ngtr$ 40 mm thick; $y-y$ buckling	$5 \cdot 5$
Rolled H-sections; > 40 mm thick; $y-y$ buckling	$8 \cdot 0$

A comparison between the various formulae given is shown in Fig. 10.16.

EXERCISES   CHAPTER 10

1. Fig. 10.17 shows a cantilever bar of steel which is 100 mm deep and 20 mm thick and which carries two loads simultaneously. Determine the maximum combined direct and bending stress in the section.

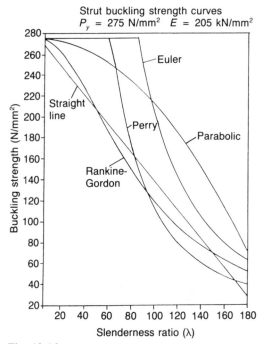

Fig. 10.16

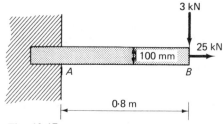

Fig. 10.17

2. A simply supported steel beam is subject to the combined application of an axial tension and a lateral load as shown in Fig. 10.18. The beam consists of a joist section having the following properties:

Depth = 152 mm; area = 2 180 mm²; second moment of area = $8 \cdot 81 \times 10^6$ mm⁴. Calculate the maximum compressive and tensile stresses acting in the beam.

3. Fig. 10.19 shows a steel bar of diameter 50 mm which is built in to a solid support at its base. Determine the magnitude of the force P which will induce in the bar a maximum stress of 150 N/mm².

4. In Fig. 10.20 the inclined member is a steel bar of diameter 25 mm which is pinned to rigid supports at A and B. Calculate the maximum combined direct and bending stress acting in the bar.

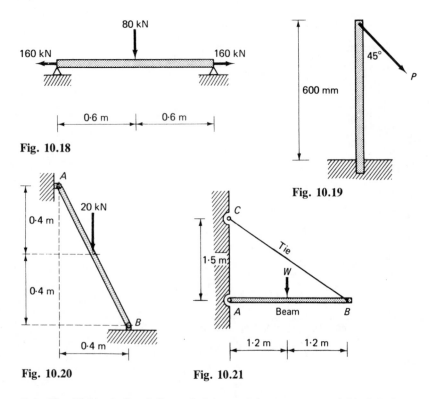

Fig. 10.18

Fig. 10.19

Fig. 10.20

Fig. 10.21

5. In Fig. 10.21, *A, B* and *C* are pin-joints and the supports are rigid. Calculate the maximum permitted value for the load *W* if the maximum combined stress in the beam is not to exceed 165 N/mm².
Properties of the beam: Area = 2 740 mm²; depth = 178 mm; second moment of area = 15·2 × 10⁶ mm⁴.

6. The member shown in Fig. 10.22 consists of a universal beam section and is pinned at both ends. Calculate the compressive and tensile stresses in the beam immediately above and below the bracket which is in the plane of the *y*−*y* axis.
Properties of the beam: Area = 5 690 mm²; depth = 352 mm; second moment of area = 120 × 10⁶ mm⁴.

7–12. Figs. 10.23–10.28 show a column of rectangular section loaded in a number of different ways. In each case, calculate the combined direct and bending stress in the extreme fibres at each corner of the section.

13. Fig. 10.29 shows the plan of a foundation slab of thickness 0·8 m which carries three column loads. Each load lies 0·8 m from the adjacent edges. Calculate the maximum bearing pressure induced onto the soil beneath the slab. Density of concrete = 2 240 kg/m³.

14. Fig. 10.30 shows the plan of a stanchion consisting of a universal column H-section and carrying an axial load of 600 kN together with an eccentric load from

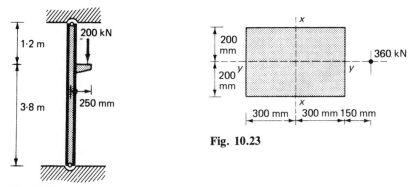

Fig. 10.22

Fig. 10.23

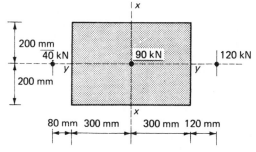

Fig. 10.24

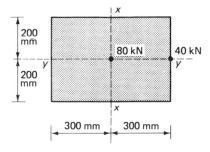

Fig. 10.25

a bracket of 150 kN. Calculate the maximum compressive and tensile stresses acting in the stanchion.

Properties of the section: Area = 11 400 mm²; depth = 260 mm; second moment of area (I_{xx}) = 143 × 10⁶ mm⁴.

15. Fig. 10.31 shows the plan of a compound stanchion consisting of two universal beam sections and carrying an axial load of 600 kN together with three loads from incoming beams of 300 kN, 300 kN and 200 kN respectively. Calculate the maximum compressive and tensile stresses acting in the stanchion.

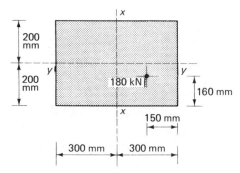

Fig. 10.26

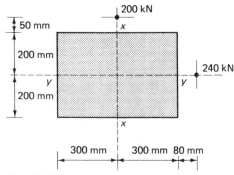

Fig. 10.27

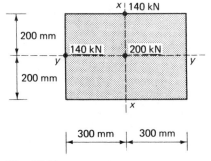

Fig. 10.28

16. A load of 120 kN is eccentrically applied to a column of circular cross-section as shown in Fig. 10.32. Calculate the value of the eccentricity e that will give (a) a maximum compressive stress of 8 N/mm^2 at A and (b) zero stress at B.

17. In the column of rectangular section shown in Fig. 10.33, the maximum combined stress must not exceed 5 N/mm^2. Calculate the maximum magnitude allowable for the load W.

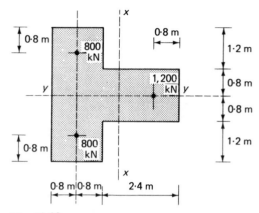

Fig. 10.29

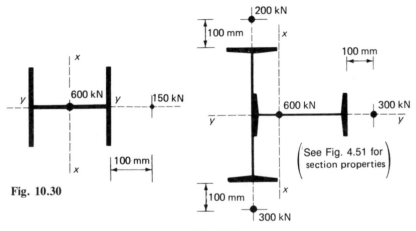

Fig. 10.30

Fig. 10.31

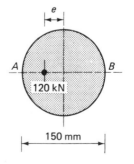

Fig. 10.32

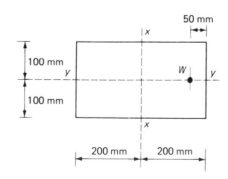

Fig. 10.33

336

18. Fig. 10.34 shows a steel tubular column of external diameter 150 mm and internal diameter 125 mm which is built in to a solid support at its base C. Calculate the maximum combined stress occurring at sections B and C due to the loading shown.

19. Fig. 10.35 shows the cross-section of a cast iron pillar. Calculate the maximum compressive stress that will result from the application of a longitudinal load of $1 \cdot 2$ MN (a) at point O and (b) at point A.

20. The vertical pillar of a pressing frame consists of a steel box section as shown in Fig. 10.36. If the maximum compressive stress developed at section $A-A$ must not exceed 120 N/mm², calculate the maximum allowable magnitude for the force P.

21. Fig. 10.37 shows a steel G-clamp upon which is exerted a clamping force of 500 N. Calculate the maximum compressive and tensile stresses acting at section $A-A$.

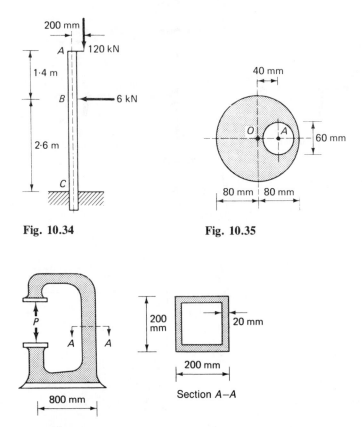

Fig. 10.34

Fig. 10.35

Fig. 10.36

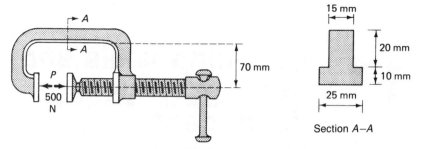

Fig. 10.37

22. Determine the Euler buckling load for a steel rod of diameter 40 mm if the length between its pin-jointed ends is 4 m.
Young's modulus = 210 kN/mm².

23. A steel tube is to be used as a strut carrying an axial load of 25 kN. If the tube has an external diameter of 50 mm and a wall thickness of 6 mm, calculate the critical buckling length (a) when both ends are pinned and (b) when both ends are fixed in direction.
Young's modulus = 210 kN/mm².

24. Calculate the safe axial load for a timber strut of section 150 × 150 mm and length 4·5 m if it is pinned at both ends and a factor of safety of 3 against buckling is required. What will be the maximum compressive stress in the timber at this safe load?
Young's modulus = 10 kN/mm².

11 Retaining walls and other gravity-dependent structures

Symbols

A	=	area of base
B	=	breadth of base
e	=	eccentricity of resultant ground reaction about centreline of base
F_s	=	factor of safety
H,h	=	height, depth
K_a	=	coefficient of active lateral earth pressure
L	=	length
l	=	sloping height of an inclined face
P	=	resultant lateral thrust
P_A	=	resultant active lateral thrust due to earth pressure
P_{AS}	=	resultant active lateral thrust due to surcharge
p	=	pressure
p_a	=	active lateral earth pressure
q	=	surcharge pressure
q_A, q_B	=	ground bearing pressures
R_S	=	sliding resistance force
V	=	resultant ground reaction
W	=	load, force
W_s	=	force due to mass of soil (weight of soil)
x	=	distance to line of ground reaction
α	=	angle between plane of rupture and face of wall
γ	=	unit weight, bulk unit weight of soil

γ_w = unit weight of water
δ = angle of wall friction
θ = angle of inclination of a sloping wall face
μ = coefficient of friction = $\tan \delta$
ϕ = angle of shearing resistance of soil and other granular materials
ρ · = density, bulk density of soil
ρ_w = density of water

11.1 Gravity-dependent structures

Any structure that is constructed on the surface of the Earth will be pulled towards the Earth's centre (i.e. downward) by the force of gravitational attraction. This downward-acting force is usually called the WEIGHT of the structure. When a structure is almost or entirely dependent upon its weight for stability, it may be termed a GRAVITY-DEPENDENT STRUCTURE. Most structures are gravity-dependent to some extent, but not all would fail if the gravity effect was to be, as it were, switched off! A truly gravity-dependent structure *would* fail under these circumstances and therefore special equilibrium conditions must be maintained in order to give a satisfactory factor of safety against failure.

In this chapter a number of structures will be considered which are subject to lateral (usually horizontal or nearly so) loading and which depend largely upon gravity for their stability. Examples will be included involving dams, retaining walls, tall chimneys and towers.

11.2 Failure criteria

Calculations are carried out to establish the minimum degree of stability that a given structure may possess with respect to critical modes of failure. Failure may be said to have occurred if one of several *ultimate limit states* is exceeded. The factors of safety against failure for each of these limit states must therefore be determined. For the type of structure under consideration here, five such limit states exist and these are:

1. Overturning.
2. Tension in the joints.
3. Bearing on the ground.
4. Sliding along the base.
5. Rotational slip.

These five apply to a *truly gravity-dependent structure*; others may be added to the list if there are additional means of support. For instance, in tower structures the tension in the holding-down bolts would have to be considered; or in the case of reinforced concrete structures, the internal design of the reinforced concrete itself would be important. The five cases are shown in Figs 11.1−11.5 where a trapezoidal retaining wall is seen at the moment of failure in each case.

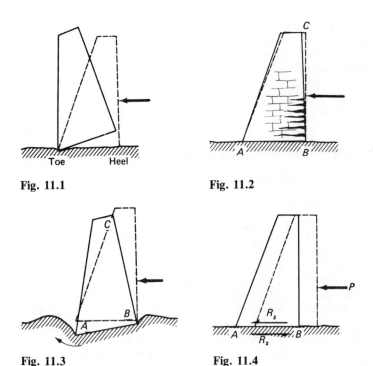

Fig. 11.1

Fig. 11.2

Fig. 11.3

Fig. 11.4

FIGURE 11.1 OVERTURNING. The wall will overturn by rotating about some point near the toe if the overturning moment of the lateral force (anticlockwise) is greater than the stabilizing moment of the weight of the wall (clockwise). This condition, however, is usually less critical than the next (Fig. 11.2), nevertheless the factor of safety against overturning should be at least *two*.

FIGURE 11.2 TENSION IN THE JOINTS. If the wall is considered as a vertical cantilever (*BC*), it will be apparent that, superimposed upon the compressive stress induced in the horizontal joints due to the weight of the wall, there is a bending stress which will be tensile along *BC*. If this bending tension exceeds the direct compression, a net tensile stress will result in the wall at and just above *B*. In brickwork or masonry walls, this tension may have the effect of opening the joints; this in turn may lead to a deterioration of the fabric of the wall.

FIGURE 11.3 BEARING ON THE GROUND. The combination of the downward-acting weight of the wall and the overturning moment will set up a bearing pressure between the base of the wall and the ground upon which it rests. In the case shown, the maximum ground-bearing pressure will occur beneath the toe *A*. Failure will occur if the applied pressure exceeds the ULTIMATE BEARING CAPACITY of the soil causing the soil at this point to be disturbed. In practice the applied pressure is not allowed to exceed the ALLOWABLE BEARING CAPACITY of the soil, which is related to the shear strength and compressibility characteristics of the soil and incorporates a factor of safety. This factor of safety should, in any event, be at least *three*. (See Table 11.3).

FIGURE 11.4 SLIDING. If the underside of the wall was smooth and it rested on a smooth surface, the wall would slide easily along its base *AB*. However, the base is usually rough and also the ground surface can develop frictional resistance. The wall will therefore only slide if the lateral thrust *P* exceeds the frictional resistance developed between the base of the wall and the soil beneath. The factor of safety against the wall sliding should not be less than *two*.

FIGURE 11.5 ROTATIONAL SLIP. The wall, together with a large amount of the retained material, will rotate about some centre *O* if the total moment of the weight of the wall (W_1) and soil mass $ABC(W_2)$ exceeds the moment of the shearing resistance developed along the circular arc *AC* (i.e. the surface of rupture or failure). This type of failure is not uncommon, particularly in clay soils. However, since the analysis of such a mechanism requires a knowledge of certain principles of soil mechanics, consideration of this mode of failure is beyond the scope of this book.

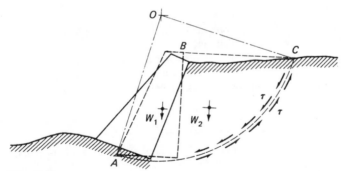

Fig. 11.5

11.3 Lateral pressure and thrust

The lateral force which must be resisted by a retaining or similar structure may result from wind pressure, from pressure due to a retained liquid, or from pressue due to a retained granular material. In addition, there may be situations in which lateral forces are transmitted to a retaining wall (for example) from adjacent structures. The obvious first step, therefore, is to calculate the magnitude and position of the resultant lateral thrust.

11.4 Wind pressure

The intensity of pressure against the side of a structure due to wind depends on a number of factors, such as the average wind velocity, the amount and frequency of gusting, the degree of exposure of the site, the amount of shelter provided by adjacent buildings, the height and shape of the structure and a few others. A good deal of complex calculation is required in order to evaluate the actual wind pressure. For the purposes required here it will be sufficient to assume that a uniform (although this is not strictly true) pressure, acting normal to the exposed surfaces of the structure, is the result of wind.

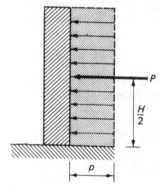

Fig. 11.6

Figure 11.6 shows a vertical wall of height H which is subject to a uniform wind pressure p. The resultant thrust per unit length of wall (P) acts through the centre of area of the pressure diagram — since this is a rectangle, the resultant strikes the wall at a height of $H/2$ above the base. The magnitude of P is given by the area of the pressure diagram:

$$P = pH$$

EXAMPLE 11.1   A vertical wall of height 3 m is subject to a uniform horizontal wind pressure of intensity 1·2 kN/m². Determine the magnitude and position of the resultant thrust per metre length of wall.

Resultant thrust $P = 1·2 \times 3$

$\qquad\qquad\quad = 3·6$ kN/m (Answer)

$\qquad\qquad\qquad$ (Acting at a height of 1·5 m above the base)

11.5 Liquid pressure

Dams and other liquid-retaining structures are required to sustain a lateral thrust from the retained liquid. Earth-retaining walls may also have to resist hydrostatic pressure due to groundwater.

It is necessary at this stage to recall two simple principles of hydrostatics:

1. The pressure at any point in a liquid acts uniformly in all directions.
2. The pressure acting at a depth h below the surface of a liquid is equal to the product of the weight per unit volume of the liquid and the depth:

$$P = \rho_w g\, h = \gamma_w h \qquad\qquad [11.1]$$

where:
ρ_w = density of the liquid

γ_w = unit weight of liquid

g = gravitational acceleration ($9 \cdot 81$ m/s^2)

From the second of these statements it will be seen that the hydrostatic pressure increases uniformly with depth. The pressure distribution diagram will therefore be a triangle.

Note: The units of *density* (ρ) may be kg/m^3 or Mg/m^3, whereas *unit weight* (γ) = ρg = $9 \cdot 81 \rho$ and will therefore have units of N/m^3 or kN/m^3.

Example: density of water, $\rho w = 1\,000$ kg/m^3
$= 1 \cdot 00$ Mg/m^3

unit weight of water, $\gamma w = 9\,810$ N/m^3
$= 9 \cdot 81$ kN/m^3

Consider the pressure distribution on the vertical surface AB in Fig. 11.7. The pressure at a depth H is $\gamma_w H$ giving the base of the diagram.

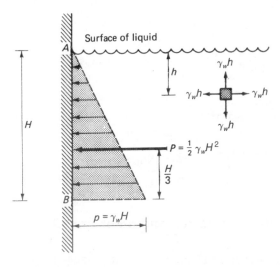

Fig. 11.7

The magnitude of the resultant thrust is given by the area of the diagram:

$$P = \tfrac{1}{2} \times \gamma_w H \times H$$
$$= \tfrac{1}{2} \gamma_w H^2 \qquad [11.2]$$

This resultant will act through the centre of area of the pressure triangle, i.e. at a height of $H/3$ above B. It is important to note that the depth is always measured from the *surface of the liquid* — see the following example.

EXAMPLE 11.2 A masonry dam of height 6 m retains water on its vertical face. Calculate the resultant hydrostatic thrust acting on a metre length of wall when: (a) the water level reaches the top of the dam (Fig. 11.8) (b) the water level is 1·8 m below the top of the dam (Fig. 11.9). Density of water = 1 000 kg/m³.

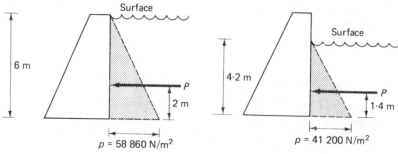

Fig. 11.8   **Fig. 11.9**

(a) Hydrostatic pressure p_w at 6 m depth $= 1\,000 \times 9 \cdot 81 \times 6$
$$= 58\,860 \text{ N/m}^2$$

Hydrostatic thrust, $P = \dfrac{58\,860 \times 6}{1\,000 \times 2} = 176 \cdot 6$ kN/m (Answer)
(acting 2 m above the base)

(b) In this case the water surface is 4·2 m above the base.
Hydrostatic pressure p_w at 4·2 m depth $= 1\,000 \times 9 \cdot 81 \times 4 \cdot 2$
$$= 41\,200 \text{ N/m}^2$$

Hydrostatic thrust, $P = \dfrac{41\,200 \times 4 \cdot 2}{1\,000 \times 2} = 86 \cdot 5$ kN/m (Answer)
(acting 1·4 m above the base)

11.6 Pressure due to retained soil and other granular materials

Since the most obvious application here is in the case of earth-retaining walls, this will be the type of case considered. The following principles and methods, however, are equally applicable to problems involving all granular materials which possess the property of internal friction (friction between adjacent grains) whilst not possessing the property of internal cohesion. Materials that may be considered under this heading include sandy and gravelly soils, coal, coke, grain and certain ores. For cohesive materials, such as clay soils, the principles involved are more complex and are therefore omitted from this work.

It should be realized at the outset that soils, unlike liquids, are not usually uniform, nor do their properties necessarily remain constant during the passage of time. Natural soils often occur in layers called strata, or in lenses, which may show radical differences in properties from one layer to another. Also, with changes in weather, drainage conditions, etc., the moisture content of the soil may vary, bringing about consequent changes in density and, more important, changes in the strength properties of the soil. Furthermore, different soils will react to these changes in different ways.

Several useful theories and methods exist which may be used to determine the lateral pressure on, say, a retaining wall. Some of these are relatively simple, while others are quite complex, and in all some degree of approximation is present. For the purposes of the work in this chapter it will be sufficient only to consider two of the simpler theories — Rankine's theory and Coulomb's wedge theory.

For further work relating to pressures and deformations in and from soil masses readers should consult an appropriate soil mechanics text (e.g. Whitlow, R. *Basic soil mechanics*, 2nd edn, Longman, 1990).

11.7 Rankine's theory of lateral earth pressure

The assumptions made in this theory are that the material is incompressible, homogeneous, granular and cohesionless, and that it possesses internal frictional resistance to movement between the grains. Thus, if a quantity of such material was tipped onto a flat surface it would flow out to form a conical heap. The material within the cone supports itself due to the internal friction between the grains. The angle that the side of the cone makes with horizontal is known as the ANGLE OF REPOSE (Fig. 11.10).

A simple experiment with dry sand will illustrate this. If, however, wet sand is used it will be seen that the slope of the pile is steeper and that it is also uneven. The angle of repose, therefore, appears to be greater, but the internal resistance of the material may not have changed very much,

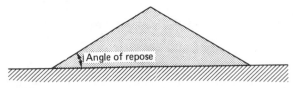

Angle of repose

Fig. 11.10

if at all. A more reliable measure of the internal friction of the soil is obtained by testing the soil in the laboratory and establishing the value of the ANGLE OF SHEARING RESISTANCE (ϕ) of the soil. The angle of shearing resistance and the angle of repose of a perfectly dry and loose sand will usually be equal. The angle of shearing resistance is the essential property upon which Rankine's theory is based.

It has been stated previously that liquid pressure acts equally in all directions (Fig. 11.7). This illustrates the fact that, since liquids do not possess internal resistance to shearing, the angle of shearing resistance for a liquid is zero. In frictional materials, such as sand, the lateral pressure exerted is somewhat less than the vertical pressure. In Rankine's theory the ratio between the lateral and vertical pressure is evaluated in terms of the angle of shearing resistance (Fig. 11.11).

In order to create a pressure against a supporting surface the soil moves slightly *towards* that surface. The pressure resisted by the supporting structure is then called the ACTIVE LATERAL EARTH PRESSURE (p_a). The ratio between the active lateral earth pressure and the vertical pressure is called the COEFFICIENT OF ACTIVE LATERAL EARTH PRESSURE (K_a).

$$\frac{p_a}{\gamma h} = K_a$$

When evaluated in terms of the angle of shearing resistance (ϕ):

$$K_a = \frac{1 - \sin\phi}{1 + \sin\phi} \qquad [11.3]$$

Then at depth H (Fig. 11.11):

$$p_a = K_a \times \gamma H$$
$$= \left(\frac{1 - \sin\phi}{1 + \sin\phi}\right) \gamma H \qquad [11.4]$$

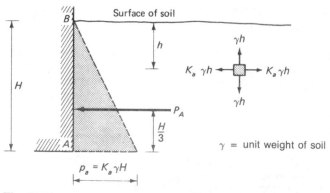

Fig. 11.11

Since p_a varies uniformly with depth in a homogeneous soil, the pressure distribution diagram for the lateral pressure on AB is a triangle.

Hence the lateral active thrust, $P_A = \frac{1}{2} \times p_a H$

$$= \frac{1}{2} K_a \gamma H^2 \qquad [11.5]$$
(acting at a height $H/3$ above A)

EXAMPLE 11.3 A masonry wall with a vertical back 4·5 m high retains earth having a unit weight of 19·6 kN/m³ and an angle of shearing resistance of 30°: Calculate the active lateral thrust acting on the wall using Rankine's theory.

The coefficient of active lateral earth pressure is calculated first:

$$K_a = \frac{1 - \sin \phi}{1 + \sin \phi} = \frac{1 - 0·5}{1 + 0·5} = \frac{1}{3}$$

At a depth of 4·5 m the active lateral pressure will be

$$p_a = K_a \gamma H$$
$$= \frac{1}{3} \times 19·6 \times 4·5 = 29·4 \text{ kN/m}$$

Active lateral thrust, $P_A = \frac{1}{2} \times 29·4 \times 4·5$

$$= 66·2 \text{ kN/m (Answer)}$$
(acting 1·5 m above the base).

11.8 Stepped wall faces

In the interests of both stability and economy, mass concrete, brickwork and masonry walls are often constructed with their front, back or both faces either stepped or inclined. The lateral pressure is assumed to act *normal* to the supporting surface and therefore, in the case of a wall having a vertical stepped back (Fig. 11.12), the direction of this pressure

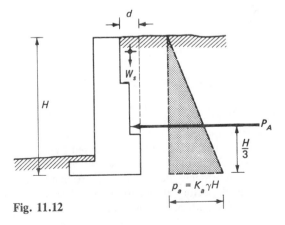

Fig. 11.12

will be horizontal. The pressure distribution is shown in Fig. 11.12 and may be considered to be the same as that against a simple vertical face of the same total depth.

An additional aid to stability is obtained from the weight (W_S) of the strip of retained material of width d which bears down on the steps. The moment of this downward-acting force may be included in the total restraining moment.

11.9 Inclined wall faces

Since the pressure acts *normal* to the supporting surface, then in the case of an inclined wall face, the pressure must act in a direction other than horizontal. As shown in Fig. 11.13, the pressure diagram is again a triangle, the area of which gives the resultant thrust P_A.

$$P_A = \tfrac{1}{2} p_a l \qquad [11.6]$$

where l = sloping height of the inclined face

This resultant thrust may be treated as a single force acting at an angle θ to horizontal. Alternatively, the horizontal and vertical components, p_h and W_s respectively, may be considered; both acting at the point of application of P_A.

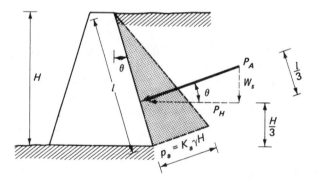

Fig. 11.13

11.10 Effect of surcharge

When a superimposed pressure is applied at the surface of the retained soil, as in Fig. 11.4(a), the active lateral pressure is increased by a proportional amount. The pressure distribution is then as shown in Fig. 11.4(b), the rectangular portion represents the lateral pressure due to the surcharge; the resultant thrust is P_{AS} acting at half the depth. The total overturning moment is therefore the sum of the moments of P_A and P_{AS}.

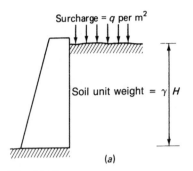

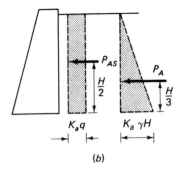

Fig. 11.14

EXAMPLE 11.4 Determine the active lateral thrust acting on the retaining wall described in Example 11.3 if there is also a surcharge of 20 kN/m² acting on the surface of the soil.

Active lateral thrust due to the surcharge:

$$P_{AS} = K_a q H$$
$$= \tfrac{1}{3} \times 20 \times 4 \cdot 5$$
$$= 30 \text{ kN/m}$$

The active lateral thrust due to the retained soil, as calculated before, is 66·2 kN/m.

Thus the TOTAL ACTIVE LATERAL THRUST $= 30 \cdot 0 + 66 \cdot 2$
$$= 96 \cdot 2 \text{ kN/m (Answer)}$$

11.11 Coulomb's wedge theory

Whereas in Rankine's theory *pressures* at a given depth below the surface are considered, in Coulomb's wedge theory the forces maintaining the stability of a wedge-shaped mass of soil lying adjacent to the wall are considered. This wedge of soil is assumed to slide down the surface of rupture as failure takes place and so bears against the face of the wall (Fig. 11.15).

An approximate solution for walls with a vertical, or nearly vertical, back may be obtained by assuming that the plane of rupture bisects the angle between the ϕ-line and a vertical plane passing through B (i.e. the face of the wall, if this is itself vertical).

Figure 11.16 illustrates the solution for a wall having a vertical back. The weight of the soil in wedge ABC acts vertically downwards through its centre of mass (G). The wedge is restrained against movement by two other forces; the wall offers a reaction P_A to the active lateral thrust and the soil beneath the plane of rupture BC offers a reaction R. These three forces acting upon the wedge intersect at a common point D. A triangle

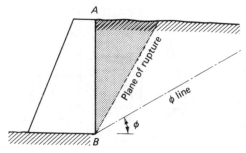

Fig. 11.15

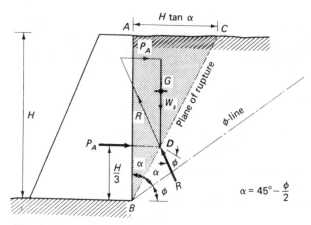

Fig. 11.16

of forces can be drawn and its solution will yield the magnitude of the resultant active lateral thrust P_A.

The procedure is as follows:

1. Evaluate angle $\alpha = 45° - \dfrac{\phi}{2}$

2. Calculate the weight of the wedge of soil ABC (per unit length of wall)

$$
\begin{aligned}
W_s &= \text{volume of wedge} \times \gamma \\
&= \text{area of triangle } ABC \times 1 \times \gamma \\
&= \tfrac{1}{2} \times H \times H \tan\alpha \times \gamma \\
&= \tfrac{1}{2} H^2 \gamma \tan\alpha
\end{aligned}
\tag{11.7}
$$

3. Calculate the active lateral thrust P_A (per unit length of wall)

$$
\begin{aligned}
P_A &= W_s \tan \alpha \\
&= \tfrac{1}{2} H^2 \gamma \tan^2\alpha
\end{aligned}
\tag{11.8}
$$

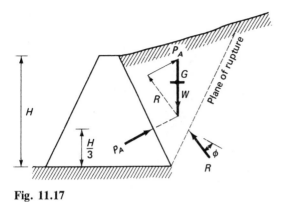

Fig. 11.17

It is interesting to note that this solution is identical with that obtained using Rankine's theory, since

$$\tan^2\alpha = \tan^2\left(45° - \frac{\phi}{2}\right) = \frac{1 - \sin\phi}{1 + \sin\phi} = K_a$$

The wedge theory is of particular value when the wall has an inclined face, or when the surface of the retained soil is inclined, Fig. 11.17. The use of squared paper and a graphical analysis make a relatively simple solution then possible.

11.12 Stability calculations

A retaining wall (or similar structure) is said to be stable if the limit states set out in Section 11.2 are not exceeded. In order to examine these criteria for a particular case, a number of basic calculations must be made. Those applicable to the more usual cases are summarized as follows, additional work may be necessary for special cases.

1. Calculation of weight and lateral thrusts
2. Determination of the position of the ground reaction
3. Determination of the ground bearing pressure
4. Determination of the factor of safety against sliding.

11.13 Determination of the position of the ground reaction

The equilibrium of a retaining wall is maintained by an upward reaction from the ground upon which it rests (Fig. 11.18). The magnitude of this reaction is given by the sum of the vertical forces acting downward.

$$V = W_1 + W_2 + \dots \text{ etc.} \qquad [11.9]$$

The position at which the ground reaction intersects the base *AB* is most important in the assessment of the stability of the wall. Since the

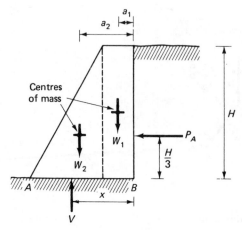

Fig. 11.18

magnitude of V is readily found, the value of x can be determined by taking moments about B.

$$Vx = P_A \frac{H}{3} + W_1a_1 + W_2a_2 \qquad [11.10]$$

from which x may be evaluated.

Where the base of the wall is stepped, the weight of the soil prisms resting on the steps may be included in this calculation, see Fig. 11.19.

$$V = W_1 + W_2 + W_s \qquad [11.11]$$

$$Vx = P_A \frac{H}{3} + W_1a_1 + W_2a_2 + W_sa_s \qquad [11.12]$$

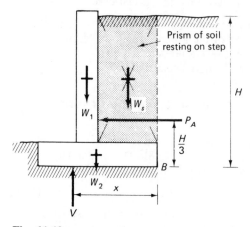

Fig. 11.19

11.14 Ground bearing pressure

The combination of the downward acting forces (weights) and the overturning moments gives rise to a reaction pressure from the ground beneath the base. The distribution of this GROUND BEARING PRESSURE and its maximum value must be determined in order to assess the factor of safety against failure of the foundation soil. From the work done in the previous chapter, it will be seen that the ground bearing pressure consists of a direct stress component and a bending stress component (Fig. 11.20).

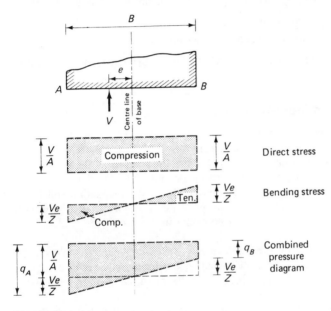

Fig. 11.20

Direct load $= V$ Direct stress $= \dfrac{V}{A}$

'Bending moment' about the centre line of the base $= Ve$ 'Bending stress' $= \dfrac{Ve}{Z}$

The ground bearing pressures (stresses) at A and B are obtained from the algebraic addition of the direct and bending stresses:

$$q_A = \frac{V}{A} + \frac{Ve}{z} \qquad\qquad [11.13]$$

$$q_B = \frac{V}{A} - \frac{Ve}{Z} \qquad\qquad [11.14]$$

where V = the resultant ground reaction
e = the eccentricity of V about the centre line of the base
A = the area of the base
= $1 \times B$ (per unit length of wall)
Z = the section modulus of the base about its centre line
$$= \frac{1 \times B^2}{6} \text{ (per unit length of wall)}$$

The factor of safety against failure of the foundation soil is given by:

$$F_s = \frac{\text{ultimate bearing capacity of the soil}}{\text{maximum ground bearing pressure}}$$

It is recommended that THE VALUE OF F_s SHOULD BE AT LEAST THREE.

11.15   The middle-third rule

The stability condition of the wall may be examined by considering a number of limiting values of the eccentricity e.

Firstly, if $e = 0$, only direct stress exists beneath the base and the pressure distribution diagram is a rectangle (Fig. 11.21). This condition would appear to be the most desirable as far as the foundation soil is concerned, since the ground bearing pressure is uniform, but it is usually only obtainable at the expense of economy in the use of the walling materials.

If e were found to be greater than $B/2$, this would indicate that the wall would overturn, since, theoretically, the reaction V would not bear against the base. It is usual to provide a factor of safety of at least two against overturning.

Next, consider a horizontal plane within the wall, just above the base AB. The distribution of pressure on this plane will be as shown in Fig. 11.22. After examining the three cases shown it will be seen that tension will occur at B when $e > B/6$. The critical value of e is therefore that

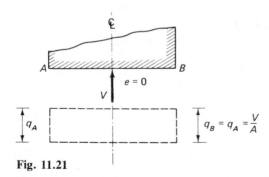

Fig. 11.21

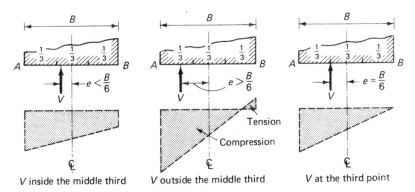

Fig. 11.22

which will give zero pressure at B, i.e. $e = B/6$, when the resultant V passes through a point on the base one-third of the way in.

It may be stated therefore that NO TENSION WILL OCCUR IN THE WALL JUST ABOVE THE BASE IF THE RESULTANT GROUND REACTION CUTS THE BASE WITHIN THE MIDDLE THIRD. This statement is known as the **middle-third rule**.

In jointed construction, such as masonry and brickwork, it is usual to use the middle-third rule as a design limit state, in order to prevent the possibility of tension occurring in the joints which could lead to cracking and subsequent deterioration of the wall. Some degree of tension can be tolerated, however, in reinforced concrete walls and systems of sheet piling. In such cases a modification to the procedure for calculating the maximum ground bearing pressure is necessary.

Since tension cannot exist between the base of the wall and the soil beneath, the pressure distribution will be as shown in Fig. 11.23. The area of the pressure diagram is equal to the ground reaction V.

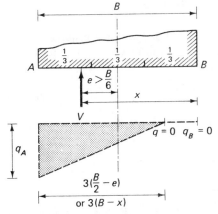

Fig. 11.23

Then
$$V = \tfrac{1}{2} \times q_A \times 3 \left(\frac{B}{2} - e \right)$$

Hence
$$q_A = \frac{2V}{3\left(\dfrac{B}{2} - e \right)} \qquad [11.15]$$

or
$$q_A = \frac{2V}{3(B - x)} \qquad [11.16]$$

11.16 Sliding

The wall will begin to slide if the resistance (R_S), developed by friction between the underside of the base and the soil beneath, is less than the horizontal thrust P — see Fig. 11.24.

The value of R_S increases uniformly with V (the total weight acting downward). If values of R_S are plotted against the corresponding values of V, a straight-line graph is produced (Fig. 11.25) the angle of slope of which is defined as the ANGLE OF WALL FRICTION (δ).

The coefficient of friction, $\mu = \tan \delta$ \qquad [11.17]

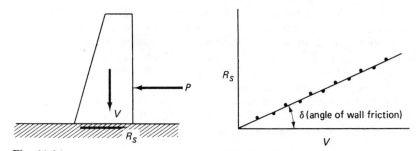

Fig. 11.24   **Fig. 11.25**

The value of δ varies with the type of soil and with the material and texture of the wall base. However, it is usual to assume a value for δ beween $0 \cdot 75 \, \phi$ and $1 \cdot 00 \, \phi$.

Factor of safety against sliding: $F_s = \dfrac{R_S}{P}$ \qquad [11.18]

THIS FACTOR OF SAFETY SHOULD BE NOT LESS THAN TWO.

WORKED EXAMPLES

EXAMPLE 11.5 The masonry dam, whose cross-section is shown in Fig. 11.26 retains water against its vertical face. Examine the stability condition of the dam

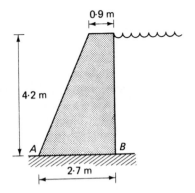

Fig. 11.26

with respect to tension in the joints, ground bearing pressure and sliding. The unit weight of the masonry is $21 \cdot 6 \ kN/m^3$.

First calculate the lateral thrust due to the retained liquid.

$$P = \tfrac{1}{2}\gamma H^2 = \tfrac{1}{2} \times 9 \cdot 81 \times 4 \cdot 2^2$$
$$= 86 \cdot 5 \ kN/m$$

Now divide the cross-section of the wall into a rectangle and triangle and calculate their respective weights W_1 and W_2 (Fig. 11.27).

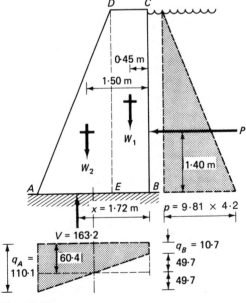

Fig. 11.27

$$W_1 = 4 \cdot 2 \times 0 \cdot 9 \times 21 \cdot 6 = 81 \cdot 6 \text{ kN/m}$$

$$W_2 = 4 \cdot 2 \times \frac{1 \cdot 8}{2} \times 21 \cdot 6 = 81 \cdot 6 \text{ kN/m}$$

and
$$V = W_1 + W_2 = 163 \cdot 2 \text{ kN/m}$$

The moment of P, W_1 and W_2 about B must be balanced by the moment of V also about B.

Then $Vx = P \times 1 \cdot 4 + W_1 \times 0 \cdot 45 + W_2 \times 1 \cdot 5$

From which the position of V may be determined.

$$x = \frac{86 \cdot 5 \times 1 \cdot 4 + 81 \cdot 6 \times 0 \cdot 45 + 81 \cdot 6 \times 1 \cdot 5}{163 \cdot 2}$$

$$= \frac{280 \cdot 2}{163 \cdot 2} = 1 \cdot 72 \text{ m}$$

This value lies between $0 \cdot 9$ m and $1 \cdot 8$ m and therefore V CUTS THE BASE WITHIN THE MIDDLE THIRD and hence no tension will develop in the joints. The eccentricity of V about the centre line will be:

$$e = 1 \cdot 72 - 1 \cdot 35$$

$$= 0 \cdot 37 \text{ m}$$

The ground bearing pressure is now found using eqns [11.13] and [11.14].

$$\frac{V}{A} = \frac{163 \cdot 2}{2 \cdot 7 \times 1 \cdot 0} \qquad = 60 \cdot 4 \text{ kN/m}^2$$

$$\frac{Ve}{Z} = \frac{163 \cdot 2 \times 0 \cdot 37 \times 6}{1 \cdot 0 \times 2 \cdot 7^2} = 49 \cdot 7 \text{ kN/m}^2$$

Then
$$q_A = 60 \cdot 4 + 49 \cdot 7 = 110 \cdot 1 \text{ kN/m}^2$$
$$q_B = 60 \cdot 4 - 49 \cdot 7 = 10 \cdot 7 \text{ kN/m}^2 \quad \text{(Answer)}$$

The allowable bearing pressure of the soil must therefore be not less than 110 kN/m². In order to prevent sliding, the frictional resistance developed between the base and the soil beneath must be at least equal to P.

Therefore the coefficient of friction required $(\mu) = \dfrac{86 \cdot 5}{163 \cdot 2}$

$$= 0 \cdot 53$$

Now the angle of wall friction (δ) required $= \arctan 0 \cdot 53$

$$= 28°$$

If it is assumed that $\delta = \phi$, then ϕ would have to be at least 28° if sliding is not to occur. (Answer)

EXAMPLE 11.6   Figure 11.28 shows a reinforced concrete retaining wall which supports a cohesionless soil having the following properties:
unit weight = 19·8 kN/m³, angle of shearing resistance = 30°. The unit weight of concrete is 24·5 kN/m³

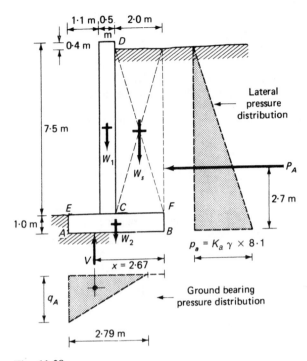

Fig. 11.28

Determine: (a) the position at which the resultant ground reaction cuts the base,
        (b) the ground bearing pressure at A and B.
        (c) the factor of safety against sliding, if the angle of wall friction is
        equal to the angle of shearing resistance.

Rankine's coefficient, $K_a = \dfrac{1 - \sin\phi}{1 + \sin\phi} = \dfrac{1 - 0\cdot5}{1 + 0\cdot5} = \dfrac{1}{3}$

Then the active thrust $P_A = \frac{1}{2} \times \frac{1}{3} \times 19\cdot8 \times 8\cdot1^2$
$$= 216\cdot5 \text{ kN/m}$$

Three vertical forces are considered: the weights of the base and the stalk of the wall (W_1 and W_2 respectively) and the weight of the soil resting on the base behind the wall (W_s).

$$
\begin{aligned}
W_1 &= 7\cdot5 \times 0\cdot5 \times 24\cdot5 &&= 91\cdot9 \text{ kN/m} \\
W_2 &= 1\cdot0 \times 3\cdot6 \times 24\cdot5 &&= 88\cdot2 \text{ kN/m} \\
W_s &= 7\cdot1 \times 2\cdot0 \times 19\cdot8 &&= \underline{281\cdot2 \text{ kN/m}} \\
\text{and } V = W_1 + W_2 + W_s &&&= \underline{\underline{461\cdot3 \text{ kN/m}}}
\end{aligned}
$$

The *position* of V is determined by taking moments about B:

$$461\cdot3x = 216\cdot5 \times 2\cdot7 + 91\cdot9 \times 2\cdot25 + 88\cdot2 \times 1\cdot8 + 281\cdot2 \times 1\cdot0$$

Therefore

$$x = \frac{585 \cdot 1 + 206 \cdot 8 + 158 \cdot 8 + 281 \cdot 2}{461 \cdot 7} = \frac{1231}{461 \cdot 3} = 2 \cdot 67 \text{ m}$$
$$\text{(from } B)\text{(Answer)}$$

Since the base is $3 \cdot 6$ m wide, its middle third lies between $1 \cdot 2$ and $2 \cdot 4$ m. In this case, therefore, the resultant V cuts the base *outside* the middle third. This will not necessarily produce adverse consequences in reinforced concrete; however, the ground bearing pressure will have to be calculated using either eqn [11.15] or [11.16].

Hence from eqn [11.16]:

$$q_A = \frac{2 \times 461 \cdot 3}{3(3 \cdot 60 - 2 \cdot 67)} = 331 \text{ kN/m}^2 \quad \text{(Answer)}$$

and
$$q_B = 0$$

Since the angle of wall friction is equal to the angle of shearing resistance, the ultimate sliding resistance that could be developed beneath the base is as follows:

$$R_s = V \tan \delta = V \tan 30° = 461 \cdot 3 \times 0 \cdot 577$$
$$= 266 \cdot 2 \text{ kN/m}$$

If this is the *only* resistance to sliding, the factor of safety will be:

$$F_{(sliding)} = \frac{\text{force resisting sliding}}{\text{force tending to cause sliding}}$$

$$= \frac{R_s}{P_A}$$

$$= \frac{266 \cdot 2}{216 \cdot 5} = 1 \cdot 23 \quad \text{(Answer)}$$

This figure is not really adequate, a value of $2 \cdot 0$ is the required minimum. Some measure of redesign would therefore appear to be necessary.

EXAMPLE 11.7 *The masonry retaining wall shown in Fig. 11.29 supports a cohesionless soil, the surface of which is inclined upwards from the top of the wall at an angle of 10°. The soil has a unit weight of 18·6 kN/m³ and an angle of shearing resistance of 32°; the unit weight of the masonry is 23·5 kN/m³. The angle of wall friction may be taken as 1·0 φ along the base and as 0·75 φ along the supporting face of the wall. Determine: (a) the position at which the resultant ground reaction cuts the base AB, (b) the maximum ground bearing pressure occurring beneath the base, (c) the factor of safety that exists against a sliding failure of the wall.*

(a) In view of the sloping ground surface, the wedge theory will be used to determine the value of the active lateral earth pressure.

The critical wedge of soil is shown in Fig. 11.30, i.e. the triangular portion BCD. The plane of rupture (BD) is assumed to bisect the angle between the face BC and the 'φ-line'.

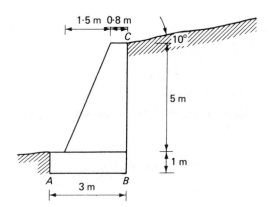

Fig. 11.29

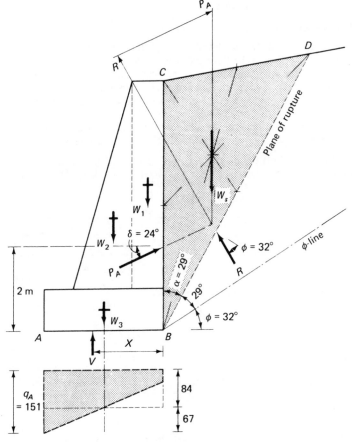

Fig. 11.30

Then
$$\alpha = 45° - \frac{\phi}{2}$$
$$= 45° - 16° = 29°$$

The diagram is drawn to scale and therefore the length CD may be scaled:
$$CD = 3·75 \text{ m}$$

The area of the wedge can either be calculated or found with the aid of squared tracing paper. A calculation will be done here:

$$\text{Area of wedge } BDC = \tfrac{1}{2} BC \ CD \sin 100°$$
$$= \tfrac{1}{2} \times 6·0 \times 3·75 \times \sin 80° \text{ (since } \sin 100° = \sin 80°)$$
$$= 11·1 \text{ m}^2$$

$$\text{Weight of wedge } BCD = \text{area} \times \gamma$$
$$= 11·1 \times 18·6$$
$$= 206 \text{ kN/m}$$

The triangle of forces is now drawn, consisting of:

1. The weight of the wedge (W_s) which acts through the centre of the wedge located at the intersection of the three medians.
2. The reaction from the wall equal to the active lateral thrust (P_A) which acts at a point one-third of the way up BC and makes an angle δ ($\delta = 0·75 \ \phi = 24°$) to horizontal (this allows for the effect of the wall friction along the face of the wall).
3. The reaction from the soil beneath the plane of rupture (R) which makes an angle ϕ to the plane of rupture. The *value* of R is not actually required.

The value of P_A is scaled off the diagram: $P_A = 101$ kN/m. The stability calculation may be set out in tabulated form to save space and keep the figures together.

Force (kN/m)		Lever arm about B	Moment about B
$P_A = 101$		$a_p = 2 \cos 24°$ $= 1·827$	185
$W_1 = 0·8 \times 5·0 \times 23·5 =$	94	0·4	38
$W_2 = \dfrac{1·5}{2} \times 5·0 \times 23·5 =$	88	1·3	114
$W_3 = 1·0 \times 3·0 \times 23·5 =$	71	1·5	107
$V = \Sigma(W) =$	253 kN/m	$\Sigma(Wa) = 444$	

Therefore
$$x = \frac{\Sigma(Wa)}{\Sigma(W)} = \frac{444}{253} = 1·76 \text{ m (Answer)}$$

Hence V lies within the middle third of the base and so no tension will be developed in the joints.

(b) Since V lies within the middle third, the ground bearing pressure diagram will be a trapezium and eqns [11.13] and [11.14] will be used.

$$\frac{V}{A} = \frac{253}{3} = 84 \text{ kN/m}^2$$

$$\frac{Ve}{Z} = \frac{253\,(1\cdot76 - 1\cdot50) \times 6}{3^2} = 42\cdot8 \text{ kN/m}^2$$

The maximum ground bearing pressure occurs at the toe A:

$$q_A = 84 + 42\cdot8 = 126\cdot8 \text{ kN/m}^2 \quad \text{(Answer)}$$

(c) Ultimate sliding resistance: $R_S = V \times \tan 32°$
$$= 253 \times 0\cdot625$$
$$= 158 \text{ kN/m}$$

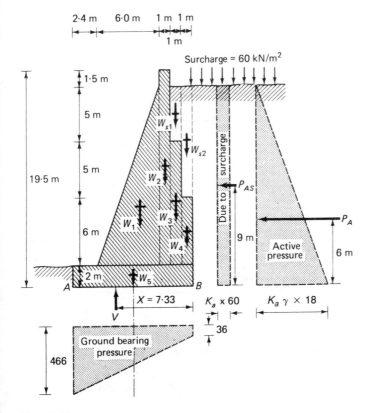

Fig. 11.31

Then the factor of safety against sliding will be:

$$F_{\text{(sliding)}} = \frac{R_S}{\text{horizontal component of } P_A}$$

$$= \frac{158}{101 \times \cos 24°}$$

$$= 1 \cdot 43 \quad \text{(Answer)}$$

EXAMPLE 11.8 *The masonry retaining wall shown in Fig. 11.31 supports a cohesionless soil applied to the surface of which there is a surcharge pressure of 60 kN/m². The soil has a unit weight of 19·1 kN/m³ and an angle of shearing resistance of 28°; the unit weight of the masonry is 24·5 kN/m³.*
Determine: (a) the position of the resultant ground reaction, (b) the ground bearing pressures at A and B, (c) the factor of safety against sliding.

(a) Rankine's coefficient, $\quad K_a = \dfrac{1 - \sin 28°}{1 + \sin 28°} = \dfrac{1 - 0·470}{1 + 0·470} = 0·361$

	Force (kN/m)			Lever arm about B	Moment about B
$P_A = 0·361 \times 19·1 \times \dfrac{18^2}{2}$		=	1 117	6·0	6 702
$P_{AS} = 0·361 \times 60 \times 18$		=	390	9·0	3 510
$W_1 = \dfrac{6}{2} \times 16 \times 24·5$		=	1 176	5·0	5 880
$W_2 = 1·0 \times 17·5 \times 24·5$		=	429	2·5	1 073
$W_3 = 1·0 \times 11·0 \times 24·5$		=	270	1·5	405
$W_4 = 1·0 \times 6·0 \times 24·5$		=	147	0·5	74
$W_5 = 11·4 \times 2·0 \times 24·5$		=	559	5·55	3 102
$W_{S_1} = 1·0 \times 5·0 \times 19·1$		=	96	1·5	144
$W_{S_2} = 1·0 \times 10·0 \times 19·1$		=	191	0·5	96
		$V = \Sigma(W) =$	2 868	$\Sigma(Wa) =$	20 985

Therefore $\qquad\qquad\qquad\qquad x = \dfrac{20\ 985}{2\ 868} = 7·32$ m (Answer)

Hence V lies within the middle third of the base and so no tension will occur in the joints.

(b) The ground bearing pressure may be determined by using eqns [11.13] and [11.14].

$$\frac{V}{A} = \frac{2868}{11·4} = 252 \text{ kN/m}^2$$

$$\frac{Ve}{Z} = \frac{2868\,(7\cdot32\,-\,5\cdot70)\,\times\,6}{11\cdot4^2} = 215\ \text{kN/m}^2$$

Then $\qquad\qquad q_A = 252 + 215 = 467\ \text{kN/m}^2$ (Answer)

and $\qquad\qquad q_B = 252 - 215 = 37\ \text{kN/m}^2$

(c) Assuming that $\delta = \phi$, then the ultimate shearing resistance along the base will be:

$$R_S = V \times \tan 28°$$
$$= 2\ 868 \times 0\cdot532 = 1\ 526\ \text{kN/m}$$

Hence $\qquad\qquad F_{(\text{sliding})} = \dfrac{R_s}{P_A + P_{AS}}$

$$= \frac{1\ 526}{1\ 117 + 390} = 1\cdot01\quad\text{(Answer)}$$

This value is much too near to $1\cdot00$ to be acceptable; it would appear that the wall, in its present form, is on the brink of failure; a modified design would now have to be considered.

Table 11.1 Typical properties of cohesionless materials

Material	Angle of shearing resistance (ϕ) (deg)	Drained bulk unit weight (γ) (kN/m³)
Gravel	35−45	16−20
Sand — loose	25−35	17−19
— compact	30−40	18−21
Organic topsoil	15−30	13−18
Coal and coke	30−45	6−9
Broken brick	35−45	11−16
Broken rock	35−45	14−21
Ashes and clinker	35−45	6−10

Table 11.2 Typical densities of wall fabric

Material	Unit weight (γ) (kN/m³)
Brickwork	18−20
Concrete — unreinforced	22−24
— reinforced	23−25
Masonry — granite	25−26
— limestone	21−26
— sandstone	22−25
— slate	25−27

Table 11.3 Recommended maximum allowable bearing capacities

Type of soil	Maximum allowable bearing capacity (kN/m^2)
Compact sands and gravels	400–600
Loose sands and gravels	150–400
Solid non-fissured rocks	600–3000
Hard clays and soft rocks	300–600
Stiff clays and sandy clays	150–300
Firm clays and sandy clays	75–150
Soft clays and silts	0–75
Fill and made-ground	variable

Notes: 1. The above values only apply where the foundation is 1 m or more wide and is founded at a depth of at least 0·6 m.
2. Ultimate bearing capacity = allowable bearing capacity × 3.

EXERCISES   CHAPTER 11

1. A vertical wall of height 2·2 m is subject to a uniform wind pressure of 800 N/m². Calculate the overturning moment acting at the base of the wall in kNm per metre run of wall.

2. Calculate the lateral thrust exerted by a depth of water of 3·5 m.

3. The water table stands at a depth of 4·2 m against the supporting face of a retaining wall. Calculate the overturning moment acting at the base of the wall due to hydrostatic pressure.

4. The resultant horizontal force exerted by water pressure against the face of a retaining wall is 450 kN/m run. Calculate the height of the water table above the base of the wall.

5. Soil having a unit weight of 18·8 kN/m³ and an angle of shearing resistance of 30° is retained by a wall with a vertical back. Calculate the overturning moment about the base of the wall due to a depth of soil of 3·6 m.

6. Fig. 11.32 shows a retaining wall which supports soil having a unit weight of 18·6 kN/m³ and an angle of shearing resistance of 30°. Using the wedge theory, determine the magnitude of the horizontal thrust acting on the wall (neglect wall friction).

7–10. Figs 11.33–11.36 show the forces acting on a metre length of wall. In each case determine the position of the ground reaction V.

11. Fig. 11.37 shows a retaining wall for which the factor of safety against overturning is to be 2. Calculate the maximum allowable value for the horizontal thrust P.

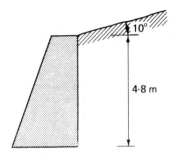

Fig. 11.32

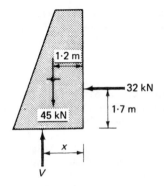

Fig. 11.33

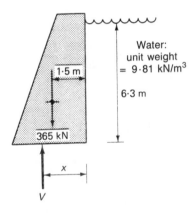

Fig. 11.34

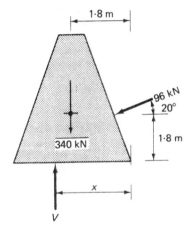

Fig. 11.35

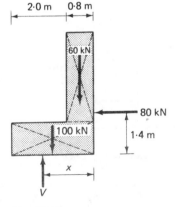

Fig. 11.36

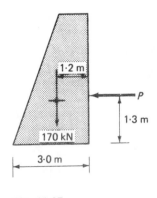

Fig. 11.37

12. A brick wall is of uniform thickness, 0·45 m, and is 4·6 m high from the ground. Determine the factor of safety against overturning if the unit weight of the brickwork is 18·6 kN/m³ and the wind pressure may be assumed to be uniform at 750 N/m².

13. Fig. 11.38 shows a brickwork wall of thickness 0·70 m which, in addition to supporting the vertical load of 120 kN, retains soil which imposes a horiztonal thrust of 16·7 kN per metre length. Calculate the factor of safety against overturning. The unit weight of the brickwork is 19·1 kN/m³.

14. Fig. 11.39 shows a water tank supported by a steel tower subject to wind loads as indicated. Each of the four tower legs is supported on a concrete slab foundation. Due to the mass of the tank, the steel framework and the concrete slabs, a total force of 350 kN acts downwards along the vertical centre-line. Due to the liquid carried in the tank a maximum force of 900 kN may act downwards also along the vertical centre-line. For the evaluation shown calculate the factors of safety against overturning for the tank-empty and tank-full conditions. Calculate also the maximum possible values for the reactions R_A and R_B.

15. A masonry dam of trapezoidal section has a vertical retaining face of height 14 m. The base is 8·4 m wide and the top is 3·0 m wide. If the maximum water level coincides with the top of the dam, examine the stability conditions with regard to tension in the joints and ground bearing pressure.
Unit weight of the masonry = 23·5 kN/m³

16. A masonry dam of trapezoidal section is 35 m high and has a base of width 20 m. The retaining face is vertical and the top is 3·5 m wide. The unit weight of the masonry is 23·5 kN/m³. Draw a diagram showing the calculated normal stress distribution beneath the base of the dam.

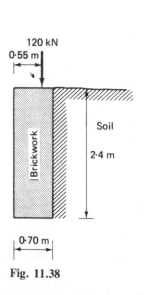

Fig. 11.38

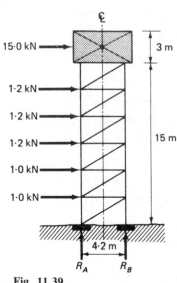

Fig. 11.39

17. Fig. 11.40 shows a mass concrete wall which retains soil against its vertical face up to the top of the wall. The unit weight of the concrete is 24 kN/m³; the soil has a unit weight of 18 kN/m³ and an angle of shearing resistance of 30°. Calculate the intensity of the bearing pressure beneath the toe A and the heel B.

18. Fig. 11.41 shows a masonry retaining wall which retains a cohesionless soil against its vertical face. The unit weight of the masonry is 22 kN/m³; the soil has a unit weight of 17·5 kN/m³ and an angle of shearing resistance of 28°.

(a) Examine the stability of the wall with respect to tension occurring in the joints.
(b) Calculate the maximum ground bearing pressure occurring beneath the base of the wall.
(c) Calculate the factor of safety against sliding assuming an angle of friction beneath the base of 24°.

19. Fig. 11.42 shows a mass concrete retaining wall which supports soil having a unit weight of 16·5 kN/m³ and an angle of shearing resistance of 30°. The unit weight of the concrete is 24 kN/m³.

(a) Calculate the maximum ground bearing pressure occurring beneath the base.
(b) Calculate the factor of safety against sliding if the angle of friction beneath the base may be taken as 0·9 ϕ.

20. Fig. 11.43 shows the section of a reinforced concrete retaining wall which supports a cohesionless soil having a unit weight of 18 kN/m³ and an angle of shearing resistance of 32°. The unit weight of the concrete is 24 kN/m³. Calculate the maximum ground bearing pressure exerted on the soil beneath the base.

21. Fig. 11.44 shows a masonry retaining wall which supports against its steeper face a cohesionless soil having a unit weight of 18 kN/m³ and an angle of shearing resistance of 30°. The unit weight of the masonry is 22 kN/m³. The angle of wall friction may be assumed to be 24° along the supporting face and 28° along the base.

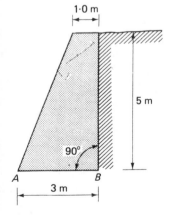

Fig. 11.40

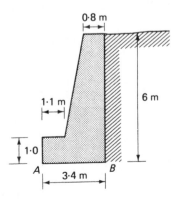

Fig. 11.41

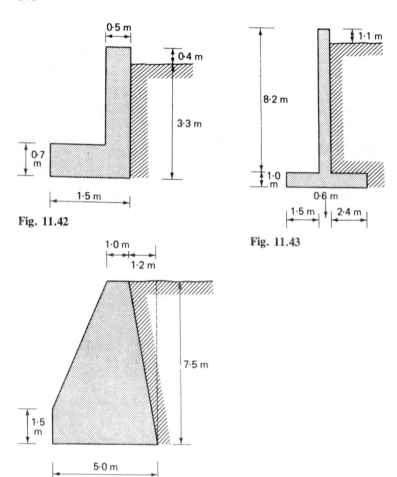

Fig. 11.42

Fig. 11.43

Fig. 11.44

(a) Determine the position of the ground reaction force and from this deduce the stability of the wall with respect to tension occurring in the joints near the base.
(b) Calculate the maximum ground bearing pressure beneath the base.
(c) Calculate the factor of safety against sliding.

22. A wall of rectangular section is required to support a vertical soil face of height $4 \cdot 8$ m. The soil has a unit weight of 17 kN/m³ and an angle of shearing resistance of 30°; the unit weight of the wall fabric is 24 kN/m³. Calculate the thickness of wall required when

(a) a factor of safety of 2 against overturning is required;
(b) the minimum condition to prevent tension occurring in the joints near the base is satisfied.

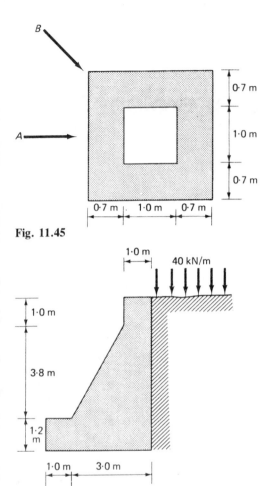

Fig. 11.45

Fig. 11.46

23. Fig. 11.45 shows a sectional plan of a brick chimney which has parallel sides and a height of 18 m. Assuming a uniform distribution of wind pressure over the full height of the chimney, calculate the minimum value of the wind pressure at which tension will occur in the joints. The unit weight of the brickwork is 20 kN/m³:

(a) when the wind is in direction A,
(b) when the wind is in direction B.

24. Fig. 11.46 shows a retaining wall which supports a cohesionless soil having a unit weight of 18·5 kN/m³ and an angle of shearing resistance of 30°. The unit weight of the fabric is 24 kN/m³. A surcharge load of 40 kN/m² is applied to the surface of the soil. Calculate the maximum ground bearing pressure occurring beneath the base of the wall.

Answers to exercises

Chapter 1

1.1 1962 N; **1.2** 97·38, 48·69, 19·48,
9·74 mm; **1.3** 7848 N; **1.4** 21·6 kN,
31·7°; **1.5** 166·5 kN, −19·9°; **1.6**
71·3 kN, 52·5°; **1.7** 18·22 kN,
−53·9°; **1.8** 34·4 kN, 26·1°; **1.9**
579·5 kN, 15°; **1.10** $P = 44·8$ kN,
$Q = 36·6$ kN; **1.11** $P = 51·8$ kN,
$Q = 73·2$ kN; **1.12** $P = 57·7$ kN,
$Q = 115·5$ kN; **1.13** $P = 57·7$ kN,
$Q = 57·7$ kN; **1.14** $P = 70·7$ kN,
$Q = 70·7$ kN; **1.15** $P = 75·0$ kN,
$Q = 125·0$ kN; **1.16** 14·9 N; **1.17**
22 kN at 1·091 m from 10 N force;
1.18 29 kN at 4·00 m from 5 N force;
1.19 55 kN at 4·41 m from LH end;
1.20 36 kN at 5·22 m from LH end;
1.21 21 kN at 3·14 m from LH end;
1.22 18 kN at 2·63 m from LH end;
1.23 18 kN at 4·22 m from LH end;
1.24 37 kN at 0·568 m from LH end;
1.25 $R_A = 11·25$ kN, $R_B = 13·75$ kN;
1.26 $R_A = 21·875$ kN,
$R_B = 18·125$ kN; **1.27** $R_A = 8·00$ kN,
$R_B = 22·00$ kN; **1.28** $R_A = 47·5$ kN,
$R_B = 82·5$ kN; **1.29** $R_A = 13·17$ kN,

$R_B = 16·83$ kN; **1.30** $R_A = 8·00$ kN,
$R_B = 4·00$ kN; **1.31** $R_A = 19·60$ kN,
$R_B = 8·40$ kN; **1.32** $R_A = 14·20$ kN,
$R_B = 11·80$ kN; **1.33** $R_A = 9·00$ kN,
$R_B = 3·00$ kN; **1.34** $R_A = 26·00$ kN,
$R_B = 19·00$ kN; **1.35** $R_A = 7·44$ kN,
$R_B = 9·06$ kN; **1.36** $R_A = 15·00$ kN,
horizontal to left, $R_B = 18·03$ kN at
33·7° to horizontal; **1.37** $R_A = 0$ kN,
$R_B = 12·0$ kN vertically upward; **1.38**
$R_A = 3 W − 100 (x − 1)$,
$R_B = 100x − 2W$; **1.39**
$H_A = 259·8$ N horizontal to left,
$R_B = 278·4$ kN at 21·05° to
horizontal; **1.40** $H_A = 31·6$ N
horizontal to right, $R_B = 16·1$ kN at
12·6° to horizontal; **1.41**
$R_A = 45·0$ kN, $R_B = 51·5$ kN at
29·1° to horizontal; **1.42**
$R_{A'} = −5·6$ kN, $R_B = 34·4$ kN at
70·8° to horizontal.

Chapter 2

2.1 to **2.16** (Figs 2.32 to 2.47) See the
following diagrams.
2.17 (Fig. 2.32) See Ans. **2.1.**

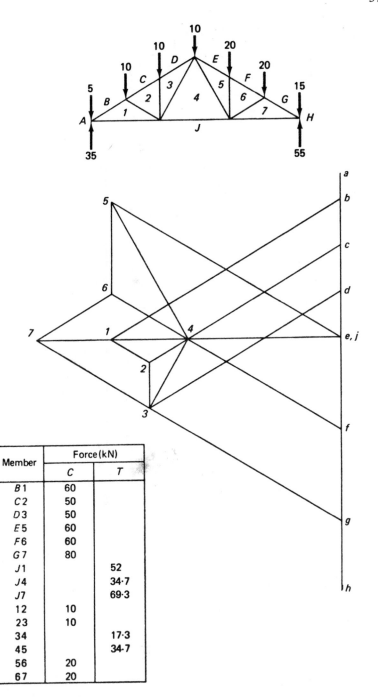

Member	Force (kN)	
	C	T
B 1	60	
C 2	50	
D 3	50	
E 5	60	
F 6	60	
G 7	80	
J 1		52
J 4		34·7
J 7		69·3
1 2	10	
2 3	10	
3 4		17·3
4 5		34·7
5 6	20	
6 7	20	

Exercise 2.1

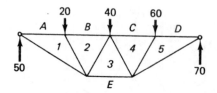

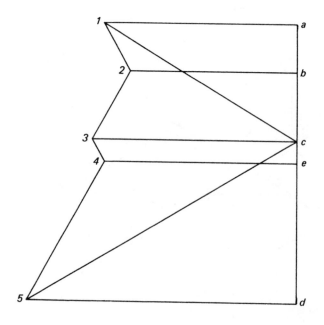

Member	Force (kN)	
	C	T
A 1	86·7	
B 2	75·2	
C 4	86·7	
D 5	121·2	
E 1		100·0
E 3		92·3
E 5		140·0
1 2	23·1	
2 3	34·7	
3 4	11·5	
4 5	69·3	

Exercise 2.2

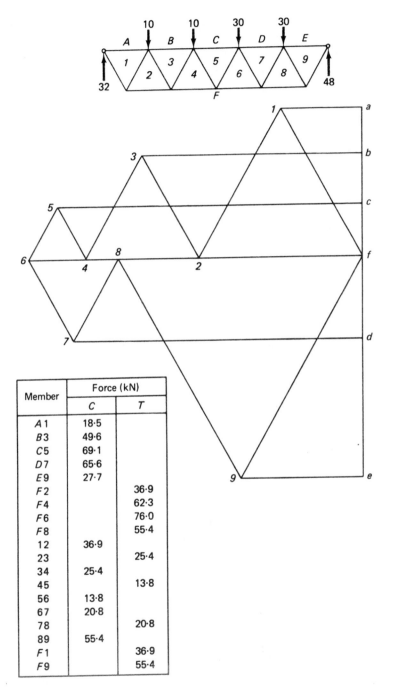

Member	Force (kN)	
	C	T
A1	18·5	
B3	49·6	
C5	69·1	
D7	65·6	
E9	27·7	
F2		36·9
F4		62·3
F6		76·0
F8		55·4
12	36·9	
23		25·4
34	25·4	
45		13·8
56	13·8	
67	20·8	
78		20·8
89	55·4	
F1		36·9
F9		55·4

Exercise 2.3

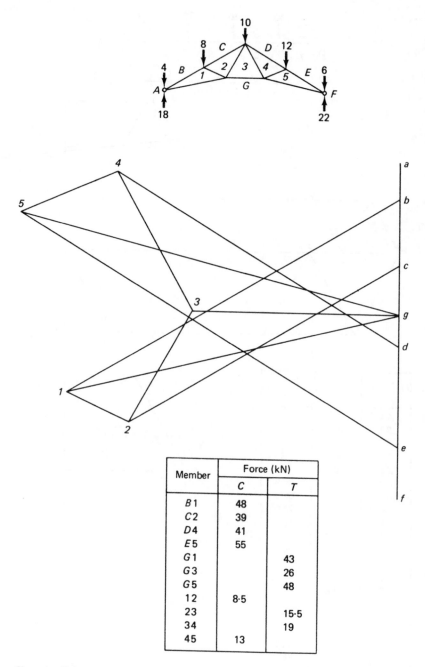

Member	Force (kN)	
	C	T
B1	48	
C2	39	
D4	41	
E5	55	
G1		43
G3		26
G5		48
12	8·5	
23		15·5
34		19
45	13	

Exercise 2.4

377

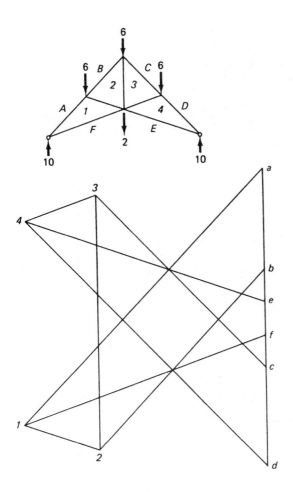

Member	Force (kN)	
	C	T
A1	21·5	
B2	14·9	
C3	14·9	
D4	21·5	
E4		15·8
F1		15·8
12	4·8	
23		15·1
34	4·8	

Exercise 2.5

378

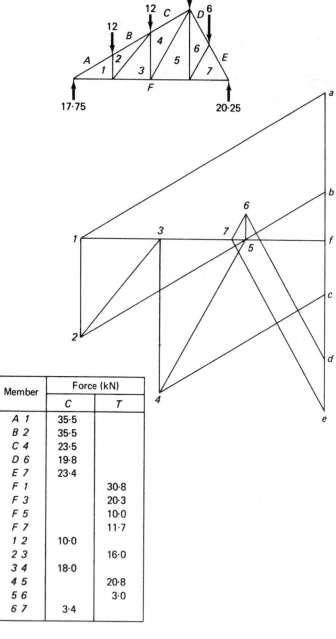

Member	Force (kN)	
	C	T
A 1	35·5	
B 2	35·5	
C 4	23·5	
D 6	19·8	
E 7	23·4	
F 1		30·8
F 3		20·3
F 5		10·0
F 7		11·7
1 2	10·0	
2 3		16·0
3 4	18·0	
4 5		20·8
5 6		3·0
6 7	3·4	

Exercise 2.6

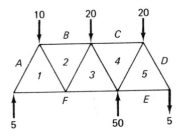

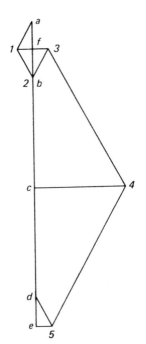

Member	Force (kN)	
	C	T
A 1	5·8	
B 2	0	
C 4		17·3
D 5		5·8
E 5	2·9	
F 1		2·9
F 3	2·9	
1 2	5·8	
2 3		5·8
3 4	29·0	
4 5	29·0	

Exercise 2.7

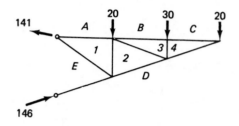

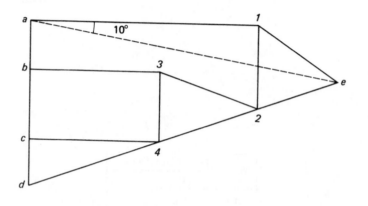

Member	Force (kN)	
	C	T
A 1		114
B 3		60
C 4		60
D 2	109	
D 4	63	
D E	146	
1 2	35	
2 3		46
3 4	30	
E 1		42

Exercise 2.8

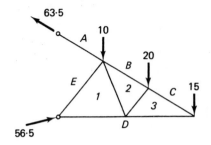

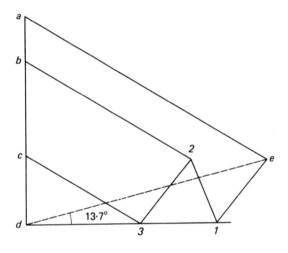

Member	Force (kN)	
	C	T
A E		63·5
B 2		43·5
C 3		30·0
D 1	43·5	
D 3	26·0	
E 1	17·3	
1 2		14·7
2 3	17·6	

Exercise 2.9

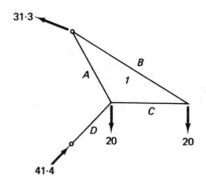

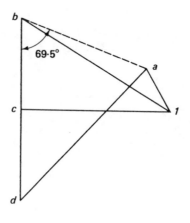

Member	Force (kN)	
	C	T
A 1	10·7	
B 1		40·0
C 1	34·6	
A D	41·4	

Exercise 2.10

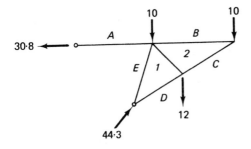

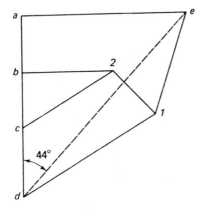

Member	Force (kN)	
	C	T
B 2		17·3
C 2	20·0	
D 1	28·8	
A E		30·8
E 1	19·6	
1 2		10·8

Exercise 2.11

384

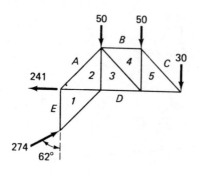

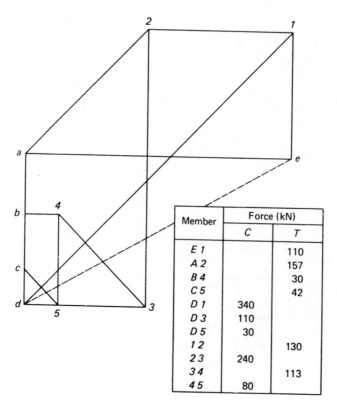

Member	Force (kN)	
	C	T
E 1		110
A 2		157
B 4		30
C 5		42
D 1	340	
D 3	110	
D 5	30	
1 2		130
2 3	240	
3 4		113
4 5	80	

Exercise 2.12

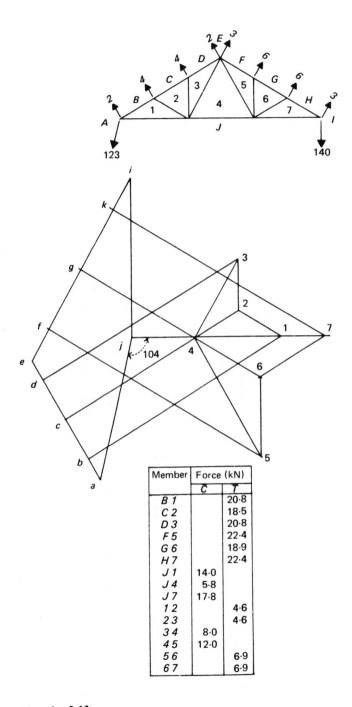

Member	Force (kN)	
	C	T
B 1		20·8
C 2		18·5
D 3		20·8
F 5		22·4
G 6		18·9
H 7		22·4
J 1	14·0	
J 4	5·8	
J 7	17·8	
1 2		4·6
2 3		4·6
3 4	8·0	
4 5	12·0	
5 6		6·9
6 7		6·9

Exercise 2.13

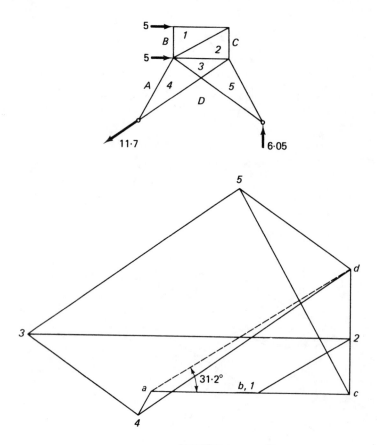

Member	Force (kN)	
	C	T
A 4		1·3
B 1	0	
C 1	5·0	
C 2	2·7	
C 5	11·4	
D 4		12·8
D 5		6·9
1 2		5·7
2 3	16·4	
3 4		6·9
3 5		12·8

Exercise 2.14

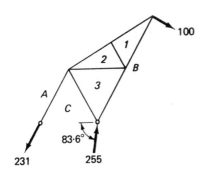

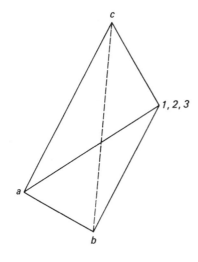

Member	Force (kN)	
	C	T
A C		231
A 1, A 2		200
B 1, B 3	173	
C 3	116	
1 2, 2 3	0	

Exercise 2.15

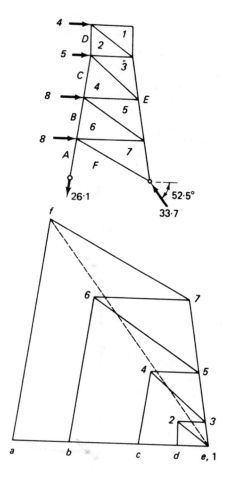

Member	Force (kN)	
	C	T
A F		26·1
B 6		17·4
C 4		8·9
D 2		3·0
E 1, E 2		0
E 3	3·0	
E 5	8·9	
E 7	17·4	
F 7	20·2	
1 2	4·9	
2 3		3·4
3 4	9·2	
4 5		6·1
5 6	15·2	
6 7		11·4

Exercise 2.16

2.18 (Fig. 2.33) See Ans. **2.2.**
2.19 (Fig. 2.35) See Ans. **2.4.**
2.20 (Fig. 2.36) See Ans. **2.5.**
2.21 (Fig. 2.41) See Ans. **2.10.**
2.22 (Fig. 2.43) See Ans. **2.12.**
2.23 (Fig. 2.44) See Ans. **2.13.**
2.24 (Fig. 2.46) See Ans. **2.15.**
2.25 (Fig. 2.33) See Ans. **2.2.**
2.26 (Fig. 2.34) See Ans. **2.3.**
2.27 (Fig. 2.35) See Ans. **2.4.**
2.28 (Fig. 2.36) See Ans. **2.5.**
2.29 (Fig. 2.37) See Ans. **2.6.**
2.30 (Fig. 2.38) See Ans. **2.7.**
2.31 (Fig. 2.39) See Ans. **2.8.**
2.32 (Fig. 2.41) See Ans. **2.10.**
2.33 (Fig. 2.43) See Ans. **2.12.**
2.34 (Fig. 2.46) See Ans. **2.15**
2.35 $17 \cdot 6$ kN; **2.36** $92 \cdot 3$ kN,
$65 \cdot 4$ kN; **2.37** $F_{AC} = 50 \cdot 0$ kN(T),
$F_{BC} = 86 \cdot 6$ kN(C),
(a) $F_{AD} = F_{AE} = 40 \cdot 0$ kN(T),
    $F_{BD} = F_{BE} = 28 \cdot 3$ kN(C),
    $F_{AB} = 24 \cdot 0$ kN(C),
    $V_B = 106 \cdot 6$ kN(up),
    $V_D = V_E = 28 \cdot 3$ kN(down);
(b) $F_{AD} = 61 \cdot 2$ kN(T)
    $F_{BD} = 43 \cdot 3$ kN(C)
    $F_{AB} = 18 \cdot 3$ kN(C),
    $F_{AE} = F_{BE} = V_E = 0$,
    $V_D = 43 \cdot 3$ kN(down),
    $V_B = 93 \cdot 3$ kN(up);
(c) $F_{AD} = 61 \cdot 2$ kN(T),
    $F_{BD} = 43 \cdot 3$ kN(C),
    $F_{AB} = 26 \cdot 8$ kN(C),
    $F_{AE} = 31 \cdot 1$ kN(T),
    $F_{BE} = 22 \cdot 0$ kN(C)
    $V_B = 101 \cdot 8$ kN(up),
    $V_D = 41 \cdot 3$ kN(down),
    $V_E = 22 \cdot 0$ kN(down).
2.38 In the two vertical members under
the loads: $F = 20$ kN(C). In all other
members: $F = 0$.

Chapter 3

3.1 $4 \cdot 33$ mm; **3.2** $139 \cdot 5$ kN; **3.3**
163 N/mm^2; **3.4** $20 \cdot 2$ mm; **3.5**
$165 \cdot 8$ N/mm^2, $65 \cdot 0$ mm; **3.6** $31 \cdot 9$ kN,
$16 \cdot 2$ mm; **3.7** 333 mm (350 mm to
nearest brick size); **3.8** $0 \cdot 48$ m; **3.9**
$1 \cdot 70$ m; **3.10** $76 \cdot 7$ kN; **3.11** $96 \cdot 5$ kN,
$156 \cdot 6$ N/mm^2; **3.12** 144 mm; **3.13** 8;
3.14 $6 \cdot 80$ N/mm^2; **3.15** $4 \cdot 12$; **3.16**
$12 \cdot 1$ mm; **3.17** $124 \cdot 3$, $1 \cdot 18$ mm; **3.18**
$0 \cdot 51$ mm; **3.19** $14 \cdot 95$ kN; **3.20**
220 kN/mm^2; **3.21** $8 \cdot 15$ kN/mm^2; **3.22**
(a) 204 kN/mm^2; (b) 284 N/mm^2; (c)
755 N/mm^2; **3.23** $8 \cdot 82$ mm; **3.24**
$0 \cdot 73$ mm; **3.25** $d_B = 25 \cdot 0$,
$\sigma_A = 172$ N/mm^2, $\sigma_B = 73 \cdot 3$ N/mm^2;
3.26 $\sigma_c = 7 \cdot 14$ N/mm^2,
$\sigma_S = 107 \cdot 1$ N/mm^2; **3.27** $1 \cdot 51$ MN,
$2 \cdot 0$ mm; **3.28** $\sigma_C = 36 \cdot 0$ N/mm^2,
$\sigma_S = 75 \cdot 8$ N/mm^2; **3.29** $283 \cdot 5$ kN;
3.30 $46 \cdot 7$ kN; **3.31** (a) $15 \cdot 55$ kN, (b)
$\sigma_S = 76 \cdot 0$ N/mm^2, $\sigma_B = 22 \cdot 1$ N/mm^2;
3.32 $68 \cdot 9$ N/mm^2; **3.33** $58 \cdot 5$ N/mm^2
(compressive); **3.34** $4 \cdot 8$ mm; **3.35**
$56 \cdot 0$ N/mm^2; **3.36** (a) $23 \cdot 2$ kN, (b)
$0 \cdot 58$ mm; **3.37** $2 \cdot 066$ Nm; **3.38**
135 N mm; **3.39** $223 \cdot 8$ Nm, $61 \cdot 7$ mm;
3.40 $83 \cdot 2$ N/mm^2; **3.41** $42 \cdot 8$ mm;
3.42 $76 \cdot 8$ kN.

Chapter 4

4.1 $\bar{y} = 0 \cdot 2122D$; **4.2** $\bar{y} = 0 \cdot 2122D$;
4.3 $\bar{x} = 32 \cdot 22$ mm, $\bar{y} = 48 \cdot 33$ mm;
4.4 $\bar{x} = 59 \cdot 17$ mm, $\bar{y} = 22 \cdot 92$ mm;
4.5 $\bar{x} = 18 \cdot 65$ mm, $\bar{y} = 21 \cdot 59$ mm;
4.6 $\bar{x} = (D^3 - d^3)/2(D^2 - d^2)$ along
the line of the centroids of the two
circles from the point of contact,
$\bar{y} = D/2$; **4.8** From B, $\bar{x} = 16 \cdot 7$ mm,
$\bar{y} = 66 \cdot 7$ mm; **4.9** From B,
$\bar{x} = 63 \cdot 4$ mm; $\bar{y} = 209 \cdot 8$ mm; **4.10**
From A, $\bar{x} = 245 \cdot 9$ mm,
$\bar{y} = 327 \cdot 6$ mm; **4.11** From A,
$x = 78 \cdot 1$ mm, $y = 312 \cdot 5$ mm; **4.12**
(a) $1\,125 \times 10^6$ mm^4, (b)
375×10^6 mm^4; **4.13**
144×10^6 mm^4, 368×10^6 mm^4;
4.14 $I_{xx} = 2 \cdot 11 \times 10^6$,
$I_{yy} = 2 \cdot 98 \times 10^6$ mm^4; **4.15**
$I_{xx} = 29 \cdot 01 \times 10^6$ mm^4,
$I_{yy} = 64 \cdot 50 \times 10^6$ mm^4; **4.16**
$I_{xx} = 29 \cdot 23 \times 10^6$ mm^4;
$I_{oo} = 29 \cdot 3 \times 10^6$ mm^4; **4.17**
$I_{xx} = 66 \cdot 5 \times 10^6$ mm^4,
$I_{yy} = 17 \cdot 7 \times 10^6$ mm^4; **4.18**
$20 \cdot 36 \times 10^6$ mm^4; **4.19**
$588 \cdot 4 \times 10^6$ mm^4; **4.35** $142 \cdot 52$ mm;
4.20 to **4.34**:

Question No.	I_{xx} (10^6mm^4)	r_{xx} (mm)	I_{yy} (10^6mm^4)	r_{yy} (mm)	$Z_{xx(A)}$ (10^6mm^3)
4.20	395	166·8	33·4	48·5	1·97
4.21	2 906	310·4	14·5	69·3	7·75
4.22	235·3	110·1	147·6	87·2	1·57
4.23	56·25	77·0	16·82	42·1	0·316
4.24	14·42	75·1	2·12	28·8	0·134
4.25	75·93	111·6	15·78	50·9	0·471
4.26	835	199·3	153·4	85·4	3·61
4.27	291·1	116·0	155·7	84·8	2·10
4.28	17·42	48·0	15·07	44·7	0·173
4.29	20·7	63·5	7·08	37·1	0·334
4.30	539·1	153·3	200·5	93·5	3·77
4.31	363·1	182·4	34·04	55·9	0·924
4.32	3 392	277·8	417·6	97·5	9·67
4.33	181·4	110·7	327·7	148·8	1·34
4.34	6 034	291·8	783·3	105·1	14·2

Chapter 5

5.23 114·9 kNm at 1·025 m to RH of B, −39·8 kNm at D; **5.24** 34·6 kNm at 1·31 m to LH of E, −16·3 kNm at B; **5.25** 182·0 kNm at 0·444 m to LH of C, −96·6 kNm at B; **5.26** 80·0 kNm at midspan; **5.27** 68·2 kNm at midspan, −3·84 kNm at B and C; **5.28** 9·0 kNm just to RH of B, −6·0 just to LH of B; **5.29** (a) $L/4$, (b) $L/4$; **5.30** (a) $L/\sqrt{8}$, (b) $L/2$; **5.31** 116 kN, −267·2 kNm; **5.32** 72 kN, −1 071·4 kNm; **5.33** 21·1 kNm at 1·56 m to RH of A; **5.34** 53·5 kNm at 1·86 m to LH of C, **5.35** 76·5 kNm at 0·56 m to RH of B; **5.36** 315·2 kNm at 2·93 m to LH of D; **5.37** 282·4 kNm at 2·02 m to RH of B; **5.38** 110·6 kNm at 0·0700 m to LH of D; **5.39** 66·8 kNm at B, −29·4 kNm at F; **5.40** 142·5 kNm at 0·992 m to LH of E, −82·75 kNm at C; **5.41** W = 143 kN, R_A = 455 kN, R_B = 686 kN; **5.42** 55·3 kNm at 1·549 m to LH of B, −25·6 kNm at C; **5.43** M_{max} = 82·1 kNm at 3·203 m to LH of C, X_c = 6·406 m from C, M at midspan = 77·0 kNm; **5.44** M at midspan = 38·5 kNm, X_{cL} = 0·829 m from B, X_{cR} = 0·892 m from E; **5.45** X_{cL} = 0·70 m and X_{cR} = 3·78 m from B; **5.46** M_{max} = 35·3 kNm at 1·817 m to RH of B, X_{cL} = 0·102 m from B, X_{cR} = 0·862 m from D;

Chapter 6

6.1 394 N/mm^2; **6.2** 468 N/mm^2; **6.3** (a) 17·8 N/mm^2, (b) 35·5 N/mm^2; **6.4** 5·25 kN/m; **6.5** 50·4 kN; **6.6** 13·36 mm, 521·9 kNm, 375 N/mm^2, 702 N/mm^2; **6.7** M_{xx}/M_{yy} = 6·25; **6.8** 8·36 m; **6.9** 8·22 N/mm^2; **6.10** 1·46 kN; **6.11** 100 mm, 62·8%; **6.12** (a) 474 kN, (b) 80 kN; **6.13** 121 N/mm^2; **6.14** 211 N/mm^2; **6.15** 213%; **6.16** 169 N/mm^2; **6.17** 1160 kN, 138 N/mm^2; **6.18** (a) 7·11 N/mm^2, (b) 6·15 N/mm^2; **6.19** 518 N/mm^2, 518 N/mm^2; **6.20** 5·69 N/mm^2, 126·1 N/mm^2; **6.21** 88·0 kN; **6.22** 34·5 mm; **6.23** 11·7 N/mm^2, 171·6 N/mm^2; **6.24** d_n = 0·477d_1, l_a = 0·841 d_1, K = 1·71; **6.25** d_n = 0·357d_1, l_a = 0·881d_1, K = 1·34; **6.26** d_n = 0·395d_1, l_a = 0·868d_1, K = 1·72; **6.27** d_n = 0·418d_1, l_a = 0·861d_1, K = 1·98; **6.28** 171·6 mm; **6.29** 234 kNm, 2874 mm^2; **6.** 294 mm, 1 355 mm^2; **6.31** 227 mm, 548 mm^2; **6.32** 0·83 N/mm^2; **6.35** (a) 78·7%, (b) 21·3%.

Chapter 7

7.1 $\theta_c = wL^3/6EI$, $y_c = -wL^4/8EI$; **7.2** $\theta_{\text{max}} = WL^2/16EI$,

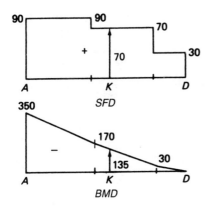

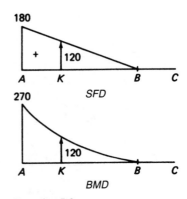

Exercise 5.1

Exercise 5.2

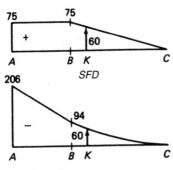

Exercise 5.3

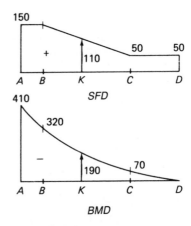

Exercise 5.4

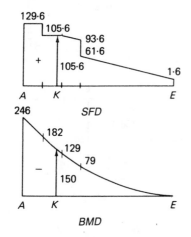

Exercise 5.5

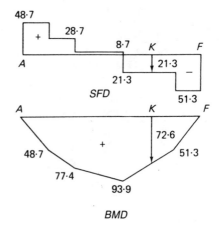

Exercise 5.6

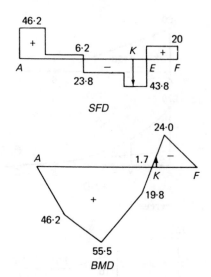

SFD

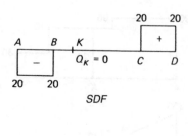

SDF

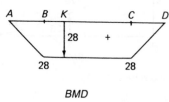

BMD

Exercise 5.8

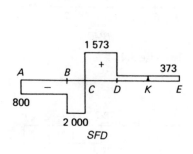

SFD

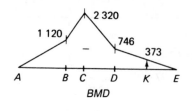

BMD

Exercise 5.7

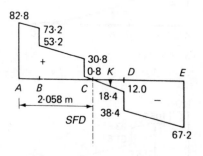

SFD

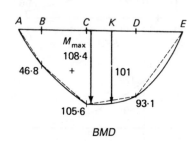

BMD

Exercise 5.9

Exercise 5.10

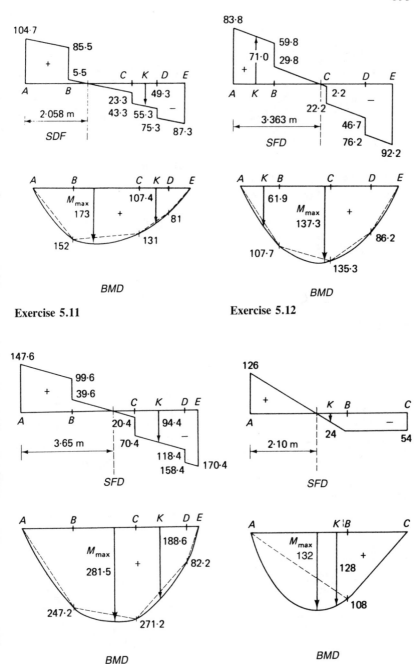

SDF

BMD

Exercise 5.11

SFD

BMD

Exercise 5.12

SFD

BMD

Exercise 5.13

SFD

BMD

Exercise 5.14

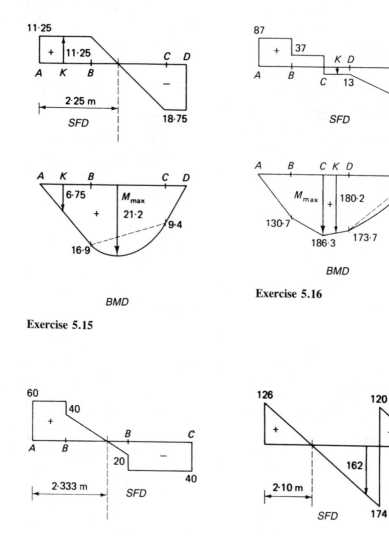

SFD

BMD

Exercise 5.15

SFD

BMD

Exercise 5.16

SFD

BMD

Exercise 5.17

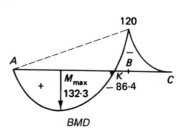

SFD

BMD

Exercise 5.18

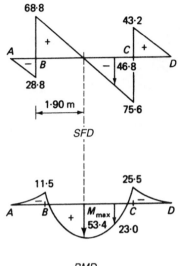

SFD

BMD

Exercise 5.19

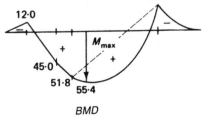

SFD

BMD

Exercise 5.20

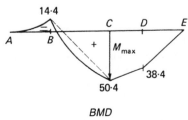

SFD

BMD

Exercise 5.21

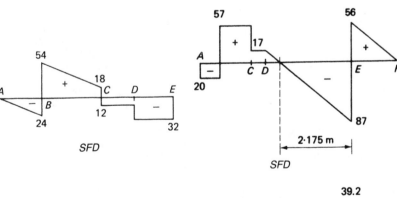

SFD

BMD

Exercise 5.22

396

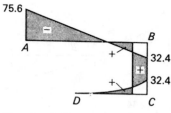

Exercise 5.47

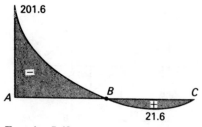

Exercise 5.48

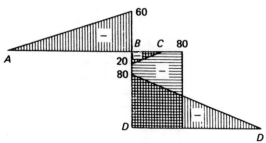

Exercise 5.49

$y_{max} = WL^3/48EI$; **7.3** $-31\cdot4$ mm;
7.4 $-43\cdot9$ mm; **7.5** $26\cdot5$ kN; **7.6**
$134\cdot3$ kN; **7.7** $-9\cdot6$ mm; **7.8**
$-10\cdot3$ mm; **7.9** $-2\cdot7$ mm; **7.10**
240 mm × 80 mm; **7.11**
$\theta_B = \theta_C = 7\cdot71 \times 10^{-3}$,
$y_b = -9\cdot26$ mm, $y_C = -13\cdot89$ mm;
7.12 $\theta_B = \theta_C = 2\cdot78 \times 10^{-3}$,
$y_B = -3\cdot75$ mm, $y_C = -5\cdot42$ mm;
7.13 $\theta_B = 3\cdot70 \times 10^{-3}$,
$\theta_C = 3\cdot80 \times 10^{-3}$, $y_B = -4\cdot17$ mm,
$y_C = -6\cdot43$ mm; **7.14**
$\theta_B = 10\cdot96 \times 10^{-3}$,
$\theta_C = 11\cdot01 \times 10^{-3}$,
$y_B = -5\cdot30$ mm, $y_C = -7\cdot96$ mm;
7.15 $12\cdot66$ kN, $-0\cdot468$ mm; **7.16**
$4\cdot94$ kN, $-0\cdot319$ mm; **7.17** $30\cdot0$ kN,
$1\cdot286 \times 10^{-3}$, $2\cdot143 \times 10^{-3}$,
$-1\cdot114$ mm; **7.18** $17\cdot1$ kN,
$-0\cdot116 \times 10^{-3}$, $0\cdot013 \times 10^{-3}$,
$-0\cdot023$ mm; **7.19** $8\cdot10$ kN,
$-0\cdot347 \times 10^{-3}$, $0\cdot347 \times 10^{-3}$,
$0\cdot21$ mm; **7.20** $9\cdot00$ kN,
$0\cdot231 \times 10^{-3}$, $0\cdot334 \times 10^{-3}$,
$-0\cdot185$ mm; **7.21** $-0\cdot58$ mm,
$-5\cdot01$ mm, $-11\cdot66$ mm; **7.22**
$y_B = -8\cdot18$ mm, $M_A = -68\cdot7$ kNm,

$M_{BC} = M_{BD} = -33\cdot9$ kNm; **7.23**
$31\cdot1$ kN, $-44\cdot5$ kNm; **7.24** $13\cdot4$ kN,
$-53\cdot0$ kNm; **7.25** 165 N/mm^2,
$-87\cdot6$ kNm, $-182\cdot4$ kNm; **7.26**
$M_A = M_B -30\cdot0$ kNm; **7.27**
$M_A = -36\cdot0$ kNm, $M_B = 22\cdot7$ kNm;
7.28 $M_A = -88\cdot88$ kNm,
$M_B = 44\cdot44$ kNm; **7.29**
$M_A = -61\cdot4$ kNm, $M_B = 60\cdot6$ kNm.

Chapter 8

8.1 $0\cdot0742$ mm; **8.2** $0\cdot291$,
$210\cdot7$ kN/mm^2; **8.3** $5\,714$ mm^3; **8.4**
21 mm^3; **8.5** $\Delta L = 0\cdot5$ mm,
$\Delta B = 0\cdot15$ mm; **8.6** $33\,000$ mm^3; **8.7**
$\sigma_n = -1\cdot25$ N/mm^2,
$\tau = -23\cdot8$ N/mm^2,
$\tau_{max} = -27\cdot5$ N/mm^2 at $45°$ to PS;
8.8 $\sigma_n = 42\cdot4$ N/mm^2,
$\tau = -13\cdot0$ N/mm^2, τ_{max} at $45°$ to Ps;
8.9 $\sigma_n = 45\cdot0$ N/mm^2,
$\tau = 26\cdot0$ N/mm^2; **8.10** (a)
$\sigma_n = 116$ N/mm^2, $\tau = 37$ N/mm^2; (b)
$\sigma_n = 90$ N/mm^2, $\tau = 60$ N/mm^2; (c)
$\sigma_n = 56$ N/mm^2, $\tau = 67$ N/mm^2; **8.11**
(a) $35\cdot8°/125\cdot8°$, $101\cdot6$ and
$38\cdot4$ N/mm^2, $\tau_{max} = 31\cdot6$ N/mm^2 at

$80 \cdot 8°/170 \cdot 8°$; (b) $16 \cdot 8°/106 \cdot 8°$, $82 \cdot 1$ and $-62 \cdot 1$ N/mm²,
$\tau_{max} = 72 \cdot 1$ N/mm² at $61 \cdot 8°/151 \cdot 8°$;
(c) $-13 \cdot 3°/-193 \cdot 3$, $109 \cdot 5$ and $-69 \cdot 5$ N/mm², $\tau_{max} = 89 \cdot 5$ N/mm² at $-59 \cdot 4°/239 \cdot 4°$; (d) $33 \cdot 7°/123 \cdot 7°$, $-35 \cdot 0$ and $-100 \cdot 0$ N/mm²,
$\tau_{max} = 32 \cdot 5$ N/mm² at $78 \cdot 7°/168 \cdot 7°$;
8.12 $104 \cdot 4$ N/mm² at $23 \cdot 3°$; **8.13** $50 \cdot 8$ and $-70 \cdot 8$ N/mm²,
$40 \cdot 3°/130 \cdot 3°$; **8.14** (a) $153 \cdot 0$ and $-53 \cdot 0$ N/mm², $14 \cdot 5°/104 \cdot 5°$; (b) $103 \cdot 0$ N/mm², $-14 \cdot 5°/-194 \cdot 5°$; (c) $69 \cdot 7$ N/mm², $-101 \cdot 1$ N/mm²; **8.15** (a) $42 \cdot 4$ N/mm², (b) $30 \cdot 8$ N/mm².

Chapter 9

9.1 491 Nm; **9.2** 230 Nm; **9.3** $26 \cdot 8$ N/mm², $2 \cdot 22$ kNm; **9.4** $97 \cdot 0$ N/mm²; **9.5** 38 mm; **9.6** 25 mm, 15 mm; **9.7** (a) $48 \cdot 3$ N/mm², (b) $60 \cdot 2$ N/mm²; **9.8** $113 \cdot 8$ N/mm², $75 \cdot 8$ N/mm²; **9.9** $97 \cdot 7$ mm; **9.10** $28 \cdot 8$ kNm; **9.11** $OD = 161$ mm $ID = 113$ mm; **9.12** $4 \cdot 01°$; **9.13** $4 \cdot 34°$; **9.14** $0 \cdot 328°$; **9.15** $65 \cdot 7$ kN/mm², $15 \cdot 1°$; **9.16** $35 \cdot 6\%$; **9.17** $76 \cdot 9$ N/mm²; **9.18** $D = 70$ mm; **9.19** $OD = 208$ mm, $8 \cdot 3°$; **9.20** 175 kW; **9.21** $37 \cdot 3$ N/mm² $22 \cdot 4$ N/mm²; **9.22** $OD = 18 \cdot 2$ mm, $ID = 10 \cdot 9$ mm; **9.23** $19 \cdot 4$ mm (say 20 mm); **9.24** 21 mm.

Chapter 10

10.1 $84 \cdot 5$ N/mm²; **10.2** 280 N/mm² (T), 134 N/mm²(C); **10.3** $2 \cdot 15$ kN; **10.4** 172 N/mm² (C); **10.5** $43 \cdot 4$ kN; **10.6** above: $f_c = 26 \cdot 4$ N/mm² and $f_t = 8 \cdot 8$ N/mm², below: $f_c = 64 \cdot 5$ N/mm² and $f_t = 46 \cdot 9$ N/mm²; **10.7** $8 \cdot 25$ N/mm² (C), $5 \cdot 25$ N/mm² (T); **10.8** $3 \cdot 08$ N/mm² (C), $1 \cdot 00$ N/mm² (T);

10.9 $1 \cdot 00$ N/mm² (C), 0; **10.10** $2 \cdot 33$ N/mm² (C), $1 \cdot 43$ N/mm² (C), $0 \cdot 07$ N/mm² (C), $0 \cdot 83$ N/mm² (T); **10.11** $8 \cdot 76$ N/mm² (C), $2 \cdot 50$ N/mm² (C), $1 \cdot 16$ N/mm² (C), $5 \cdot 10$ N/mm² (T); **10.12** $5 \cdot 50$ N/mm² (C), $2 \cdot 00$ N/mm² (C), $2 \cdot 00$ N/mm² (C), $1 \cdot 50$ N/mm² (T); **10.13** 445 kN/m²; **10.14** $97 \cdot 2$ N/mm² (C),. $13 \cdot 4$ (T); **10.15** 214 N/mm², no tensile stress developed N/mm²; **10.16** (a) $3 \cdot 34$ mm, (b)$18 \cdot 75$ mm; **10.17** $123 \cdot 1$ kN; **10.18** At B, 194 N/mm² (C) and 150 N/mm² (T), at C, $82 \cdot 4$ N/mm² (C) and $38 \cdot 0$ N/mm² (T); **10.19** (a) $90 \cdot 8$ N/mm², (b) $219 \cdot 5$ N/mm²; **10.20** $495 \cdot 5$ kN; **10.21** $15 \cdot 46$ N/mm² (C), $13 \cdot 71$ N/mm² (T); **10.22** $15 \cdot 5$ kN; **10.23** (a) $4 \cdot 12$ m, (b) $8 \cdot 24$ m; **10.24** $68 \cdot 5$ kN; $3 \cdot 04$ N/mm².

Chapter 11

11.1 $1 \cdot 94$ kNm; **11.2** $60 \cdot 1$ kN/m; **11.3** $121 \cdot 1$ kNm/m; **11.4** $9 \cdot 58$ m; **11.5** $48 \cdot 7$ kNm/m; **11.6** 80 kN/m; **11.7** $2 \cdot 41$ m; **11.8** $2 \cdot 62$ m; **11.9** $2 \cdot 34$ m; **11.10** $1 \cdot 73$ m; **11.11** $117 \cdot 7$ kN; **11.12** $1 \cdot 09$; **11.13** $5 \cdot 78$; **11.14** $F_{empty} = 2 \cdot 53$, $F_{full} = 9 \cdot 02$, $R_{max} = 694$ kN; **11.15** $e = 1 \cdot 255$ m (inside middle-third), $q_{max} = 423$ kN/m², $q_{min} = 23$ kN/m²; **11.16** $e = 4 \cdot 09$ m (outside middle-third), $q_{max} = 1 \ 090$ kN/m² at toe, $q = 0$ at $17 \cdot 73$ m from two; **11.17** 96 kN/m² 64 kN/m²; **11.18** (a) $e = 0 \cdot 326$ m (inside middle-third), (a) 114 kN/m², (c) $0 \cdot 96$; **11.19** (a) 81 kN/m², (b) $1 \cdot 04$; **11.20** 184 kN/m²; **11.21** (a) $e = 0 \cdot 134$ m (inside middle-third), (b) 144 kN/m², (c) $1 \cdot 98$; **11.22** (a) $1 \cdot 90$ m, (b) $1 \cdot 17$ m; **11.23** (a) 470 N/m², (b) 470 N/m²; **11.24** 180 kN/m².

Index